Mechanics of Offshore Pipelines

To the memory of our fathers,

Kyriakos C. Kyriakides
and
Dr. Antonio Corona.

Mechanics of Offshore Pipelines
Vol. I Buckling and Collapse

by

Stelios Kyriakides
The University of Texas at Austin

and

Edmundo Corona
University of Notre Dame

ELSEVIER

Amsterdam • Boston • Heidelberg • London • New York • Oxford
Paris • San Diego • San Francisco • Singapore • Sydney • Tokyo

Elsevier
Linacre House, Jordan Hill, Oxford OX2 8DP, UK
30 Corporate Drive, Suite 400, Burlington, MA 01803

First edition 2007

Copyright © 2007 Elsevier BV. All rights reserved

British Library Cataloguing in Publication Data
Kyriakides, S.
  Mechanics of offshore pipelines
  Vol. 1: Buckling and collapse   1. Petroleum pipelines   2. Natural gas pipelines
  3. Underwater pipelines
  I. Title   II. Corona, Edmundo
  665.5′44

Library of Congress Number: 2007924001

ISBN: 978-0-08-046732-0

For information on all Academic Press publications visit our web site
at books.elsevier.com

Typeset by Charon Tec Ltd (A Macmillan Company), Chennai, India
www.charontec.com

Printed and bound by CPI Group (UK) Ltd, Croydon, CR0 4YY

Transferred to Digital Print 2012

Contents

Preface

Although offshore oil and gas production was conducted throughout the entire 20th century, the industry's modern importance and vibrancy did not start until the early 1970s, when the North Sea became a major producer. Since then, the expansion of the offshore oil industry into many parts of the world has been both continuous and rapid. This growth has been coupled with a gradual movement to increasingly deeper waters, where today production has reached approximately 7,000 ft (2,130 m) while exploration is proceeding as deep as 10,000 ft (3,050 m).

Pipelines, and more generally long tubular structures, are major oil and gas industry tools used in exploration, drilling, production, and transmission. Installing and operating tubular structures in deep waters places unique demands on them. The high pressures and elevated temperatures of the oil wells, the high ambient external pressures, the large forces involved during installation, and generally the hostility of the environment can result in a large number of limit states that must be addressed.

The technical challenges of the field have spawned significant research and development efforts in a broad range of areas. For example, they include improvements in line grade steels and innovations in the manufacture of tubular products; development of several pipeline installation methods and lay-vessels; establishment of the mechanical behavior and the modes of failure of long tubes and pipes under various loads; investigation of new problems in fluid–structure and soil-structure interactions, issues of corrosion, welding, fatigue, and many others. The field is so broad that covering the progress achieved would require several books and monographs. The goal of this book series is to present developments in the area of mechanical behavior of pipelines and long tubular structures in general used in this industry. The approach is problem oriented, encompassing the main limit states encountered in practice. The background of each problem and scenarios where it may be encountered are first outlined. Each problem is then analyzed experimentally and analytically, first in simple terms and then in more advanced ways. Each discussion finishes with design recommendations.

Many of the problems discussed have been brought about by the demands of this new frontier, while others are classical problems of inelastic structural mechanics recast to the needs of the new applications. A prime example of the "new problem" category is the problem of buckle propagation under external pressure, together with its many related issues. Propagating collapse and its catastrophic effects were first brought to light around 1975 by researchers in the offshore pipeline industry. Since then, the underlying mechanism has been shown to be shared by several other structures and a variety of material systems. This wider applicability of the underlying mechanism has in turn attracted the attention of researchers from several different fields. By contrast, the problem of plastic buckling and collapse of a long cylinder under compression has had a 50-year life span, and can be categorized as a classical problem. However, the design needs of modern off-shore as well as onshore pipelines have generated new investigations that have produced new relevant insights into the problem.

I came to this field in 1976 through my work with Chuck Babcock at Caltech on the propagating buckle and its arrest. Although we were able at the time to establish from experiments the mechanics underlying the problem, neither the numerical tools nor the computational facilities available then were adequate for the needs of the problem. During the next 20 years, improvements in both of these enabled researchers to investigate many aspects of this problem in great detail. Thirty years of exposure to the offshore industry has enabled me to get involved in many other interesting engineering problems, often at their genesis. Significant parts of the contents of the book come from either investigations carried out by my research group at the University of Texas at Austin, or from less formal arrangements with the industry. I am indebted to the industry for their support through Joint Industry Projects as well as through individual contracts. The support by the US Office of Naval Research of part of the work is also acknowledged with thanks. The writing of this book was supported in part by the College of Engineering at the University of Texas at Austin.

The book is aimed at the practicing professional, but can also serve as a graduate level text for inelastic design of tubular structures. Each chapter deals with a specific mechanical problem that is analyzed independently, for the most part, of others. Entry-level graduate school background in nonlinear structural and solid mechanics, and working knowledge of issues in structural stability and numerical methods should make going through the analytical developments easier. A class in plasticity would enhance the brief chapter on metal plasticity at the end of Volume I. The mathematics used is at a level that should enable graduates from 5-year engineering programs to follow the formulations presented in detail.

The style of the work is one I have always found both necessary and most rewarding: investigate the problem through a combination of experiments and analysis. Solution of relatively new, challenging problems requires understanding that can only come from careful experiments. Having gained an understanding of the underlying physics, one proceeds to analysis. Because of the nonlinear nature of most of the problems addressed, analysis is often supplemented by numerical modeling. The results of these combined efforts can subsequently be distilled into more concise procedures and guidelines for the design engineer. Wherever possible, numerical procedures are outlined in terms of custom formulations to help the reader gain a deeper understanding into each problem. The insight gained from the experiments presented, coupled with results from complementary numerical efforts, should enable the interested reader to either reproduce the solution procedures and results or solve similar problems using a nonlinear finite element code. The experimental results included in each chapter should be sufficient to evaluate or calibrate new numerical models. The experimental set-ups and procedures used are described in sufficient detail to enable the execution of new experiments should the need arise.

Volume I addresses problems of buckling and collapse of long inelastic cylinders under various loads encountered in the offshore arena. Several of the solutions are also directly applicable to land pipelines. The fundamental nature of most of the problems makes them also applicable to other fields. Examples include tubular components in nuclear reactors and power plants, aerospace structures, automotive and civil engineering structures, naval vehicles and structures, and others. Volume I is co-authored with Edmundo Corona, with whom I have cooperated for more than 20 years. He was instrumental in the original work on bending and combined bending and external pressure, as well as in the writing and upgrading of various versions of the computer program BEPTICO. I am indebted to Edmundo for his long-term interactions, friendship and wise counsel.

Volume II, presently under preparation, deals with various aspects of the problem of buckle propagation and arrest in offshore pipelines and related problems. A third volume (III) that will include subjects such as fracture mechanics of pipelines, burst under internal pressure, cyclic loading and ratcheting, upheaval buckling, reeling, and other special problems is also being planned.

Many of the results presented have originated from studies performed over the years in cooperation with several outstanding graduate students. Their contributions are reflected in the references after each chapter. I would like to thank in particular Theodoro Netto, Heedo Yun, Gwo-Tarng Ju and Ken Liechti for their input in some of the chapters. Thanks are also due to Liang-Hai Lee, who reproduced several numerical results using modern finite element codes. Additional help received as well as photographs and drawings provided by individuals are acknowledged in the text. Special thanks go to Rob Cook for his help in generating several of the figures and Jason Kyriakides for proofreading the book. Much of the artwork and CAD figures were patiently produced by Jason Kyriakides. His work truly enriches the text and for this I am very grateful. Despite the generous input we have received from all mentioned and our best personal efforts, typos and some, minor errors may remain. The authors would be grateful if readers let us know of any they come across.

Over the years I have had the pleasure and privilege of interacting with many university colleagues and many researchers and engineers from industry. Although too many to list by name here, I would like to acknowledge interactions with university colleagues Professors Charles Babcock (deceased), John Hutchinson, Nicolas Triantafyllidis, Theodoro Netto, Andrew Palmer and Chris Calladine. I am grateful to Carl Langner and Serghios Barbas for keeping my compass directed in the practical direction through numerous exchanges over many years. Other industrial colleagues with whom I have interacted include Ralf Peek, Pedro Vargas, David McKeehan, Frans Klever, Roberto Bruschi, Raghu Madhavan, Duane DeGeer, Gwo-Tarng Ju, Heedo Yun and Richard Verley. I also thank my colleagues in the Department of Aerospace Engineering and Engineering Mechanics at the University of Texas at Austin for providing a fertile atmosphere for learning and intellectual growth for my graduate students and myself. Many others not mentioned have contributed to this body of work. I assure them that I am no less thankful for their contributions.

Finally, I am grateful to David Sleeman for shepherding this book project on behalf of Elsevier. The production of this book was coordinated by Debbie Clark to whom I am thankful for her patience and determination to meet the requirements of the authors.

Stelios Kyriakides
Austin, Texas, USA

1

Introduction

Pipelines constitute a major means of gathering and transporting a variety of commodities, ranging from oil, natural gas and chemicals to water, sewage, etc. Their use has expanded with time because they are more energy efficient than competing means of transportation. Although they require significant initial investment, they have a lifespan of up to 40 years and require relatively minor maintenance. Oil and natural gas pipelines are usually the longest, often crossing country and sometimes continental borders. Such major projects require investments of billions of dollars, and in addition can be contentious. Local resistance is often tied to environmental and safety issues. International and geopolitical complications arise from tensions either between countries along the route, or between them and the country of origin or destination. Furthermore, in today's climate of tight energy supply, pipelines are often entangled in geopolitical affairs. Consequently, such pipeline projects require long and careful planning and negotiations with many interest groups. One example is the *Trans-Alaska* pipeline, a 48-inch, 800-mile crude oil land pipeline that connects Prudhoe Bay on Alaska's North Slopes to the Valdez terminal in southern Alaska (see Table 1.1). The line crosses arctic regions that are environmentally sensitive as well as earthquake prone. As a result, the project took a decade of planning, engineering work and political negotiations. It was completed in 1977 at a cost of $10 BIL. A second example that has been discussed since the 1980s is the *Medgaz* pipeline, which will bring Algerian natural gas to Spain (Table 1.1). It is finally under construction and due to be completed in 2009. Primarily because of political reasons, the route chosen has the pipeline crossing the Mediterranean Sea in water depths exceeding 7,000 ft (2,130 m). Another very politicized major project is the pipeline connecting Baku on the Caspian Sea in Azerbaijan to Ceyhan on the Mediterranean coast of Turkey, passing through Tbilisi in Georgia (Table 1.1). The 1,094-mile (1,760 km) long, 42-inch pipeline was an international effort to give the energy-rich, land-locked countries of the Caspian Sea region an alternate route for exporting oil to ones going through Russia or Iran. It was in planning for 10 years before it was completed in 2005 at a cost of $3.6 BIL.

In some cases, moving transportation pipelines offshore has made them less contentious. Examples of offshore transport pipelines are the *Trans-Mediterranean* gas pipelines joining Tunisia to Sicily and then to mainland Italy; the *Blue Stream* gas pipeline connecting Russia to Turkey through the Black Sea (see Table 1.1); the *Algeria to Spain* pipeline mentioned above; the 1,100 km gas pipeline from *Oman to India* that was seriously considered in the mid-1990s and others.

During the last 30 years, significant reserves have been located offshore in places like the North Sea, the Gulf of Mexico, the Persian Gulf, offshore Brazil, West Africa,

1

Table 1.1 Examples of major land and offshore pipelines and some operational details.

	Trans-Alaska	Trans-Mediterranean	Blue Stream	MEDGAZ
Type	Land	Land/Offshore	Land/Offshore	Land/Offshore
Product	Crude Oil	Gas	Gas	Gas
From–To	Prudhoe Bay, Alaska – Valdez, Alaska	1. Hassi R'mel Oil Fields – Cape Bon, Algeria 2. Cape Bon, Algeria – Mazarra del Vallo, Sicily 3. Mazarra del Vallo, Sicily – Mortelle, Sicily 4. Mortelle, Sicily – Favazzina, Italy 5. Favazzina – Bologna, Italy	1. Stavropol, Russia – Dzhubga, Russia 2. Dzhubga, Russia – Samsun, Turkey 3. Samsun, Turkey – Ankara, Turkey	1. Hassi R'mel Oil Fields, Algeria – Beni Saf, Algeria 2. Beni Saf, Algeria – Almeria, Spain 3. Almeria, Spain – Albacete, Spain
Length	1287 km	1. 920 km 2. 160 km 3. 352 km 4. 44 km 5. 1051 km	1. 370 km 2. 774 km 3. 501 km	1. 550 km (Algerian onshore) 2. 200 km (offshore) 3. 300 km (Spanish onshore)
Capacity	Actual in 2006 ~800 000 bbl/d	30 BIL Nm³/a	16 BIL Nm³/a	16 BIL Nm³/a
Internal Pressure	1180 psi		1. 75 bar, 100 bar 2. 250 bar 3. 75 bar	
Operation Date	1977	1983	2003	2009
D × t	48 × 0.462 in and 0.562 in	1. 48 in 2. 20 in (3) 3. 48 in 4. 10 in, 20 in (3)	1. 56 in, 48 in 2. 24 × 1.254 in (2) 3. 48 × 0.563 in	2. 24 × 1.170 in
Grade	X60, X65, X70	2. X65	2. X65	2. X70
Maximum Water Depth	Not Applicable	2. 1970 ft (600 m) 4. 1181 ft (360 m)	2. 7054 ft (2150 m)	2. 7087 ft (2160 m)
Special Features	• Crosses arctic and earthquake-prone areas. • Designed to withstand 8.5 Richter scale earthquake	• Deepest offshore pipeline, 1982 • Crosses uneven sea bed and earthquake-prone areas	• Deepest offshore pipeline, 2003	–
Cost	$10 BIL	$3 BIL	$2.5 BIL	2. $0.75 BIL

	Baku-Ceyhan	Mardi Gras	Independence Trail	Jansz & Gorgon Projects
Type	Land	Land/Offshore	Land/Offshore	Land/Offshore
Product	Crude Oil	Oil/Gas	Gas	Gas
From–To	Baku, Azerbaijan – Ceyhan, Turkey	1. Na Kika/Thunderhorse-MP260 2. Thunder Horse-SP89E 3. SP89E-LOOP, LA 4. Holstein/Mad Dog/Atlantis-SS332 and Beaumont 5. Holstein/Mad Dog/Atlantis-SS332 and Neptune	Independence Hub, Mississippi Canyon Block 920, Gulf of Mexico – Tennessee Gas Pipeline, West Delta Block 68, Gulf of Mexico	1. Jansz Field to Barrow Island – Northwest Australia 2. Gorgon Gas Fields, to Barrow Island – Northwest Australia
Length	1 760 km	1. Okeanos (Gas) ~100 mi 2. Proteus (Oil) ~70 mi 3. Endymion (Oil) ~90 mi 4. Caesar (Oil) ~115 mi 5. Cleopatra (Gas) 115 mi	137 mi	1. Jansz ~180 km 2. Gorgon ~70 km
Capacity	160 Mbbl/d	1. 500 MMscf/d 2. 580 Mbbl/d 3. 750 Mbbl/d 4. 450 Mbbl/d 5. 500 MMscf/d	850 MMscf/d	825 MMscf/d
Operation Date	2006	2006	2007	2010
$D \times t$	1. 42 in 2. 36 in	1. 24 in + 20 in × 0.941–1.00 in 2. 28 × 1.088–1.210 in + 24 × 1.095–1.200 in 3. 30 in 4. 28 × 1.045–1.148 in + 20 × 0.971–1.227 in 5. 20 × 0.750–0.880 in + 16 × 0.665–0.898 in	24 in	1. 24 × 1.236 in + 28 × 1.173 in 2. 30 × 1.327 in
Grade	Not Applicable	X65	X65	X65
Maximum Water Depth	Not Applicable	4 300–7 300 ft (1 310–2 225 m)	8 050 ft (2 450 m)	4 430 ft (1 350 m)
Special Features		One of the most complex deepwater projects	When finished, world's deepest pipeline and deepest production	
Cost	$3.6 BIL	$1 BIL	$0.28 BIL	Not available

Malaysia, Indonesia, Northwest Australia, and other places. The first reserves at significant distance from the shore such as projects in the North Sea, offshore California and the Gulf of Mexico were found in the early 1970s in water depths of a few hundred feet. Since then, offshore activity has grown rapidly, and gradually moved to increasingly deeper waters (see current and future deepwater basins in the map of the world shown in Figure 1.1). Today, production has reached approximately 7,000 ft (2,130 m), while exploration is proceeding at 11,000 ft (3,350 m). Offshore exploration and production operations invariably involve long tubular structures. These include casing used in oil wells, hydraulic tubing used for drilling and production, high pressure flow lines used to gather oil from wells to a central point such as a production platform, risers used to bring oil from the sea floor to the surface, offshore transport pipelines and others.

With thousands of miles of pipelines in operation, onshore pipeline engineering is reasonably well established. Traditional design can, however, be challenged when a pipeline is crossing an earthquake-prone area, when it is constructed in arctic regions, or by the recent advent and use of new high-strength steels. By contrast, the construction and operation of pipelines and other tubular structures offshore is a younger field with unique design needs and challenges that gave birth to new technologies. Examples include: new underwater surveying technologies; a variety of methods and associated vessels for installing pipelines in increasingly deeper waters; new corrosion and sour hydrocarbon protection methods and insulation materials; underwater welding and inspection methods and facilities; seafloor trench plowing methods; remotely-controlled vehicles (ROVs) and equipment for deepwater operations; pigging and repair methods and facilities, and many others (see Chapter 2).

Major land oil and gas pipelines usually have diameters in the range of 36 to 64 inches. The major design load is the internal pressure of the flow. As a result, they have diameter-to-thickness ratios (D/t) that range from 40 to 80. By contrast, offshore pipelines usually have diameters that are under 36 inches (see Table 1.1, diameters larger than 36 are used in shallower waters). The primary loading is often external pressure, which can lead to collapse. Installation as well as other operational loads can also influence the design and dictate that the pipe material exhibit significant ductility. The combined effect is that offshore pipelines are much thicker, with D/t values ranging from 50 for shallow waters down to 15 for deep waters, and even lower for high pressure flow lines. To meet these differences between land and offshore pipelines, new steel alloys as well as improved methods of manufacturing had to be developed. The advances include transition to low carbon steel, microalloying, and thermo-mechanical control processing (TMCP), all of which contribute to grain size reduction, increase in toughness and lowering of the transition temperature. Other innovations include the introduction of continuous casting methods for plate and round billets, advances in plate rolling, advances in hot forming of seamless pipe and improvements in cold forming of seam-welded pipe (see Chapter 3).

Offshore pipelines are designed to safely sustain first the installation loads, second the operational loads and third to survive various off-design conditions. Each of these provides several design scenarios. For example, in a typical S-lay installation, the following loading conditions are often addressed (see Section 2.2): the first is encountered as the line departs the lay vessel and enters the water. The second involves the usually long, relatively straight suspended section. The third involves the zone where the line bends into a horizontal configuration in relatively deep water. In a fourth, the line bends and stretches to conform to the surface of the sea floor. Other installation design scenarios involve abandonment

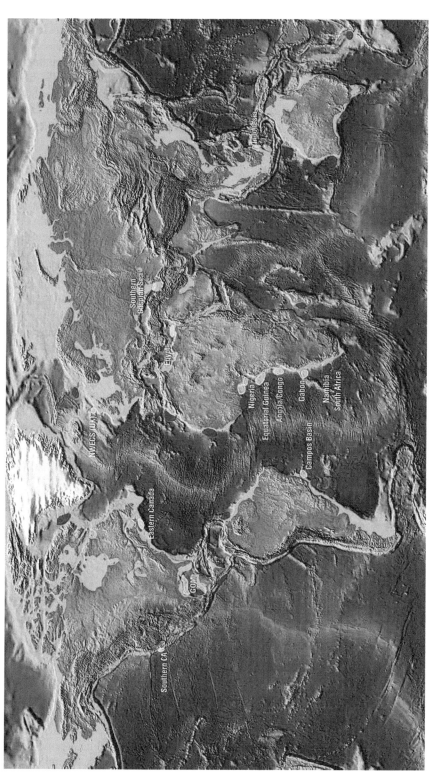

Figure 1.1 World map showing locations of current and future deepwater areas/basins with confirmed hydrocarbon reserves: ▨ Current, ■ Frontier/Future, ■ Former (reproduced with permission from "2006 Deepwater Solutions and Records for Concept Solutions," *Offshore Magazine and Mustang Engineering*).

of the line, dewatering in case of flooding, and recovery and repair procedures in case of an accident. During operation, the pipeline must safely sustain the ambient external pressure and survive in generally hostile environments. In addition, the hydrocarbons it carries can be at relatively high internal pressures and elevated temperatures. Lifetime off-design events that can impact design include the development of a propagating buckle; the rare possibility of a propagating crack for gas lines; accidental impact by a foreign object; the development of a leak and/or flooding; the survivability of the line in earthquakes, sea storms, mudslides, etc.

The majority of these loading conditions are such that traditional stress-based design is not feasible. Instead, the design of offshore pipelines is based on *limit states (limit state design)*. In other words, in many of the design scenarios plastic deformation is allowed, but the engineer must ensure that the structure is safely away from either excessive deformation or failure that are defined as limit states. The many design scenarios are characterized by an unusually large number of limit states, ranging from collapse under external pressure to burst under internal pressure and many others. The limit states can affect the manufacture of the pipe, the installation, and the long-term operation of the structure on the sea floor. This three-volume book series touches upon the main limit states encountered in practice, describes them, analyzes them experimentally and analytically, first in simple terms and then in more advanced ways, and finishes with design recommendations. The book series is aimed at the practicing professional, but it can also serve as a text for inelastic design of tubular structures in graduate classes.

Offshore pipeline engineering is a vast field and as a result impossible to cover in one book. A good demonstration of the breadth of this technology can, for example, be found in the recent book by Palmer and King [1.1]. The book covers subjects like route selection, construction, welding, corrosion resistance, inspection techniques, repairs, etc. Another viewpoint on the subject appears in a book edited by Braestrup [1.2]. This book takes a broader and more practical approach to design and includes subjects like risk and safety, a more detailed look at installation, control and documentation, operation, etc. By contrast, the present three-volume series is more narrowly focused on the mechanical behavior and limit states of offshore pipelines. Volume 1 deals with some of the main local buckling and collapse limit states that can be encountered during installation and operation of the pipelines. Volume 2 covers the problem of a propagating buckle and its arrest. Volume 3 covers additional mechanical behavior problems such as the mechanics of pipeline reeling, upheaval and lateral buckling, burst, fracture mechanics of pipelines, cyclic loading and ratcheting, etc.

1.1 OFFSHORE PIPELINE DESIGN CONSIDERATIONS

The construction of an offshore pipeline involves several engineering disciplines. Once the need for a new pipeline has been established, the project starts with the design engineer, who usually selects the diameter, wall thickness, steel grade, the method of manufacture (e.g., seamless, UOE, ERW – see Chapter 3) and the method of installation (Chapter 2). A survey of the installation site is usually commissioned and the route is selected [1.1, 1.2]. The water depth, the relief of the sea floor and the route selected define loadings that must be sustained, and as such influence the selection of the basic pipeline parameters as well as the installation method chosen for the project. Some of the most severe loading conditions seen by a pipeline often occur during installation (Section 2.2). Consequently, the installation method itself can also influence the design parameters of the pipeline. Simultaneously, of course, material and installation costs are factors that also enter into the design decisions.

Once the basic pipe parameters are specified, they are submitted to pipe mills (e.g., API Specs. 5L [1.3]). The purchaser, who interacts with the mill technical manager, is usually a person with metallurgical background, as are most of the technical people of the pipe mill. Pipe specifications include dimensional tolerances such as ovality, wall eccentricity and straightness; minimum specified yield stress and maximum yield-to-ultimate ratio; ductility, toughness and mechanical characteristics of the material such as transition temperature; weld characteristics for seam-welded pipe, special corrosion resistance characteristics, in some cases stress–strain response characteristics; etc. [1.3]. Often, the pipe mill technical manager will be in contact with the plate or round billet mill responsible for the selection of the steel alloy, its casting and its rolling into a plate or billet from which the pipe is later made. Tight specifications or specifications that do not necessarily impact the project can complicate the manufacture of the pipe, increase the cost and, in extreme cases, render it impossible to manufacture. Communication between the designer, the purchaser and the pipe mill and steel mill technical representatives helps each to understand the others' constraints, reduces the cost, and in the end results in the best pipe for the project. To facilitate this communication, Chapter 3 presents outlines on steel making, plate and round billet rolling, and of the main manufacturing methods of offshore pipelines. Manufacturing limitations and information on typical geometric tolerances, material property variations, etc. are included.

Once manufactured, the pipe is shipped to a land site close to the offshore project, and is transferred in smaller quantities to the installation site by barge (or to the fabrication and loading site for reel vessels). The overall installation is administered by a project engineer who stays in close communication with the design engineer of the pipeline. Pipe lengths are welded on the lay-vessel, X-rayed, and coated as required. The actual installation process is overseen by a third engineer, who along with the captain operates the vessel. Onboard welding, the tensioning of the pipeline on its way to the sea floor, the position and length of the stinger, the path of the vessel and its speed are some of the major issues that the operator must decide upon and oversee. The operation places some requirements on the pipe, such as straightness and ovality, that make their way in the pipe specifications. Although the operator is far removed from the pipe mill, some understanding of the actual installation limitations, if communicated early enough to the design engineer, can help avoid overly tight specifications and their impact on cost. A descriptive outline of the major offshore production facilities that act as the termination points of flowlines that bring in hydrocarbons from offshore wells is given in Chapter 2. The hydrocarbons are processed and sent off via export oil and gas pipelines either to shore or to another facility offshore. The chapter also outlines the major offshore pipeline installation methods, and the general setting that these tubulars operate in.

Chapters 4 to 12 deal individually with various limit states that a pipeline designer must consider. In particular, limit states associated with local buckling and collapse of the pipeline that can result from loads such as external and internal pressure, bending, tension, compression and combinations of these are discussed. The nature of the structures is such that the limit states are in the main inelastic. To help the reader navigate through the book, Chapter 13 presents elements of metal plasticity. Guidance on measurements and testing methods recommended for qualifying offshore pipelines appears in the Appendices. The physics of each limit state is described through experiments. Subsequently, each problem is analyzed in sufficient detail for the reader to understand and, if necessary, reproduce the calculations for a new application. In most cases, the formulations of custom analyses, such as those of the numerical model BEPTICO, are presented. Often, models developed in a nonlinear finite element code can be used instead, provided the required problem parameters are incorporated. The chapters include sensitivity studies of the limit states to key parameters and to material and geometric imperfections. The introduction of some of these imperfections in standard FE codes often requires customization.

Several of the limit states discussed in this book are addressed in offshore pipeline standards such as API RP 1111 [1.4], DNV OS-F101 [1.5], ISO 13623 [1.6] and British Standard BS 8010 [1.7] through simplified design formulae. By their nature, simplified formulae cannot always adequately capture the effect of key variables such as geometric imperfections, residual stresses or unique material properties such as anisotropy, the shape of the stress–strain curve, etc., all of which can have a significant impact on the ability of the pipe to withstand the applied loads. In many cases, the simplified formulae provided in these Standards can be overly conservative. In other cases, simplified approaches can lead to unconservative and potentially unsafe designs. Analytical and numerical models, such as the ones described in this book, should provide a more dependable means of optimizing the design while maintaining adequate safety. Today's offshore pipelines are major structures with costs that run in the hundreds of millions $US (see Table 1.1); thus they require and deserve to be designed with the care and attention that modern engineering tools can provide.

The oil and gas industry has its roots in the USA. As a consequence, most of the pipeline terminology, specifications and definitions have originated from the *American Petroleum Institute* (API). For example, the size of the pipe is specified by the outside diameter in units of inches. Standard diameters are listed in [1.3]. For every standard pipe diameter there are several standard wall thicknesses. Similarly, pipeline steels are designated as Grade A, B and X that define manufacturing and product specifications such as chemistry, fracture toughness, ultimate strength, etc. [1.3]. To help the inexperienced reader with pipeline terminology, a Glossary section is included in Appendix G.

1.2 BUCKLING AND COLLAPSE OF STRUCTURES

Instability is one of the major factors that limit the extent to which structures can be loaded or deformed. Several of the pipeline limit states discussed in this book are associated with buckling and localized collapse instabilities. In this section, we discuss some of the instability events and define the terminology used. Interested readers can learn more about the fundamentals of structural stability from Refs. [1.8–1.12]. The literature on plastic instabilities is less developed, but Refs. [1.10, 1.13, 1.14] should be helpful. The references at the end of the chapter dealing with buckling and collapse should help clarify any issues left unresolved by what is presented.

The most familiar example of structural buckling is the column. A column loaded within its design specifications is straight and carries axial compression by undergoing membrane deformations. In this regime, its stiffness is proportional to its axial rigidity. On reaching a critical load, a bent configuration becomes energetically preferable. The stiffness of

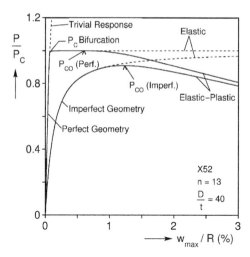

Figure 1.2 Nonlinear pressure–maximum displacement responses of a pipe under external pressure. Pipe buckles elastically but collapses due to inelastic action. Included are the perfect and imperfect structure responses.

the bent configuration is nearly zero, rendering the column structurally useless. *Buckling* describes the process of switching from the straight, stiff configuration to the bent one that has very small stiffness. The load at which this switch takes place is the *critical buckling load*. For an ideally straight column the buckling load is established through a linearized bifurcation analysis. Before the critical load, the straight configuration is the only solution to the problem. At the critical load, two solutions become possible: the straight, which is unstable, and the bent one. *Bifurcation* is the term that describes the instantaneous existence of two solutions at the critical point. Tracking the equilibrium of the buckled configuration is a nonlinear process that often requires numerical solution procedures.

Practical structures are manufactured to specified tolerances, and as a result always deviate to some degree from an ideally perfect geometric shape. In the presence of even small geometric imperfections, bifurcation does not materialize. Instead, the response of the structure approaches the bifurcated solution from below. Thus, for the column example, as the load approaches the critical value, the buckling mode is excited, the column progressively bends, and in the process, it undergoes a significant reduction in stiffness. The closer the critical load is approached, the bigger the reduction in stiffness, which once more renders the column structurally useless.

The geometries and materials used in pipelines are such that the prevalent instabilities are influenced by the plastic characteristics of the material. Material nonlinearity can complicate instabilities and can make their effect more catastrophic. The effect of inelastic material behavior will be illustrated by considering the buckling of a long tube or pipe under external pressure. It is important to note, however, that the behavior can differ with geometry and loading. We first consider a relatively thin pipe ($D/t = 40$) with a monotonically increasing stress–strain response (see Figure C.2 and nomenclature in Appendix G). The pressure–maximum radial deflection ($P - w_{max}$) response of the perfect geometry, shown in Figure 1.2, bifurcates at a pressure P_C that is low enough for the material to still be linearly elastic. The buckling mode is in the form of a uniform ovalization of the pipe cross section. Drawn in the figure is the *postbuckling* response, first assuming the material

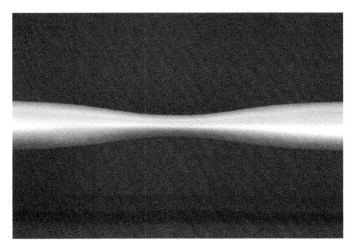

Figure 1.3   Localized collapse of a steel pipe caused by external pressure.

is linearly elastic (dashed line), and second for the elastic–plastic material (solid line). In the case of the elastic material, the ovalization grows with little additional pressure. (Note that the pressure must increase somewhat in order to ovalize the tube significantly [1.15–1.17].) By contrast, when plasticity is allowed, the combined effect of bending and membrane stresses plasticizes the material at the four most deformed locations on the cross section soon after the bifurcation. This further reduces the stiffness of the structure and causes what appears in the figure: a gradual reduction in pressure with deformation. A structure with a negative stiffness collapses on attainment of the pressure maximum. (For example, in the sea the pressure is constant at the location of buckling and consequently dynamic collapse is unavoidable.) Thus, the local pressure maximum, identified with in the figure with a caret "∧", is considered as the *collapse pressure* of the pipe (P_{CO}). An additional complication is that in the process of collapsing, the deformation switches once more, this time from uniform ovalization to *localized* ovalization. Thus, at least initially, collapse is limited to only a characteristic length of a few pipe diameters as shown in Figure 1.3. Tracking the localization requires a more complex, three-dimensional nonlinear formulation of the problem. Since the key variable of interest is the collapse pressure, such calculations are usually not necessary. In this case, the difference between P_C and P_{CO} is insignificant. In summary, inelastic action has changed the buckled structure from one with low stiffness to one that collapses catastrophically.

We next consider the corresponding solution to the more realistic problem of a pipe with small initial geometric imperfections. For external pressure loading, the most appropriate imperfection is initial ovality. Shown in Figure 1.2 are responses for initial ovality of 0.1% (see definition in Appendix G). Once again, the elastic material is drawn with a dashed line and the elastic–plastic material with a solid line. The response of the imperfect structure deviates from that of the perfect one quite early and approaches the postbuckling response from below. When the material is assumed linearly elastic, the structure loses stiffness but the response remains monotonically increasing. When inelastic action is included, the combined effect of bending and membrane stresses causes yielding at some point. This reduces the stiffness of the structure, separates the responses of the two

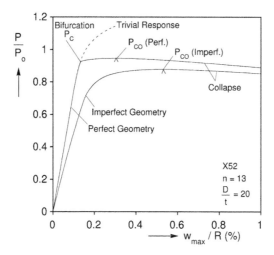

Figure 1.4   Nonlinear pressure–maximum displacement responses of a pipe under external pressure. Pipe buckles in the plastic range. Included are the perfect and imperfect structure responses.

materials, and eventually results in a pressure maximum. The pressure maximum represents the collapse pressure of the imperfect structure, which is seen to be lower than that of the perfectly circular pipe. Indeed, P_{CO} depends strongly on the value of initial ovality. Such a structure is said to be *imperfection sensitive*. In summary, when the structure stays elastic it is not imperfection sensitive. By contrast, when inelasticity sets in it becomes imperfection sensitive.

Pipes used in deep water applications, or tubulars carrying high enough internal pressure, have relatively low D/t values, so that buckling takes place at stresses that are beyond the elastic limit of the material. Such an instability is referred to as *plastic buckling*. Figure 1.4 shows results for an X52 pipe with $D/t = 20$. For this problem, the trivial response represents the pipe and material response under hydrostatic pressure loading (i.e., a biaxial state of stress). As such, it follows the nonlinearity of the stress–strain response, which starts at a pressure of about $0.8P_o$ ($P_o \equiv$ yield pressure). At pressure P_C the geometrically perfect structure bifurcates into a uniform-ovality buckling mode. Because of the material nonlinearity, linearized plastic bifurcation formulations are cast in *incremental* or instantaneous form [1.13]. The instantaneous inelastic material moduli for the state of stress at the bifurcation point are used in the calculation. The postbuckling response is followed by an appropriate elastic–plastic nonlinear analysis. An interesting difference with the elastic case is that here bifurcation occurs at an increasing pressure. The pressure increases from P_C, reaches a maximum value, P_{CO}, and then takes a downward trajectory indicating the collapse of the structure. For most plastic buckling problems (including the column), P_{CO} is higher than P_C. In this case, the difference is only about 2.3%. As was the case for the thinner pipe, collapse localizes soon after the pressure maximum (see Figure 1.3).

Results for a pipe with a uniform initial ovality of 0.1% are also included in the figure. The postbuckling response of the perfect case is approached from below. A maximum pressure is reached that represents the collapse pressure of the imperfect structure. This collapse pressure is strongly dependent on the value of initial ovality used. Thus, this structure is also categorized as imperfection sensitive.

Plastic deformations are path dependent, and as a result flow-type theories of plasticity must be used in both prebuckling and postbuckling models of plastic instabilities. However, 50 years of experience has taught us that use of flow theories with smooth yield surfaces in bifurcation calculations invariably results in overprediction of critical states. By contrast, deformation theory of plasticity, a total theory that is essentially a nonlinear elasticity theory, has been shown to yield bifurcation–buckling results that are much closer to experimental results. Plausible explanations for this unsatisfactory state of affairs are given in Section 13.3. The approach taken throughout this book is the following: a flow theory of plasticity is used in all prebuckling and postbuckling calculations, while deformation theory is used strictly for plastic bifurcation calculations using the flow theory prebuckling state of stress as input. An alternative is to use the *Corner Theory* of plasticity that combines the two [1.18]. In most practical cases this model when suitably calibrated will yield the same results as the combined models adopted in this book.

Given the size of actual pipelines, full scale testing is difficult and can be prohibitively expensive. The approach taken in most of the studies presented in this volume is to use scaling in order to reduce the size of the specimens tested. Dimensional analysis considerations easily show that a key geometric non-dimensional variable is the diameter-to-thickness ratio (D/t). Therefore, experimental results obtained using small scale tubes and pipes are transferable to actual large diameter structures with the same D/t. Regarding material properties, the elastic modulus/yield stress ratio and the hardening exponent are often key non-dimensional variables. When scaling is adopted in analyzing structural instabilities, it is also important to differentiate between ones that are elastic and those that occur in the plastic range of the material.

Parameters that may not be scaleable exist and must be addressed individually. Examples include welds and heat-affected zones, material voids and microcracks, special material anisotropies and residual stresses introduced by the manufacturing process, and a few others. Some of these will be given special attention in this book.

1.3 BUCKLE PROPAGATION IN OFFSHORE PIPELINES

The pipeline buckling phenomena discussed in this volume all lead to local collapse of the structure. Local collapse can also occur in zones of the pipeline that are weakened by denting due to impact by a foreign object or by wear caused by corrosion or erosion. Such a local compromise of the geometric integrity of an offshore pipeline can have additional consequences with far more catastrophic results. Local collapse can initiate a more global instability where, driven by the external pressure, the collapse propagates along the pipeline, often at high velocity. The result is catastrophic flattening of the structure, as shown in Figure 1.5 [1.19]. The collapse only stops if it encounters a physical obstacle that resists the flattening, or when it reaches a water depth with low enough pressure so that propagation cannot be sustained. The phenomenon is known as a *propagating buckle*. The lowest pressure that can sustain such a buckle in propagation is known as the *propagation pressure* $(P_P$ [1.15, 1.16]). For typical offshore pipelines, the propagation pressure is only a small fraction of the collapse pressure of an intact pipe $(P_P \sim 0.15–0.25 P_{CO})$. The big difference between these two critical pressures, the potential of catastrophic failure of miles of structure in a rather short time, and the potential of damage to the environment

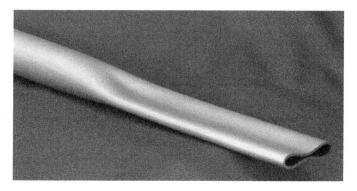

Figure 1.5   The transition between collapsed and intact sections of a pipe that developed a propagating buckle.

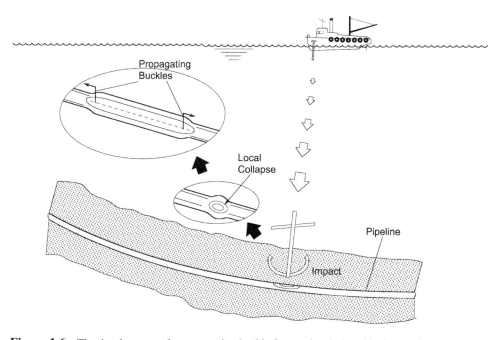

Figure 1.6   The development of a propagating buckle from a dent induced by impact by an anchor.

have made the problem a very important one that must be addressed when designing offshore pipelines.

A scenario that can lead to the initiation of a propagating buckle is shown schematically in Figure 1.6. A pipeline resting on the seabed is impacted by a foreign object and is dented. This, for example, can be an anchor or a piece of equipment dropped from a vessel operating in the area. A dent reduces the capacity of the pipe to resist external pressure leading to local collapse. The local collapse, in turn, initiates a collapse that propagates at high velocity, quickly flattening the line. Scenarios leading to local buckling and collapse of pipelines during installation are outlined in Section 2.2. Such damages can initiate propagating buckles with similarly catastrophic results.

Because the propagation pressure is so much lower than the collapse pressure, increasing the pipe wall thickness so that buckle propagation is precluded is an impractical option. Instead, pipelines are designed on the collapse pressure, and the line is protected against catastrophic collapse by a propagating buckle by the periodic addition of buckle arrestors along their length. Buckle arrestors are devices that locally stiffen the pipe circumferentially in order to prevent the downstream propagation of a buckle. Thus, in the event a propagating buckle is initiated, collapse is limited to the length of pipe between two arrestors, which typically is several hundred feet. The various aspects of buckle propagation, propagating buckle arrest and other related problems will be discussed in Volume 2 of this book series.

REFERENCES

1.1. Palmer, A.C. and King, R.A. (2004). *Subsea Pipeline Engineering*. PennWell. Tulsa, Oklahoma.
1.2. Braestrup, M.W. Ed. (2005). *Design and Installation of Marine Pipelines*. Blackwell Science. Fairfield, New Jersey.
1.3. API Specifications 5L: Specifications for Line Pipe, 43rd Edition, 2004. American Petroleum Institute.
1.4. API Recommended Practice 1111: Design, Construction, Operation and Maintenance of Offshore Hydrocarbon Pipelines (Limit State Design), 3rd Edition, July 1999.
1.5. Offshore Standard OS-F101: Submarine Pipeline Systems, January 2000, Det Norske Veritas.
1.6. International Standard ISO 13623: Petroleum and Natural Gas Industries – Pipeline Transportation Systems, 1st Edition 2000-04-15.
1.7. British Standard PD 8010-2:2004, Code of Practice for Pipelines. Part 2. Subsea Pipelines.
1.8. Timoshenko, S.P. and Gere, J.M. (1961). *Theory of Elastic Stability*, 2nd Edition, McGraw-Hill. New York.
1.9. Brush, D.O. and Almroth, B.O. (1975). *Buckling of Bars, Plates and Shells*. McGraw-Hill. New York.
1.10. Bazant, Z.P. and Cedonin, L. (1991). *Stability of Structures*. Oxford University Press. New York.
1.11. Singer, J., Arbocz, J. and Weller, T. (1998). *Buckling Experiments: Experimental Methods in Buckling of Thin-Walled Structures. Vol. 1. Basic Concepts, Columns, Beams and Plates*. John Wiley & Sons, Ltd. Chichester, UK.
1.12. Singer, J., Arbocz, J. and Weller, T. (2002). *Buckling Experiments: Experimental Methods in Buckling of Thin-Walled Structures. Vol. 2. Shell, Built-up Structures, Composites and Additional Topics*. John Wiley & Sons, Ltd. New York.
1.13. Hutchinson, J.W. (1974). Plastic buckling. In *Advances in Applied Mechanics*, Vol. 14, Ed. C.S. Yih. Academic Press, New York, pp. 67–144.
1.14. Nguyen, Q.-S. (2000). *Stability and Nonlinear Solid Mechanics*. John Wiley & Sons, Ltd. Chichester, UK.
1.15. Kyriakides, S., Yeh, M.K. and Roach, D. (1984). On the determination of the propagation pressure of long circular tubes. *ASME J. Pressure Vessel Technol.* **106**, 150–159.
1.16. Dyau, J.-Y. and Kyriakides, S. (1993). On the propagation pressure of long cylindrical shells under external pressure. *Intl J. Mech. Sci.* **35**, 675–713.
1.17. Dyau, J-Y. and Kyriakides, S. (1993). On the localization and collapse in cylindrical shells under external pressure. *Intl J. Solid. Struct.* **30**, 463–482.
1.18. Christoffersen, J. and Hutchinson, J.W. (1979). A class of phenomenological corner theories of plasticity. *J. Mech. Phys. Solid.* **27**, 465–487.
1.19. Kyriakides, S. (1993). Propagating instabilities in structures. In *Advances in Applied Mechanics*, Vol. 30, Ed. J.W. Hutchinson and T.Y. Wu. Academic Press, Boston, MA, pp. 67–189.

2
Offshore Facilities and Pipeline Installation Methods

Offshore operations associated with oil and gas exploration and production have blossomed during the last 30 years. The needs of such operations have generated a unique new engineering discipline, offshore engineering, that has pulled ideas and people from most of the traditional engineering disciplines such as petroleum engineering, ocean engineering, naval and marine engineering, civil engineering, mechanical engineering, and aeronautical engineering. This new and vibrant engineering discipline is responsible for the design, construction and operation of a plethora of new structures and vessels used in offshore operations. These include exploration vessels, drilling ships, several types of offshore platforms, well completion tubulars and equipment, flowlines, pipelines and risers as well as a variety of support vessels ranging from barges with heavy cranes to J-lay and pipe reeling ships. Section 2.1 introduces some of the major types of offshore structures used in production, as they are either the termination points of flowlines and risers that bring hydrocarbons from subsea wells, or are the starting points of export pipelines that take processed hydrocarbons to other offshore facilities or to shore. The interaction of these tubular structures with platforms and other production facilities results in challenging loading scenarios that will be introduced. Each type of production facility is first described in general, followed by a more detailed description of a representative facility of the class. In addition to describing platforms, some of the associated subsea and topside facilities, equipment, and structures that comprise an offshore development are introduced.

The installation of pipelines often generates complex loading conditions that can influence their design. The major methods of installing offshore pipelines are outlined in Section 2.2. Types of pipelay vessels used in each method are described in general and representative examples of each class are portrayed in more detail. The selection of the vessels is based more on their familiarity to the authors and ease of access of information about them rather than their prominence in the industry. The loads generated on the pipeline by each installation method and corresponding possible limit states are briefly introduced. A newcomer to offshore and deepwater oil and gas exploration and production should find the descriptive book of Leffler *et al.* on the subject [2.1] a useful starting point for acquiring a general understanding of such operations as well as a resource for the terminology used in this industry (see also [2.2, 2.3]).

2.1 OFFSHORE PLATFORMS AND RELATED PRODUCTION SYSTEMS

The development of offshore fields usually involves the construction of a structure at a central location that acts as the gathering and control center for several wells that can be a number of miles away. Such a platform plays several roles. For example, with the advent of directional drilling, modern platforms are also used to drill into a field. The drill holes spread out radially from the platform, often extending several miles to reach a particular location in the reservoir. The platform houses the hydraulic equipment that provides power to operate various valves located in *trees* sitting on top of wells on the sea floor. It also holds hydraulic equipment that controls the well flow and, when required, high pressure pumps used to pump water or gas back into the formation to enhance oil recovery. High pressure *flowlines* bring the hydrocarbons from the wells to the platform. Hydrocarbons come out of the ground as a high pressure mixture of oil, gas, water and solids. Platform-based equipment processes the mixture lowering the pressure, separating oil and gas, and extracting water, sand and other substances. Processing must be conducted in a carefully controlled manner. For example, lowering the pressure of a gas-rich flow by choking drops its temperature, turning the gas into a liquid, or worse, into a snow-like *hydrate*. Hydrates can plug a flowline, as they consist of methane trapped in icy crystals. Accordingly, provisions must be taken to heat up the flow at the point of choking. Processed crude oil and gas leave the platform via *export lines*, which connect it either to a shore terminal or to another offshore facility where they are loaded onto tankers. There exist three categories of such production facilities: platforms that are fixed to the seabed, moored and tethered floating systems, and subsea systems [2.1–2.4]. Each category is outlined in the following. The major types of platforms that operate today in water depths 1000 ft (300 m) and beyond can be seen in Figure 2.1.

2.1.1 Fixed Platforms

Although there are many different designs of fixed platforms, they can be placed in three broad categories: *Gravity platforms* are usually constructed out of reinforced concrete, and rely largely on their massive weight to keep them fixed on the seabed. Traditional frame-type *jacket platforms* are constructed out of steel tubular components and are kept in place by piles driven deep into the sea floor. *Compliant towers* are similarly made of tubular components but have a narrower base and typically a nearly uniform cross section. Rather than the relatively stiff design of jackets, compliant towers are designed to deform more but to have longer resonant periods than storm wave periods.

a. Gravity Platforms

Gravity platforms are made out of reinforced concrete and are usually quite large structures. Unlike fixed steel jacket platforms that are kept in place by massive piles, many gravity platforms simply rest on the sea floor, stabilized by their own weight. When the foundation is soft, they are secured to the seabed by cookie-cutter skirts that penetrate the sea floor. A number of large hydrocarbon storage cells are built into their base, while three or four tubular columns rise above the sea surface to support a deck. Figure 2.2 shows a schematic of Statoil's *Gullfaks C* gravity platform, which was installed in the North

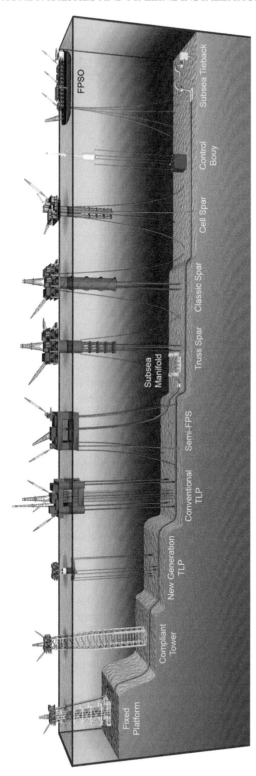

Figure 2.1 Major offshore platform concepts used in water depths beyond 1,000 ft (300 m) (reproduced with permission from "2006 Deepwater Solutions and Records for Concept Selection," courtesy *Offshore Magazine and Mustang Engineering*).

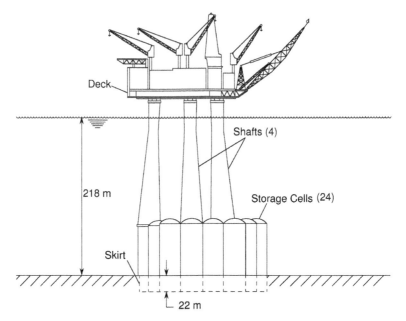

Figure 2.2   Schematic showing a four-leg gravity platform (after the Gullfaks C platform [2.5, 2.6]).

Sea in a water depth of 715 ft (218 m) in 1989, at a total cost of $2 BIL [2.5, 2.6]. The platform has a submerged weight of 500,000 t, and its base covers an area of 172,000 ft^2 (16,000 m^2). It was installed on a rather soft soil foundation, and had to be stabilized by the addition of a skirt that is 72 ft (22 m) deep. All outgoing and incoming piping passes through the four hollow shafts. Because of their large scale, gravity platforms become impractical for water depths beyond 1,000 ft (300 m) due to excessive material needs.

b. Jacket Platforms

A *Jacket* is a frustum-shaped frame structure made of tubular steel members as shown schematically in Figure 2.3 (see also Figure 2.1). This construction provides the lowest resistance to waves and currents, reducing the loads on the structure. Most commonly they have four legs, but jackets with three, six, or eight legs also exist. The jacket is secured to the sea floor by tubular piles that run either through the legs or, for larger structures, through sleeves connected to the jacket legs. In the former case, the piles run to the top of the legs where they are connected to them by plates. In the latter case, the pile-sleeve annulus is grouted. Larger platforms require several large diameter piles at each leg that can run as deep as 400 ft (120 m) into the sea floor. The design and installation of such piles is a special foundation engineering discipline [2.4].

Shell's *Bullwinkle* platform is an illustrative example of a major jacket structure. It was installed in 1988 in the Green Canyon Block 65 field, about 160 mi SSW of New Orleans [2.7–2.9]. The water depth is 1,353 ft (412 m), while the structure overall height is 1,615 ft (492 m), making it the tallest offshore platform at the time of installation. The jacket base is 408 × 487 ft (124 × 148 m), its top is 140 × 162 ft (42.7 × 49.4 m) and the

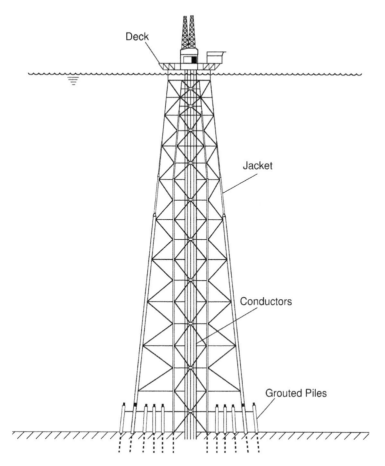

Figure 2.3 Simplified schematic of a jacket fixed platform (after the Bullwinkle platform [2.7, 2.8]).

deck 185 × 205 ft (56.4 × 62.5 m). The jacket has a weight of 50,000 ton while the deck weight exceeds 2,000 ton. The platform is secured to the sea floor by twenty-eight 84-inch piles driven into the seabed to penetration depths of up to 437 ft (133 m). The piles pass through sleeves connected to each corner of the jacket and are grouted to them once they reach full penetration [2.4]. The overall cost of the project was $500 MIL.

Slots for 60 wells were provided at the base of the Bullwinkle platform. The wells are connected to the deck using *conductors* (Figure 2.3). These are pipes that often penetrate the sea floor to a certain depth. They contain the drilling tools during the drilling phase, and on completion, the production tubing. In this case the conductors have diameters of 26 in. During production, in addition to containing the production tubes (*risers*), they also carry part of the weight of the tubing and casing that is hung in the well (see Figure 7.1).

Flowlines and export pipelines are often connected to the decks of jacket platforms by pulling them through a preinstalled tube, known as a *J-tube*, that is secured on the side of the jacket, as shown in Figure 2.4. The diameter of the J-tube is somewhat larger than that of the pipeline and has a characteristic elbow at the bottom to facilitate the incoming riser. The pipeline is first laid on the sea floor, is then attached to a wire rope and pulled

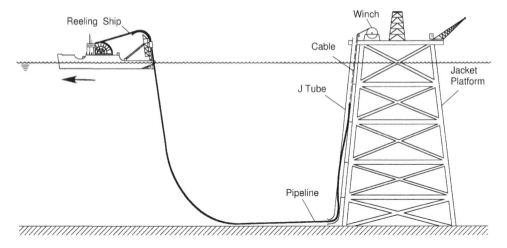

Figure 2.4   Schematic representation of a J-tube riser installed by reeling.

up the J-tube with a topside winch [2.10]. The procedure is best facilitated when the line is pulled directly from the lay vessel, as shown in the figure, because this minimizes the length of pipe that has to be dragged on the seabed. As the pipe passes through the J-tube it undergoes plastic bending in the presence of some tension at nearly the highest operational external pressure [2.11] (e.g., the J-tube radius on the Cognac platform was 140 ft (42.7 m) resulting in a maximum bending strain in the 12.75 in pipeline of 0.38%). In view of these relatively high combined loads, measures must be taken in the design of such a pipeline to ensure that it does not buckle or collapse [2.12]. The line straightens as it moves up the tube, and the pressure gradually reduces.

c. Compliant Towers

The cost of a jacket increases significantly with depth due to the increase in the size of the base and the required increase in the wall thickness of the tubular components to resist the external pressure. Consequently, jackets are generally considered uneconomical for water depths beyond about 1,500 ft (460 m). The preferred alternative is the *compliant tower*, a slender, nearly uniform cross section frame structure constructed also of steel tubular members (see Figure 2.1). Figure 2.5 shows a schematic of the Amerada Hess/Oryx *Baldpate* compliant tower that was installed in 1998 in the Gulf of Mexico in 1,648 ft (502 m) of water (1,900 ft – 579 m – tall from the tip of the flare to the sea floor) [2.13, 2.14]. The tower comprises a 351 ft (107 m) tall base section that at the sea floor has a cross section of 140 × 140 ft (42.6 m) that tapers to the tower section of 90 × 90 ft (27.4 m). It carries a 180 × 140 ft (54.4 × 42.6 m) deck, which, when fully loaded with equipment and a drilling rig, weighs 9,800 ton. The structure is fixed using three 83 in (2,108 mm) diameter piles with a penetration depth of 428 ft (130 m) at each leg. Additional flexibility is provided to the tower by allowing an articulation point at a depth of 1,143 ft (348 m). The combined effect of the top-heavy structure, its slender construction and the added flexibility due to articulation make the sway natural period over 30 s compared to hurricane sea periods of 15 s (at 30 s periods the spectral wave energy is negligible [2.15]). Thus, the waves tend to pass through the structure with relatively small reaction. Compliant towers

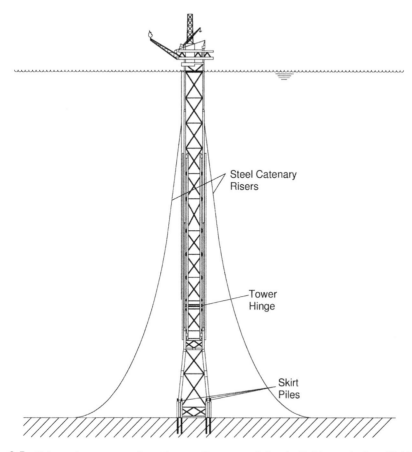

Figure 2.5   Schematic representation of a compliant tower (after the Baldpate platform [2.13, 2.14]).

are competitive as a design concept from about 1,000 ft (305 m) down to 3,000 ft (914 m). In shallower waters, they tend to be too stiff with lower natural periods, and the concept does not work. In waters deeper than 3,000 ft (914 m), the wall thickness of the tubular members required to resist the external pressure tends to be too high for the concept to be practical.

The Baldpate tower has a nine-slot drilling template at its base, which was used to drill several wells. The wells connect to the platform with 4 in high pressure (10,000 psi – 690 bar) tubes. Processed oil and gas are sent to another offshore location via 16 and 12 in export pipelines, respectively, that operate at about 2,000 psi (138 bar). The two pipelines connect to the platform through *steel catenary risers* as shown in Figure 2.5 [2.16]. In order to minimize the effects of the tower motion on the risers, they connect to it at about 300 ft (91 m) below sea level. A catenary riser is a simple continuation of the pipeline up to the platform deck. It naturally acquires the catenary shape seen in the figure. In the lower bent region, the riser is under the combined effect of external (when empty) or internal pressure (during operation) and bending. Consequently, it must be designed against buckling and collapse as well as against burst. An additional concern is the effect that cyclic loading, induced to the riser by the tower motion, will have on the fatigue life of the structure overall and in particular that of its girth welds.

2.1.2 Floating and Tethered Platforms

The maximum water depth at which compliant towers are viewed as practical is 3,000 ft (914 m). Thus, as industry moved to deeper waters, new concepts involving floating systems such as *Tension Leg Platforms*, several types of *Floating Production Systems*, *Spars* and *Floating Production, Storage and Offloading* facilities were developed (see Figure 2.1).

a. Tension Leg Platforms (TLPs)

Tension leg platforms consist of floating hulls usually made of buoyant columns and pontoons. Vertical tendons, usually steel pipes that are anchored to the sea floor, hold the hull below its natural floatation level, keeping the tendons in tension and the hull in place. TLPs were first used in shallow waters in the North Sea in the mid-1980s (e.g., Hutton TLP in 482 ft – 147 m, installed in 1984). Since then, however, their use has gradually expanded to moderately deep and deep waters. Today they are considered a viable option in water depths of up to 5,000 ft (1,525 m).

Shell's *Auger* TLP, operating in a water depth of 2,860 ft (872 m), is a prime example of this type of facility [2.17, 2.18]. It was installed in 1994 in the Garden Banks area 214 mi SW of New Orleans at a total project cost of $1.2 BIL. Its hull consists of four 74 ft (22.6 m) diameter circular cylindrical columns connected with four 28 × 35 ft (8.5 × 10.7 m) box pontoons (centers of columns 264 × 214 ft (80.5 × 65.2 m), see Figure 2.6). A 300 × 300 ft (91 m^2) deck rests on the hull and supports a drilling rig, production facilities, accommodation modules, and more. The hull is held in place by twelve tendons that are approximately 2,780 ft long (847 m). Three tendons connect to each column at the top and to special foundation templates directly below on the sea floor through flex connectors. Each template in turn transfers load to four 72-inch piles driven to a penetration of 374 ft (113 m).

The tendons consist of 26-inch, X60 grade *UOE* pipe with a wall thickness of 1.3 in (33 mm) [2.19]. Each consists of 13 sections connected by mechanical connectors. The design of such tendons is typically based first on fatigue of the welds (more than 10^9 wave-induced load cycles) and second on the capacity of the lower sections to resist collapse due to the external pressure and the time-varying axial tension. The evaluation of the collapse performance of UOE pipe for this project uncovered for the first time that this manufacturing process tends to degrade the compressive yield stress in the circumferential direction [2.20]. As a direct consequence, the collapse pressure of UOE pipe is lower than that of seamless pipe of the same geometry and grade. This subject is discussed in detail in Chapter 5.

The project made provision for 32 platform-drilled wells arranged in an oval pattern within the footprint of the TLP [2.21]. A unique mooring system was used to move the drilling rig, which is fixed on the deck directly above each well. Once completed, the wells were connected to the deck by 9.625-inch *production risers* made of seamless pipe. Each is supported at the well bay by a special tensioning system that allows some differential movement between the TLP and the riser while maintaining the tension nearly constant.

Such production risers are designed to withstand combined tension, bending and external pressure when not operational and internal pressure during production.

Oil and gas are exported to other offshore locations by two X52, 12.75 in diameter 0.562 in (14.3 mm) wall seamless pipelines [2.22]. The oil line is 72 mi (116 km) long and the gas line 36 mi (58 km). At the platform, both operate at a pressure of 2,160 psi (149 bar). Three factors played a role in deciding the pipe wall thickness: the external pressure, the operating internal pressure and the need for sufficient pipeline submerged weight to provide sea bottom stability of the line. The maximum water depth was such that the external pressure exceeded the propagation pressure of the pipeline. Consequently, *integral buckle arrestors* were installed at intervals of 1,200 ft (366 m) to all pipeline sections deeper than 400 ft (122 m) (see [2.23, 2.24]). The shallower parts of the lines (52 mi – 84 km – of the oil and 16 mi – 26 km – of the gas lines) were installed by the S-lay method. The deeper 20 mi (32 km) sections of both were installed by the J-lay method, using the J-lay tower and associated facilities (newly installed at the time) on McDermott's Derrick Barge No. 50 [2.25] (see also Section 2.2.2). The pipeline sections installed by J-lay had buckle arrestors/collars every four joints. The collars were used to hold the line at the barge while the next section was welded to it. In addition, they were designed to act as lifting points in the event of a propagating buckle, to retrieve and repair the damaged section between two arrestors.

The two pipelines connect to the TLP by catenary risers (Figure 2.6). The risers are of the same grade steel and diameter as the pipelines, but their wall thicknesses were increased to 0.688 in (17.5 mm). The design [2.26] was based on the same design criteria as those of the pipelines, plus the bending experienced near the sea floor and girth weld fatigue considerations. The risers are connected to one of the pontoons via a flexible joint that allows $\pm 14°$ rotation in order to accommodate motion of the hull in hurricane weather conditions.

b. Spar Platforms

A spar typically consists of a long circular cylindrical hull that floats vertically and carries a dry deck on top [2.27] (Figure 2.1). Typical lengths can reach 700 ft (213 m) and diameters range between 60 and 150 ft (18–45 m). Spars are held in place by mooring lines anchored or piled to the sea floor. The spar is usually ballasted so that its center of gravity is well below the center of buoyancy, making it very stable to toppling motions. Additionally, its huge mass results in very little vertical motion. Flow induced vibrations due to currents are reduced by the addition of strakes (spoilers) to the spar. The hull geometry can vary by the addition of a truss structure in the lower part (*truss spar*) or the replacement of the single cylinder by a cluster of smaller diameter cylinders that are easier to construct (*cell spar*) [2.28]. Present use has reached just under 6,000 ft (1,830 m), but the concept is relatively insensitive to water depth and has been proposed for use in waters as deep as 10,000 ft (3,050 m) [2.29].

The *Hoover–Diana* project is one of the largest and most complex operated from a spar platform [2.27]. The platform serves two fields, Hoover and Diana about 16 mi (25.7 km) apart, located approximately 160 mi (257 km) south of Galveston, Texas. Co-owned by ExxonMobil and BP, it was installed in the winter of 1999/2000 in the Alaminos Canyon above the Hoover field in a water depth of 4,850 ft (1,478 m) (see Figure 2.7). The total project cost was approximately $1.6 BIL. The hull is 705 ft (215 m) long, has a diameter of 120 ft (36.6 m) and weighs 35,000 tons (Figure 2.8). It is held in place by 12 mooring

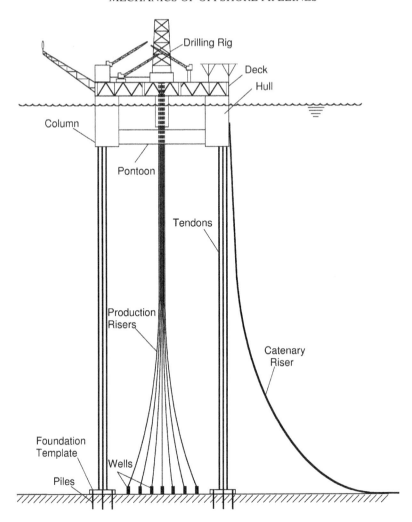

Figure 2.6   Schematic of a tension leg platform [2.17, 2.18].

lines anchored to the sea floor by *suction piles* [2.30]. These are large diameter (~20 ft–6 m) cylindrical structures that are usually only a few diameters long. They are closed at the top and are installed by self-weight, augmented by lowering the pressure of the enclosed cavity. The ones used for the Hoover platform were 21 ft (6.4 m) in diameter and 105 ft (32 m) long, constituting an industry record at the time of installation. The mooring lines consist of 6,600 ft (2,011 m) of wire rope and 1,000 ft (300 m) of chain. A three-deck multifunctional structure sits atop of the spar. The top deck supports a drilling rig and its facilities, the second deck houses production facilities and the third living quarter modules, etc.

The Diana field is operated from five wells that were pre-drilled by a deepwater drilling ship. The wells are serviced by manifolds that connect to the platform through two 10 in and one 6 in flowlines (total length of about 50 mi, see Figure 2.7). Steel catenary risers terminate the flowlines at the platform, where the hydrocarbons are processed and treated.

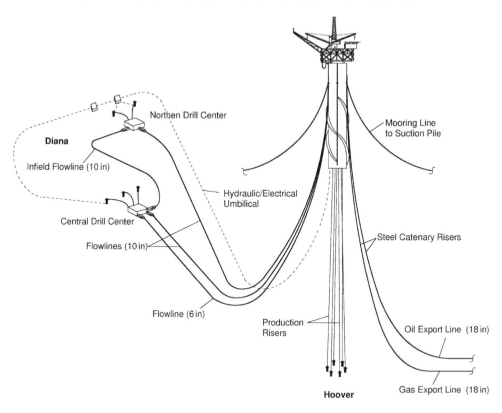

Figure 2.7 The Hoover–Diana development and its flowline and pipeline systems (figure not to scale) [2.27].

Despite the significant water depth (4,600–4,900 ft), the flowline sections in the vicinity of the drill centers were installed by the S-lay method using the vessel Lorelay. The main parts of the three flowlines that connect the drilling centers to the platform were installed by J-lay using Saipem's S-7000, as were their catenary risers. Unlike most other J-lay facilities, its J-lay tower uses track-type tensioners to hold the line while a quad-joint (160 ft – 55 m) is welded to it. Consequently collars are not required (see Section 2.2.2). Thus integral buckle arrestors used were installed at 640 ft (220 m) spacing [2.23, 2.24]. Internal pressure governed the design of the flowlines and as a result the 10 in lines had D/t of 17.2 and the 6 in 13.3. In both cases the catenary risers were made even thicker.

The Hoover field has six wells that were drilled from the platform (horizontal open-hole completion [2.31]) and connect to the production deck via production risers as shown in Figure 2.7 [2.32]. Two 18 in export pipelines, one for gas and the other for oil, take the processed products to other offshore locations (Figure 2.7). The oil pipeline has an initial section 33 mi (53 km) long where it connects to a 42-mile (68 km) 20 in line that terminates at the Galveston block. The gas line is 4 mi (6 km) long, connecting to an 82-mile (132 km) 20 in line that goes to High Island Offshore System. The pipelines were installed by S-lay, while the terminations at the platform were installed by J-lay. Both have integral buckle arrestors at 640 ft (220 m) spacing. These pipelines had variable wall thickness with D/t values ranging from 22.2 in deeper water to 35.6 in the shallower parts.

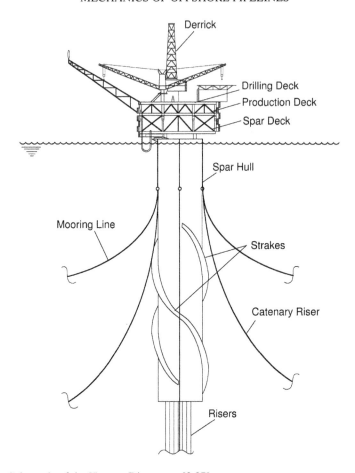

Figure 2.8 Schematic of the Hoover–Diana spar [2.27].

c. Semi-Submersible Floating Production Systems (FPSs)

Another floating platform concept is the *Floating Production System* (see Figure 2.1). Most commonly an FPS consists of a TLP-like buoyant hull that stays moored in place by long catenary-shaped mooring lines, as shown in Figure 2.9. The mooring lines usually consist of large diameter spiral strand wire with heavy chain sections close to the mudline. They connect to piles located a distance 1.5–2 times the water depth away from the hull. The hull usually consists of pontoons and columns and carries a deck with the same general structure and equipment as spars and TLPs. An FPS can serve one particular field or act as gathering point (*host*) for several fields in a larger area. Petrobras led the development of FPSs, using them as hosts in a variety of moderately deep and deepwater fields in the Campos and other basins offshore Brazil [2.1]. Today they are considered to be applicable in ultra-deep waters.

The *Na Kika* development, located in the Mississippi Canyon in the Gulf of Mexico approximately 140 mi (225 km) SE of New Orleans, uses an FPS as the host production facility. Installed in 2003 in a water depth of 6,350 ft (1,936 m), the platform serves six fields with a total of 12 wells [2.33, 2.34]. The fields, which produce both oil and

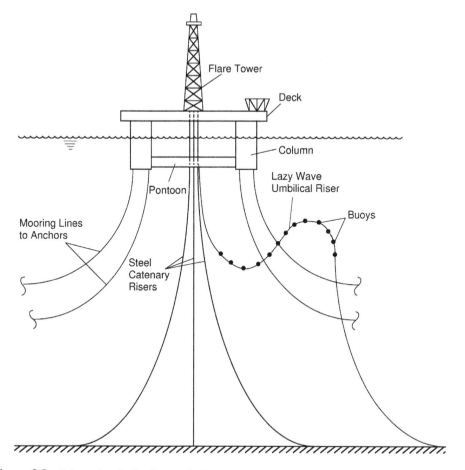

Figure 2.9   Schematic of a floating production system.

gas, were evaluated as being small to medium size, and consequently none of them could have supported a major production facility on their own. Five of the fields are spread over a distance of 25 mi (40 km) in water depths that range from 5,800 to 7,000 ft (1,768–2,134 m). The platform is located approximately mid-way between the two furthest wells. The wells connect to the platform with three flowline loops. The FPS and the five fields are co-owned by Shell and BP, who together invested $1.4 BIL towards this development. The wells were drilled using deepwater drilling facilities and are serviced via subsea systems. A sixth gas field, the Coulomb field, located 27 mi (43.5 km) from the platform is co-owned by Shell and Petrobras and was developed independently. It connects to the FPS with a single flowline.

The Na Kika semi-submersible hull is a standard four-column structure connecting to horizontal pontoons at the bottom and to the deck at the top (Figure 2.9). The columns are 56.4 ft (17.2 m) square and are set 210 ft (64 m) apart. The pontoons have a 40.7W × 34.6H ft (12.4 × 10.5 m) cross section [2.35]. The steel hull weighs 20,000 ton and has a displacement of 64,000 ton. It is kept on location by 16 semi-taut catenary mooring lines that connect to each of the four columns. The overall lengths of the mooring

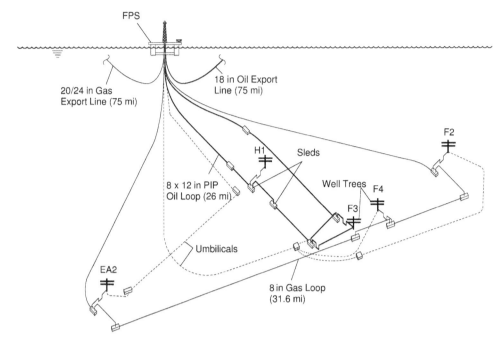

Figure 2.10   Layout of the Na Kika South Side fields [2.37].

lines range between 9,575 and 10,950 ft (2,918–3,337 m) [2.36]. The main parts consist
of 4.5 in (114 mm) steel spiral strand wire, but connect to 4.5 in chain sections that are
about 1,200 ft (365 m) long at the lower end and 450 ft (137 m) long at the platform level.
The mooring lines are anchored to 14 ft (4.3 m) diameter, 78 ft (23.8 m) long suction piles.
They connect to the piles at padeyes located 54 ft (16.5 m) below the seabed.

The project involves 12 deepwater wells spread over a relatively large area. Some
wells produce mainly oil at relatively low temperatures (~130°F) and pressures, while
others produce mainly gas. These conditions placed unusually stringent design demands
on the subsea flowline system that brings the oil and gas to the FPS [2.37]. Three loops
connect the wells in the Na Kika development. Two of the loops are insulated pipe-in-
pipe systems. Insulation is necessary because the flowline legs are 10–15 mi long and
the oil will otherwise cool down, causing paraffin to solidify on the pipe inner wall and
eventually clog the line [2.1]. One 8-inch uninsulated loop carries gas from three wells.
A separate 8-inch flowline brings gas from the two wells of the Coulomb field 27 mi
(43.5 km) away. All flowlines terminate at the FPS with catenary steel risers.

The layout of the wells, flowlines, umbilicals and sleds of one of the developments
(South Side) of the Na Kika project is shown in Figure 2.10 [2.37]. The oil loop is a pipe-
in-pipe system that consists of an 8-inch inner pipe that carries the hydrocarbons and a
12-inch outer one that insulates it. It was installed by reeling using the Deep Blue vessel
(see Section 2.2.3). Typically, polymeric centralizers, placed periodically in the annulus
between them, keep the two lines concentric. The 26 mi (42 km) long loop connects wells
H1 and F3. Each well is connected with a *jumper* to *sleds*. A sled is a structure that
is anchored to the sea floor and acts as the starting point for a flowline. A jumper is

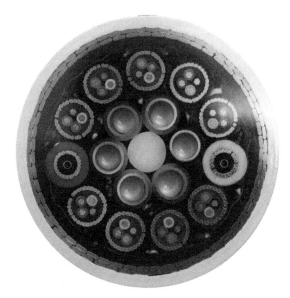

Figure 2.11 Cross section of an umbilical (courtesy Nexans).

a relatively short section of prefabricated rigid or flexible pipe that connects the well to the sled. It is usually installed by a tethered drop from a barge guided by an ROV. Separating the flowline from the well allows independent installation of wellheads, trees and flowlines. In other projects, the sled is replaced by a *manifold* that receives flows from several wells, mixes them, and sends them on to the host by a single flow line.

The design flow pressure of the oil loop is 7,300 psi (503 bar) at the wellhead and 5,900 psi (407 bar) at the platform. The internal pressure, including its variability up the riser, governs the wall thickness of the inner pipe. By contrast, the outer pipe design is governed mainly by external pressure and combined pressure and bending experienced in the sagbend of the riser that can lead to collapse. The line is protected against the potential occurrence of a propagating buckle by periodic placement of clamped buckle arrestors [2.38] installed at 800 ft (244 m) spacing. The 8-inch gas loop connects fields F2, F4 and EA4 to the host platform (Figure 2.10). It is 31.6 mi long (50.9 km) and operates at 8,400 psi (579 bar) at the wellhead and 7,550 psi (521 bar) at the surface. In this case, the design was governed by internal pressure considerations. It was again installed by reeling and has clamped buckle arrestors with the same spacing as the oil loop 800 ft (244 m).

The two flowline loops terminate at the top of the platform with steel catenary risers that were also installed by reeling. The risers connect at flex joints that allow the ends to rotate when the FPS moves in a storm. Fairings were placed over part of the length of the risers to suppress vortex-induced vibrations.

Umbilicals are conduits that connect facilities on the host platform with subsea facilities, in this case trees and sleds. They provide electrical, hydraulic, chemical injection and fiber optic connections for operating the subsea facilities. The cross section of an umbilical, encompassing several carbon steel and duplex stainless steel service tubes and cables covered by a polyethylene sleeve is shown in Figure 2.11. Typical overall diameters are in the range of 4.33–5.94 in (110–150 mm); 77 mi (124 km) of umbilicals were used in the Na Kika development [2.39]. They were installed off a large reel very much like a pipeline and connect to the FPS directly. Projected storm motions of the host were found to result in unacceptable compression in the sagbend as well as in cyclic bending, which would lower the design fatigue life of the umbilicals. This was remedied by the use of a *lazy wave catenary* near the platform to isolate the lower part of the catenary from the topside motions (see Figure 2.9). Buoys are placed on the conduit at regular intervals providing the required lift to generate the "lazy wave" shape. This and other similar techniques (e.g., *steep wave*, *lazy S*, *steep S* [2.1]) are also used for the same purpose on steel catenary risers.

A third oil flowline loop, with a total length of 22 mi (35 km), connects the remaining five wells of the Na Kika development. It consists of a 10×16 in pipe-in-pipe system. Because of its size and weight, this line was installed by J-lay using the vessel *Balder*. In this installation method, the outer pipe is held at the vessel in a vertical position while four pre-welded joints are welded to both the inner and outer pipes (see Section 2.2.2). During this operation it must be ensured that the inner pipe does not slide down the outer one, and buckle spirally. This was done by adding shear plugs at regular intervals in the annulus between the two pipes. During installation the pipe hangs from the vessel from collars that are attached to the end of each quad-joint. The collars also serve as buckle arrestors.

Processed oil and gas are respectively exported via 18 and 20/24 in export pipelines that leave the platform via catenary risers (Figure 2.10). The 75 mi (120 km) long oil line goes to Shell's Main Pass 69 in shallow water. Its maximum capacity is 140 Mbbl/d at a discharge pressure of 2,400 psi (166 bar) that drops to 100 psi (6.9 bar) at the terminal. The pipeline design is unique in that its deeper part was installed while flooded using the deepwater S-lay vessel *Solitaire* [2.40, 2.41]. Wet installation is very rare, especially in deep waters because, among other reasons, of the very high tension that must be applied at the lay vessel. In this case, the tension exceeded 900 kips (410 t), causing some plastic deformation to the pipeline on the special deepwater stinger of the Solitaire [2.42]. The wet lay option was exercised because the pipeline will remain filled with crude during its entire lifetime. Consequently, the loading is the difference between the external and internal pressure, which enabled the use of a 0.562 in (14.3 mm) wall thickness in the deeper part. A conventional design that would have required the pipeline to withstand the external pressure at the maximum water depth would have required a wall thickness of 0.840 in (21.3 mm). Slip-on type buckle arrestors were installed on this pipeline at 800 ft (244 m) intervals (see [2.24, 2.43]). The part of the pipeline in less than 800 ft (244 m) of water has a wall thickness of 0.500 in (12.7 mm), and was installed empty by conventional S-lay using a different vessel. The pipeline connects to the FPS with a steel catenary riser that was installed by J-lay.

The gas export pipeline is part of the Okeanos pipeline, which is part of the larger Mardi Gras transportation system (see Figure 2.12). The pipeline is nearly 75 mi long (120 km), with all but 5 mi of it being 24 in, and the section closer to the Na Kika host being 20 in. The 20 in was preferred as it simplified the design of the catenary riser with which it connects to the FPS. This pipeline was also installed by deepwater S-lay, but because of its size it was installed empty in order to keep the tension within the capacity of the *Solitaire*. Consequently, its design is governed by external pressure collapse considerations. Integral buckle arrestors were installed at 800 ft (244 m) intervals [2.23, 2.24]. High bending and tension at the stinger increased the risk of accidental occurrence of a buckle and possibly a *wet buckle* near the top (a wet buckle is one that causes fracturing accompanied by flooding of the pipeline). The pipeline could not be allowed to flood, as the tensioner capacity would have been exceeded, resulting in loss of the line. This risk was mitigated

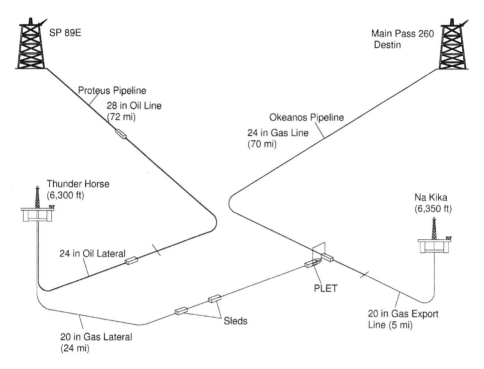

Figure 2.12 Mississippi Canyon pipeline system part of Mardi Gras development [2.95].

by hanging a SMART™ plug inside the pipeline at a water depth of 650 ft (200 m) [2.37]. The tethered plug was kept stationary during installation and would have sealed the pipeline in case flooding occurred above it. As it turned out the pipeline did not flood and this precautionary step did not have to be activated. As with the oil line, this pipeline terminated at the FPS with a steel catenary riser that was installed by J-lay using the vessel *Balder* (see Section 2.2.2).

In summary, the Na Kika development had about 150 mi (240 km) of export pipelines, 100 mi (160 km) of flow lines, 11 catenary risers and several sleds and jumpers. The total cost of these facilities exceeded $350 MIL (not including the Coulomb flowline), which represents about 25% of the overall project cost [2.33, 2.37]. This provides a sense of the importance of such tubular structures in the development of deepwater projects.

d. Floating Production, Storage and Offloading Facilities (FPSOs)

FPSOs are large tanker-like ships that are either custom-built or converted to receive, process and store oil production from several wells in a development [2.1–2.3]. They do not store gas, and differ from other platforms in that they do not usually house drilling rigs and related facilities. Produced gas is either sent back down for reinjection into the reservoir or sent to another gas-receiving host facility. Some FPSOs can store up to 2 MMbbl of oil, which is collected by shuttle tankers that visit at least once a week. FPSOs are typically used in remote areas where no pipeline infrastructure exists or is possible to develop, in areas where local conditions make connection to a coastal facility risky (offshore West Africa has the biggest concentration of FPSOs), and in areas where harsh

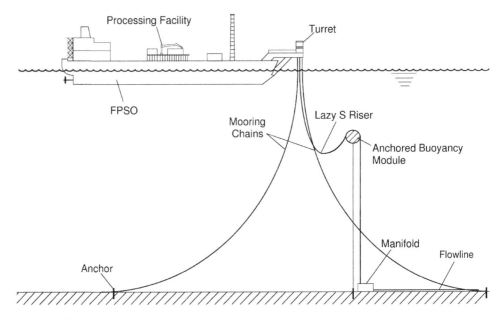

Figure 2.13 Schematic of a typical floating, production, storage and offloading system.

weather conditions do not facilitate land connection (e.g., Newfoundland [2.44], northern North Sea [2.1]).

A problem with keeping a long ship stationary is that weather conditions often require it to weathervane into the prevailing wind twisting risers connected to it in the process. Twisting of the risers is usually avoided by connecting them to a revolving turret. The turret is built either into the hull [2.45] or is cantilevered off the bow of the ship as shown in Figure 2.13 [2.46–2.48]. It connects to mooring lines anchored to the seafloor and becomes a stationary point around which the ship can weathervane. The turret contains swivel stacks that allow electrical and fluid flow continuity during the rotation of the ship. The oil is processed and stored on board, while separated water and gas are sent back to the reservoir. Some turret designs allow disconnection from their lower part and abandonment when severe weather conditions dictate departure of the FPSO [2.45]. The turret mooring lines are similar to those described in the two previous sections. The risers can be simple steel catenary risers or, when excessive motions are expected, systems like the *lazy S riser* shown in Figure 2.13. This type of riser isolates the main part of the suspended structure from topside motions (unlike the lazy wave riser that is supported by discrete buoys, the lazy S riser is supported by one large buoyancy module).

The *Girassol* development offshore Angola makes use of a permanently moored FPSO for production of oil from 23 wells spread over an area 6.2×8.7 mi (10×14 km). The field is located in 4,600 ft (1,400 m) of water, 93 mi (150 km) offshore Angola approximately 130 mi (210 km) NW of the capital Luanda [2.49]. The massive field (production ~200 Mbbl/d) is served by a newly constructed FPSO that is 984 ft long, 197 ft wide (300×60 m) and has a displacement of 396,000 t [2.50]. Unlike most FPSOs, this ship is kept at a fixed heading with sixteen 8,630 ft (2,630 m) long mooring lines that connect to 14.8 ft diameter 56.8 ft long (4.5×17.3 m) suction piles [2.51]. The FPSO supports oil

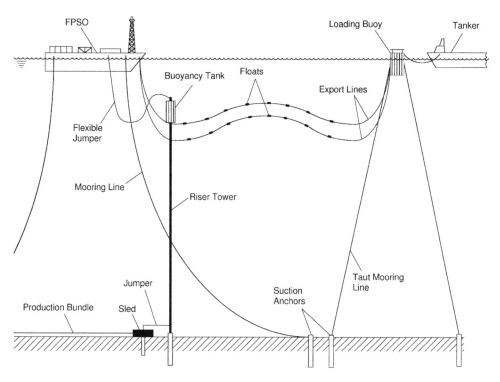

Figure 2.14   Schematic showing the Girassol FPSO and its offloading buoy [2.50, 2.52].

processing equipment, a large water injection plant, a sulfate removal facility and living quarters for 140 personnel. It provides storage for up to 2 MMbbl, while all gas produced is re-injected into the reservoir. Oil is transferred to transport tankers from an offloading buoy located 1.2 mi (2 km) from the FPSO. A consortium led by TotalFinaElf (operator), which owns the *Girassol* field, invested $2.8 BIL towards its development.

The 23 wells connect via manifolds to 11 production bundles that comprise 5 production loops. The bundles connect to three riser towers that take the hydrocarbons to the FPSO. The production bundles consist of two 8 in, 3,915 psi (270 bar) flowlines and a 2 in service line. The hydrocarbon temperature must be kept above 40°C (initial temperature ~65°C) to avoid paraffin deposition. Consequently, the bundles were insulated with foam and placed inside a 30 in carrier pipe that provides mechanical protection (filled with water [2.52]). The bundles, with lengths ranging between 2,300 and 9,800 ft (700–3,000 m), were assembled on shore and *bottom-towed* to *Girassol* (see Section 2.2.4).

Three freestanding insulated riser towers take the hydrocarbons to the FPSO [2.53]. These are bundles arranged around a 22 in core pipe. They consist of four 8 in production risers, four 3 in gas lift lines, two 2 in service lines and two 8 in injection risers (used for either gas or water). The bundles are surrounded by a high pressure glass syntactic foam with pure amine hardener for insulation and buoyancy, making the outside diameter 59 in (1.5 m). The riser towers were assembled as 4,110 ft (1,250 m) long single units on shore and were transported to the site by the *Controlled Depth Tow* method (Section 2.2.4) [2.53]. Here they were upended and stabbed onto suction piles via flex joints (Figure 2.14). Since the towers are only nearly neutrally buoyant, 131 ft long, 26 ft diameter (40 × 8 m) buoyancy modules are connected to their tops. The production and injection lines within each tower are connected to the FPSO by flexible jumpers, while the service and gas lift tubes via bundles of super duplex stainless steel tubes.

Oil is exported via two 16 in pipelines that connect to an offloading buoy 1.2 mi (2 km) away from the FPSO. The buoy, which measures 62 ft (19 m) in diameter and 33 ft (10 m) in height, has a taut-leg mooring system using nine mainly polyester rope mooring lines. The nearly 6,550 ft (2,000 m) long mooring lines spread out radially and connect to suction piles. The export lines are hung from the Girassol using flex joints, one from the port and the second from the starboard side. They connect to the buoy by flex joints also. The suspended sections between them form W-shapes as shown in Figure 2.14 [2.54]. The upper line is 7,875 ft (2,400 m) long and goes down to water depths ranging between 1,445 and 1,740 ft (440–530 m). The lower line is 9,020 ft (2,750 m) long, with the W-shape being in water depth of 2,070–2,430 ft (630–740 m). Cylindrical floats 9.8 ft (3 m) long, 4.9 ft (1.5 m) outer diameter were strapped periodically around the central 1,476 ft (450 m) long sections of the two lines to provide the buoyancy required to form the W-configuration. The pipelines are X65 grade and their main parts have a wall thickness of 0.563 in (14.3 mm). Fatigue considerations dictated that the 984 ft (300 m) long sections that connect to the buoy be made 1 in thick (25.4 mm). The two lines were installed by J-lay using the SaiBOS FDS vessel (see Section 2.2.2).

2.2   OFFSHORE PIPELINE INSTALLATION METHODS

The installation of pipelines and flowlines and their connection to platforms constitute some of the most challenging offshore operations. The level of engineering sophistication and effort required, as well as the size and cost of the various types of installation vessels used have reached such a level that pipelaying has developed into an engineering discipline of its own accord. The most commonly used installation processes and associated vessels are outlined in this section.

2.2.1   S-Lay

S-lay refers to an installation method in which the pipeline starts in a horizontal position on the vessel and acquires a characteristic S-shape on the way to the seabed, as shown in Figure 2.15 [2.55]. The first role of the vessel is to act as a work platform for assembling the line and for storing incoming pipe lengths. Usually, a linearly-arranged series of stations (firing line) weld 40 to 80 ft (12–24 m) lengths to the free end of the line. The welds are X-rayed and coated and the vessel moves forward, paying the line into the sea. The line leaves at the stern of the vessel via a sloping ramp (see Figure 2.16(b)). At the end of the ramp it comes in contact with a long boom-like curved structure known as a *stinger*. The stinger is an open-frame structure that supports the line on v-shaped rollers, providing a controlled-shape transition from the horizontal to the inclined suspended section. Older stingers were rigid, whereas modern ones are articulated, involving several segments that are connected via hinges. The stinger shape is prescribed by setting the segments at chosen angles. Stinger lengths vary with water depth and the submerged weight of the line, but in conventional S-lay they can reach 330 ft (100 m). The suspended length of pipeline is held by tensioners that are usually located on the ramp. Most commonly these involve

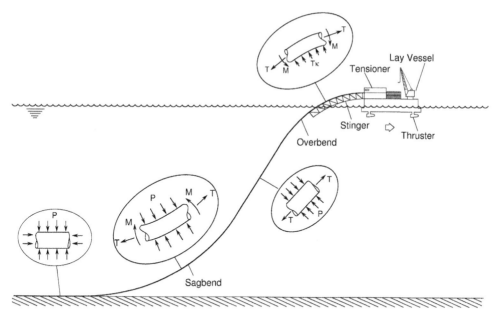

Figure 2.15 Schematic representation of S-lay pipeline installation and associated pipeline loadings.

v-shaped Caterpillar tracks with rubber pads that press on to the surface of the pipe from the top and bottom. The pipe is paid over the stinger by rotation of the tracks. In this setup, the section of pipeline on the stinger experiences bending combined with relatively high tension (see Chapter 10). Too short a stinger can result in excessive bending at the end of the stinger, which can buckle the pipeline (see Figure 8.4). Such a buckle can in turn cause fracturing and flooding of the line (*wet buckle*). Flooding of the pipeline can make it too heavy to hold by the tensioners, which can result in loss of the line to the seabed.

The upper curved part of the pipeline is known as *overbend* (upper generator convex). The line leaves the stinger at a chosen angle. Further down, it straightens and then gradually bends in the opposite direction, as shown in Figure 2.15. Often the maximum curvature occurs closer to the seabed in the *sagbend* region (upper generator concave), which is nearly at the maximum water depth. Thus it must be ensured that the combined bending and pressure loads can be safely sustained. The curvature in the sagbend is controlled by the tension applied at the top. Sudden movement of the ship or loss of tension for whatever reason can result in excessive bending, local buckling and collapse (see Chapter 9). Local collapse, in turn, has the potential of initiating a propagating buckle as shown schematically in Figure 2.17. Soon after the sagbend, the line touches the seabed and conforms to its relief. If the seabed is relatively flat, the pipeline can be considered to be under *hydrostatic* external pressure loading while empty. Its design is often based on avoiding collapse under this type of pressure loading (Chapter 4).

One of the main roles of the lay-vessel is to provide the tension that holds the suspended line and controls its shape. In older lay barges, the tension is reacted by several long mooring lines connected to anchors. The mooring lines are attached to winches, and the barge moves forward by winding in the mooring lines. This is a delicate operation

(a)

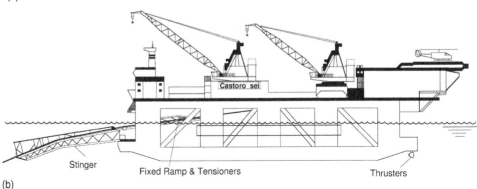

Stinger   Fixed Ramp & Tensioners   Thrusters

(b)

Figure 2.16   Saipem's Castoro Sei semi-submersible S-lay vessel: (a) photograph and (b) schematic
[2.56] (photograph courtesy Saipem).

essential to keeping the position and direction of the lay barge in accordance with the
planned route. The loss of a mooring anchor during such an operation can cause sudden
yawing or drifting of the barge, which in turn can result in buckling of the pipe at the
end of the stinger due to excessive bending. More modern S-lay vessels used in deeper
waters use *dynamic positioning* to control their position. This is achieved by thrusters
(shrouded propellers that can be freely directed) that are computer-controlled using GPS.
Dynamic positioning requires significantly more power but it increases the efficiency of
the lay operation (laying speed is as high as 4 mi/d for pipe of diameters up to 30 in).
The Castoro Sei, shown in Figure 2.16, is one of the larger dynamically positioned S-lay
vessels [2.56]. Its semi-submersible main structure is 499 ft (152 m) long, 231 ft (70.5 m)

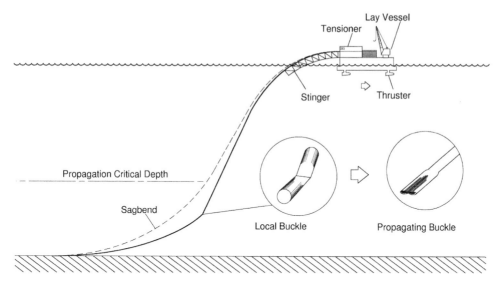

Figure 2.17 Schematic showing the initiation of a propagating buckle in a pipeline from a local bending buckle during S-lay installation.

wide and has four 37 t azimuthal thrusters. It has an articulated stinger and three 110 t tensioners.

The long suspended section of the pipeline behaves more like a cable rather than a beam, and thus its length as well as the sagbend curvature are mainly governed by the water depth, the submerged weight of the line and by the tension applied at the barge. Although modern S-lay vessels can apply very significant tensions, this comes at a significant cost to the operation. Consequently, the great majority of pipelines are installed empty in order to reduce the installation tension. The philosophy of the installation design is first to avoid buckling failures either in the overbend or the sagbend, and second to keep the pipeline in the elastic regime. The curvature in the sagbend is controlled mainly by tension. However, excessive tension can be detrimental to the section over the stinger, perhaps plasticizing the pipe. In some cases, high lay tension can also increase the cost of the operation by requiring a larger installation vessel. In general, plastic deformation on either the stinger or the sagbend is avoided, as it can cause excessive ovalization to the pipe cross section and spiraling of the pipeline on the seabed. Overall, the installation parameters are optimized to take all of these issues as well as material and installation costs into account.

Traditionally, S-lay has been the main pipe installation method for water depths up to 3,300 ft (1,000 m). More recently, S-lay water depth has been nearly doubled by the design and installation of longer articulated stingers on dynamically-positioned vessels with high tension capacities. Figure 2.18 schematically shows the deepwater stinger of the Allseas vessel *Solitaire* [2.40, 2.57]. The stinger is 460 ft (140 m) long and the tension capacity is 1,930 kips (875 t). This vessel was used to install the deepwater part of the 18 in Na Kika export oil pipeline [2.37] in a water depth that reached 6,350 ft (1,935 m). The pipeline was installed flooded, and the stinger was set to a radius of 330 ft (100 m). The applied tension reached values around 900 kips (410 t). The same vessel was also used to install most of the 24 in Okeanos gas pipeline that takes gas from Na Kika and Thunder Horse to Main Pass 260 in shallower water (see Figure 2.12).

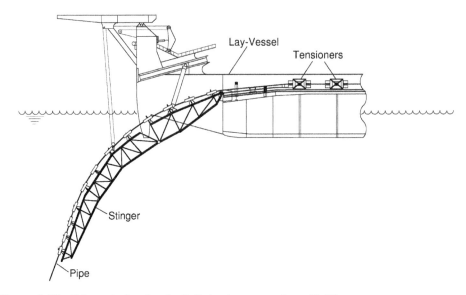

Figure 2.18   Schematic showing the Solitaire deepwater stinger [2.40].

2.2.2 J-Lay

As the water depth increases, the suspended length in conventional S-lay increases, and as a result the tension that must be applied by the lay vessel goes up. In addition, the required stinger length increases and its shape becomes more complex [2.55, 2.57–2.59]. These tough requirements are avoided by dropping the condition that the line start in a horizontal position. *J-lay* is an alternative installation method in which the pipeline leaves the vessel from a nearly vertical position, as shown schematically in Figure 2.19 (actual tower angles vary between 0° and 15° from the vertical) [2.57–2.59]. On the way down to the seabed, it acquires the characteristic J-shape from which the name J-lay is derived. The first effect of the J-configuration is that the suspended length is reduced by comparison to S-lay. In this case, the role of the tension is to support the shorter suspended length and to control the line curvature in the sagbend. A consequent second effect is a reduced tension requirement from the vessel and a significant reduction in the required thruster power [2.58].

   In J-lay there is usually only one welding and one inspection station. For this reason, longer pipe sections are used in order to increase the efficiency of the operation. These usually consist of four to six 40 ft (12 m) sections that are pre-welded on shore. Each multiple length section is then raised to the tower (Figure 2.20), aligned with the suspended pipe, welded to it, inspected and coated. The long section is then lowered into the water while the vessel moves forward, installing a corresponding length to the seabed. A short support structure (stinger) below the holding point guides the direction of the line close to the water surface. Since the touchdown point is not that far behind the vessel, the positioning of the pipeline can be more precise. Better vessel control also results from the fact that only a short length of the line close to the surface is exposed to wave motion. An additional advantage is that the lower tension in the line on the seabed translates into shorter *free spans* [2.57, 2.59].

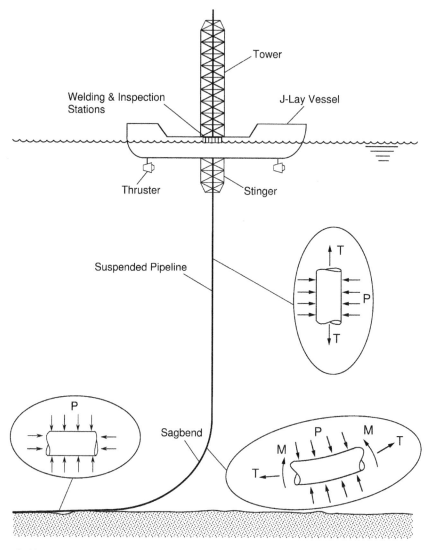

Figure 2.19   Schematic representation of J-lay pipeline installation and associated pipeline loading.

J-lay is somewhat slower than traditional S-lay, but it has been projected to be capable of installing pipelines down to 11,000 ft (3,350 m) of water [2.58]. The loads experienced by the pipe during such deepwater J-lay are as illustrated schematically in Figure 2.19. High tension and relatively small external pressure close to the surface of the sea, progressively increasing pressure and decreasing tension further down the long suspended section, high external pressure and bending in the sagbend, and essentially hydrostatic pressure on a flat seabed. Each of these loadings must be designed for. In addition, in such deep waters, the possibility of accidentally initiating a propagating buckle cannot be overlooked, so installation of buckle arrestors is usually obligatory.

J-lay was initially an outgrowth of offshore drilling, where it is common practice to hang long tubulars vertically from a ship. The first dedicated J-lay installation facility was

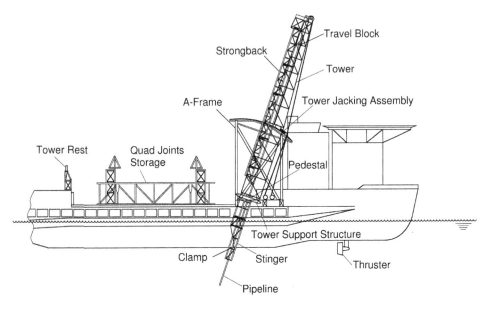

Figure 2.20   Schematic showing DP 50 and its J-lay tower [2.25].

installed on McDermott's dynamically positioned Derrick Barge (DB) No. 50 in the early 1990s. DB 50 was used in 1993 to install the 20 mi (32 km) deepwater sections of the two 12.75 in pipelines that connect with catenary risers to Shell's Auger TLP platform [2.22, 2.25]. A 200 ft (61 m) tower hangs off the starboard side of the large barge, supported by an A-frame (see Figure 2.20). The tower is pinned to a base structure that enables it to have an adjustable inclination to the vertical in the range of 0 to 20°. Below the base structure is a stinger that extends 50 ft (15 m) into the water. The stinger supports three retractable full-encirclement roller assemblies (each retracts to allow thicker sections such as collars to pass through). The system is designed to handle one quadruple pipe joint at a time. Each 167 ft (51 m) joint has a collar welded at the upper end, which is used to hold the suspended pipeline but also serves as a buckle arrestor [2.22–2.24]. The pipeline is held in place by a pedestal located on the tower support structure that engages it just below the collar. In the case of the Auger pipelines, the collars were 2 in (51 mm) wide and extended 1.625 in (41.3 mm) beyond the outer surface of the 12.75 in pipe.

The quad sections are welded on shore. Each is raised from the horizontal to the vertical position via a strongback. The pipe section is aligned to the hanging line, welded to it and the weld is inspected and coated using single stations located at the foot of the tower. The collar at the top of the new joint is then engaged from below by a travel block connected to a tensioning device that can support the weight of the whole pipeline. The bottom pedestal is retracted and the pipeline is gradually lowered into the sea while the barge moves forward. Despite the novelty of the facility and the single welding station employed, the speed of installation in the Auger project reached 1.9 mi/d (3.1 km/d) which compared with a best performance of 2.26 mi/d (3.63 km/d) achieved by S-lay in the shallower sections of the two pipelines [2.22].

Several additional J-lay facilities have been developed since 1998. Two of the largest are Saipem's S-7000 [2.58, 2.60, 2.61] and Herema's Balder [2.62, 2.63]. Both are large semi-submersible crane vessels to which J-lay facilities were added. A third facility is Saipem's SaiBOS Field Development Ship (FDS) [2.54, 2.61]. The *S-7000*, one of the largest dynamically-positioned semi-submersible crane vessels in the industry, is equipped with a removable 427 ft (130 m) J-lay tower at its stern (see Figure 2.21). The tower can be oriented up to 20° off vertical and handles quadruple pipe joints (160 ft – 49 m) that are welded on shore. The hanging line is held in place by three track-type tensioners, each with a capacity of 386 kips (175 t), while two 1,100 kips (500 t) friction clamps are present for enhanced safety. The quad joint is lifted to the upper part of the tower by an elevator. In the tower, it is lowered and aligned with the end of the hanging pipeline and is subsequently welded to it, inspected and coated. The line is then gradually lowered while the vessel moves forward. S-7000 was first used to install the main parts of the Hoover–Diana Flowlines, as well as all its steel catenary risers [2.27, 2.61]. It was then used to install the two Blue Stream 24 in pipelines that crossed the Black Sea, in water depth that reached 7,054 ft (2,150 m) (see Table 1.1). Despite the size of the pipe and the very deep water involved, these two lines were installed flooded [2.64].

The *Balder* is another large dynamically-positioned semi-submersible crane vessel converted in 2001 into a Deepwater Construction Vessel (DCV) by the addition of a J-lay tower. The tower, which is located on Balder's port side, handles 240 ft (73 m) hex joints, each with a holding collar welded to its top end [2.62]. The installation operation is similar to that of the DB 50 in that the pipeline hangs from a collar clamped at the base of the tower. A hex joint is lifted into the tower, is aligned and welded to it and the weld is inspected and coated. Subsequently, the upper collar is engaged by a 2,300 kips (1,050 t) capacity hoist, the lower collar is released and the pipeline is lowered down by 240 ft (73 m) while the vessel moves forward. DCV Balder has been used extensively to install the deepwater sections of the various pipelines and catenary risers in the Mardi Gras development (see Section 2.3).

The *SaiBOS Field Development Ship* (FDS), equipped with a J-lay tower and crane, is a smaller but more versatile ship than the S-7000 or Balder DCV (see Figure 2.22). This ship is dynamically positioned and can handle up to 22 in pipelines installed in maximum water depth of 8,000 ft (2,500 m) [2.61].* Its tower is operated at angles ranging from −6° to 45° to the vertical. The hanging pipe is held by a friction-type clamp (Hang Off Clamp) located below the tower. A fixed stinger below the clamp contains rollers for controlling the curvature of the pipeline during the lay operation. A four-joint section of pipe is placed horizontally on a loading arm hinged to the bottom of the tower. The loading arm then lifts the quad joint to the J-lay tower, where it is grabbed by line-up clamps, lowered to the assembly station, and aligned with the hanging pipeline for welding. Once welded, inspected and coated, the load is transferred to a traveling friction clamp mounted on the tower. The Hang Off Clamp opens and the quad joint is lowered through a stinger while the ship moves forward. The tension capacity of the facilities is 970 kips (440 t). Amongst other projects, SaiBOS FDS was used to install the two 16 in export pipelines of the Girassol project that go from the FPSO to the offloading buoy [2.54], as described in Section 2.1.2(d).

---

* We are grateful to Kimon Ardavanis of Saipem for information he provided on SaiBOS FDS.

(a)

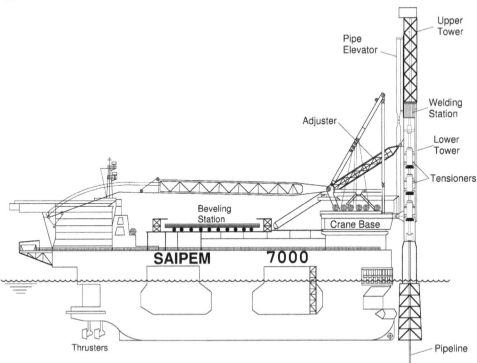

(b)

Figure 2.21   The Saipem 7000 semi-submersible crane vessel with a J-lay tower at its stern: (a) photograph and (b) schematic [2.60] (photograph courtesy Saipem).

Figure 2.22   Photograph showing Saipem's SaiBOS Field Development Ship (presently Saipem FDS, courtesy Saipem).

2.2.3 Reeling

One of the most efficient installation methods for offshore pipelines is the *Reel Vessel Method* [2.55]. In this method a section of pipe, usually several miles long, is wound onto a large diameter reel that is mounted onto a sea-going vessel while docked at a home base. The vessel travels to the installation site and installs the pipe by gradually unspooling the line. Existing reel vessels can lay pipe at speeds of up to two knots. The continuity of the method and transfer of most of the fabrication processes (assembly, welding, inspection, coating) on-shore result in significant reductions in installation time and overall cost of such projects.

Some of the first pipelines laid from a reel date back to World War II. In 1944, 25 mi long 3-inch lines were wrapped onto floating spools and towed by tug boats across the English Channel, where the lines were unspooled [2.65]. No effort was made to straighten the lines as they left the spools. The operation known as PLUTO (Pipe Line Under the Ocean) was a quick way of bringing fuel from England to the allied forces in Normandy.

Although reeling was used periodically to install small diameter flowlines in the years following the war, the first major reel vessel able to lay 12.75 in pipe was Fluor Corporation's RB-2. This vessel, built in 1970, is a 275 × 80 ft (84 × 24 m) flatbed barge equipped with a horizontal reel with a 20 ft (6.1 m) radius hub and a system for straightening the pipe as it is unwound from the reel. The vessel and the reeling patents were purchased by Santa Fe International in 1973, who renamed it *Chickasaw* [2.65]. Over the years the Chickasaw has installed a plethora of pipelines and flowlines primarily in the Gulf of Mexico. It has undergone several upgrades and remains operational in the fleet of Global Industries (see Figure 2.23; [2.66]). The upgrades include the addition of dynamic positioning [2.67], increased tension capacity (180 kips – 82 t), the addition of a longer stinger, etc. (major pipelay characteristics listed in Table 2.1). With a capacity of reeling 2,500 t of pipe (e.g., 11.5 mi of 8-inch pipe), today the vessel remains competitive for installation of flowlines and smaller diameter pipelines in shallow and moderately deep waters [e.g., 2.68].

The next major development in reeling technology was Santa Fe's design and construction of a sea-going ship equipped with a reel named *Apache* (see Figure 2.24(a)) [2.65, 2.69]. In this reelship, the reel is vertically oriented, as shown in Figure 2.24 [2.70]. The reel is 22 ft (6.5 m) wide, with a hub radius of 27 ft (8.23 m), flanges with 82 ft (25 m) diameter, and has capacity of 2,000 t of pipe. Thus, the reel can carry for example 19.7 mi of 8-inch pipe. The vessel is capable of handling up to 16-inch pipe. The pipe straightens as it comes off the reel and is paid into the sea by going over a special ramp, shown in Figure 2.24(b). The pipe bends again over the overbend of the ramp (radius of 333 ft – 10 m), straightens further down the ramp, and is reverse-bent by a special straightener. The ramp is a 105 × 30 ft (32 × 9 m) structure with a level wind mounted on it. The level wind carries all pipe-handling equipment such as the overbend track, straightener and tensioner. As the pipe is unwound, the level wind moves from side to side between the reel flanges, keeping the pipeline aligned with the equipment. The track-type tensioner with a capacity of 72 t is used to hold the suspended section of pipeline. Additional tension (∼90 t) can be reacted by the reel. The ramp inclination can be adjusted between 18° and 60° (72° with special additions) to the horizontal, allowing installation in different water depths.

In view of the reel dimensions, the spooling and unspooling processes induce bending curvatures to the pipe that are well into the plastic range of the material. For example, in the case of the 27 ft radius of the Apache reel, a 12-inch pipe bends to maximum strain of 1.93%, and a 16-inch pipe to 2.41% strain. It is thus imperative that the wall thickness as well as the mechanical properties of the pipe be chosen such that local buckling by bending is avoided (see Chapter 8). The possibility of such local buckling is further reduced by applying some level of tension during both the winding and the unwinding of the pipe on the reel. Additional plastic bending cycles are introduced by the other operational characteristics of the vessel. Figure 2.25 shows the moment-curvature history seen by the pipe during installation schematically. The pipe is first plastically deformed to curvature κ_1 during the winding onto the reel (0–1). During unwinding, the pipe straightens (1–2) due to the tension, and bends again to curvature κ_3 as it goes over the overbend on the ramp. Downstream of the overbend, the pipe straightens once more (3–4) and finally is reverse bent in the straightener (4–5), so that on unloading it ends up at zero moment and curvature (approximately).

Such bending loading histories have an effect on the geometry of the pipe and possibly on its fatigue life. Despite the fact that the process is designed to avoid bending buckles,

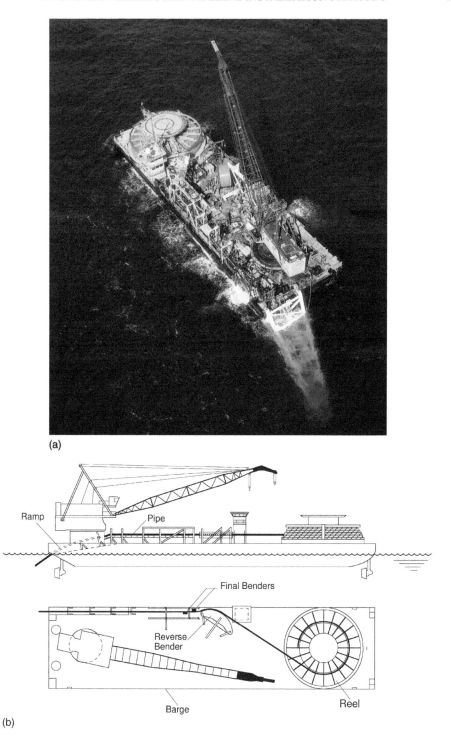

(a)

(b)

Figure 2.23   Global Industries' Chickasaw reel pipelay barge: (a) photograph and (b) schematic (photograph courtesy Global Industries).

Table 2.1 Basic characteristics of reels on commercial reeling vessels.

Specs.	Apache	Chickasaw	Deep Blue	Hercules	Skandi Navica	Seven Oceans
Reel Type	Vertical	Horizontal	Vertical (2)	Horizontal	Vertical	Vertical
Reel Radius (ID, m)	8.23	6.1 (7.2)	9.75	9	7.5	9
Flange Radius (m)	12.5	12.2		17.5	12.5	14
Reel Width (m)	6.5	3.35		7	6.7	10
Ramp Radius (m)	10	*	9	*	–	9
Pipe Capacity (t)	2,000	2,500	2,500 × 2	6,500	2,500	3,500
Pipe Diameters (in)	4–16	2–12.75	4–18	4–18	4–16	4–16
Tension/Reel (t)	84–128	–		–	100	100
Tension/Tensioner (t)	72	82	275 × 2	544	37	400
Date of Operation	1979	1970	2001	2001	2001	2007

*Pipe reverse bent to approximately the yield curvature.

the repeated excursions into the plastic range tend to ovalize the pipe cross section, cause some permanent elongation, and result in changes to the mechanical properties of the material. All of these are exacerbated when tension is applied simultaneously to bending (see Chapter 10). Induced ovalization can alter the subsequent performance of the pipeline by reducing its capacity to resist external pressure, a load of prime importance to moderately deep and deepwater lines (see Chapter 4). The induced elongation can adversely influence defects in girth welds, potentially leading to fatigue cracking. Such fatigue life reduction is a particularly acute issue for steel catenary risers [2.71, 2.72]. The mechanics of reeling and unreeling will be discussed in a chapter dedicated to the subject in Vol. 3 of this book series.

Despite these challenges, the Apache has been able to travel and install pipelines in all parts of the world with offshore oil and gas production. At each location, a home base is established, where pipeline stalks on the order of 1 km long are prefabricated and stored. The vessel travels back and forth between the base and the installation site, loading and installing pipeline until the project is completed. Over the years, the vessel has had several upgrades (e.g., [2.73]), and has successfully reeled insulated pipes as well as pipe-in-pipe systems (e.g., [2.74, 2.75]). Despite the recent introduction of newer and larger reeling vessels, the Apache remains a strong competitor for laying larger pipes in moderately deep waters, and for smaller pipe diameters in deep water (operated today by Technip). It is limited, however, by the 72° maximum inclination of its ramp and by limitations on the ship thrust. As the water depth increases, the installation angle must be increased in order to keep the vessel-applied tension and forward thrust within reasonable levels.

Several new reeling vessels have been developed in the last few years. The main reeling characteristics of a select number are listed in Table 2.1. One of the most prominent of the new reeling vessels is Technip's *Deep Blue* multipurpose construction vessel [2.76], which was introduced in 2001 (see Figure 2.26). The Deep Blue is a 678 ft (206 m) vessel equipped with two steel pipeline reels, a flexible pipe reel, and large cranes. The characteristics of the reels given in Table 2.1 enable reeling pipe of up to 18 in diameter, and allow loading up to 5,500 t of pipe [2.76, 2.77]. The ship has a 16.5 × 16.5 ft (5 × 5 m) moonpool in the middle, to which the ramp is fitted. During installation, the pipe passes over a 30 ft (9 m) radius sheave located at the top of the ramp, straightens as it travels down the ramp, and is reverse-bent by the straightener, ending up at zero curvature and

(a)

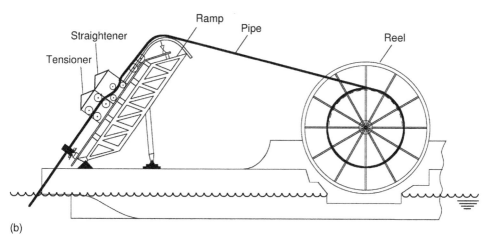

(b)

Figure 2.24 Technip's Apache reelship: (a) photograph and (b) schematic (photograph courtesy Technip).

moment (approximately). The hanging line is supported by two track-type tensioners with a total tension capacity of 550 t. Due to its design, the ramp inclination can vary anywhere from 90° to 58° to the horizontal. With these modifications, the Deep Blue can install 18-inch pipelines in water depths up to 8,200 ft (2,500 m). A J-lay ramp is also available for installation of larger diameter pipes of up to 30 in. The vessel has participated in several major deepwater projects, including Na Kika [2.37], Marlin [2.77], Devils Tower [2.78], Matterhorn [2.79], K2 [2.80] and others.

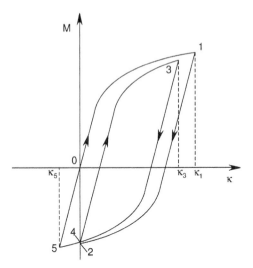

Figure 2.25 Moment-curvature history seen by pipe during reeling and unreeling on the Apache.

Global Industries also developed a new reeling vessel, the *Hercules* [2.81, 2.82]. This is a large dynamically positioned derrick barge (482 × 140 ft − 147 × 43 m) equipped with a large crane and pipeline installation facilities (see Figure 2.27). The vessel is versatile in that it can be used either in S-lay or in reel lay mode, in shallow as well as deep waters. Like in the Chickasaw, the Hercules reel is horizontally oriented but differs in that it is removable. It has a hub radius of 28.5 ft (9 m), a pipe capacity of 6,500 t, and can reel pipe 4–18 inches in diameter (Table 2.1). The straightener is similar to that of the Chickasaw, with the pipe being paid into the sea over a stinger. The tensioner capacity is 1,200 kips (544 t). Since its construction in 2001, the vessel has participated in several pipeline installation projects in the Gulf of Mexico, such as Petronius, Zia and Gunnison.

The reeling characteristics of two additional reeling vessels operated by Subsea 7 are listed in Table 2.1 [2.83]. The first is *Skandi Navica* which has reeling facilities transferred from DSND's *Fennica* [2.84]. It has worked extensively offshore Brazil. The second, named *Seven Oceans*, is a larger ship that will be introduced in 2007.

2.2.4 Towing

Another method of constructing and installing offshore pipelines is by towing them to the site. A section of pipeline is constructed onshore and is then towed to the installation site using one of the methods shown in Figure 2.28. An advantage of the technique is that welding, inspection and testing are conducted onshore before installation. Towing is ideal for shorter pipeline sections and shore approaches [2.59, 2.85–2.87], as well as for bundles and some risers [2.53, 2.88]. In the *Surface Tow* and *Near Surface Tow* (Figure 2.28(a)) methods, the pipeline is made buoyant by the periodic addition of buoys, so that it floats just below the surface of the sea. It is then towed out to location by a tugboat, while a trailing tug keeps the line taut. Once on location, the pipeline is lowered to the seabed by flooding the buoys in a controlled manner. Cross-currents and waves can be problematic,

(a)

(b)

Figure 2.26 Technip's Deep Blue reel and construction vessel: (a) photograph and (b) schematic (photograph courtesy Technip).

leading to fatigue and in some cases unstable oscillations of the trailing end [2.89, 2.90]. As a result, this method is mainly used in shallow waters.

In the *Controlled Depth Tow* [2.59, 2.85, 2.91], the pipeline is suspended between two tugboats well below the surface of the sea, as shown in Figure 2.28(b). In this method the effect of surface waves is reduced, although they still affect the tugboats and through them, the line. The pipeline is usually buoyant, so it is weighted down by the addition of chains. Tension keeps the curvature of the suspended section within acceptable limits.

(a)

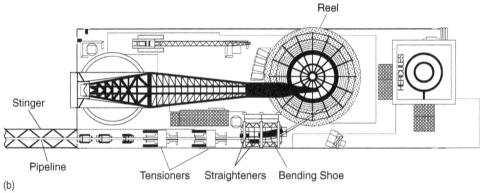

(b)

Figure 2.27 Global Industries' Hercules reel and construction vessel: (a) photograph and (b) schematic (photograph courtesy Global Industries).

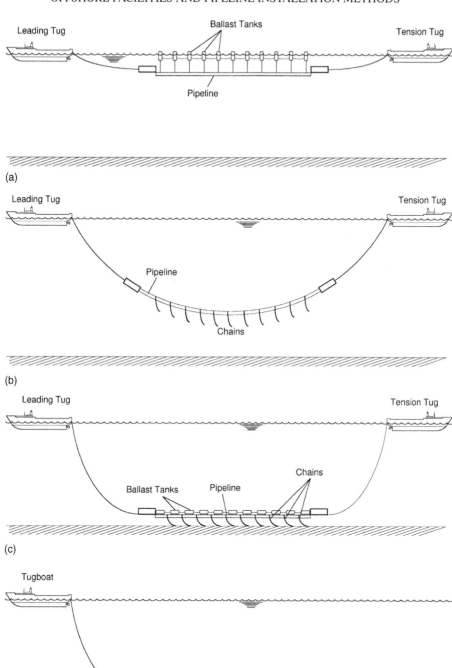

Figure 2.28 Schematics of tow installation methods: (a) surface tow, (b) controlled depth tow, (c) off-bottom tow and (d) bottom tow.

In the *Off-Bottom Tow* method, the pipeline is weighted down by chains and is held by the tugboats just above the seabed with the chains dragging on it (Figure 2.28(c)). In this manner, the effects of surface waves and currents is reduced even further.

In the *Bottom Tow* method, the pipeline is in contact with the seabed as shown in Figure 2.28(d) and a tugboat pulls it along the chosen route. In the case of heavier lines, pontoons can be added to reduce the frictional force that must be overcome. In this method, the pipeline is less susceptible to currents and waves and concerns about fatigue, from which all towing methods suffer to some degree, are reduced. One must ensure, however, that the coating does not get damaged and that the route is free of obstacles that can damage the line.

Examples of recent use of the tow methods to install deepwater production bundles and production risers come from the Girassol project off the coast of Angola [2.52]. The construction of the production bundles described in Section 2.1.2(d) took place on shore. They were encased in 30 in carrier pipes and had lengths that varied between 2,300 and 9,840 ft (700–3,000 m). They were bottom-towed to the site as described in [2.52]. Three production riser bundles approximately 4,110 ft (1,250 m) long were similarly assembled on shore and were towed to the site by the Controlled Depth Tow method as described in [2.53], where they were upended and inserted into suction piles.

2.3 THE MARDI GRAS PROJECT

The *Mardi Gras* project is one of the major new deepwater developments in the Gulf of Mexico, consisting of five deepwater prospects. It is also one of the largest and most complex offshore projects to date that has pushed the technology in several new directions. It utilizes two of the floating platform concepts discussed in Section 2.1 and has a 485 mi (780 km) pipeline transportation system, installed by the latest S-lay and J-lay technology. For these reasons, it deserves special mention in this chapter. Mardi Gras encompasses the *Holstein*, *Mad Dog* and *Atlantis* fields in the Green Canyon and the *Na Kika* and *Thunder Horse* fields in the southeastern Mississippi Canyon in the Gulf of Mexico. Each is owned by different partners and was developed somewhat independently between 2001 and 2006, but BP is the operator of the overall project [2.92]. The Holstein host was installed in 2004 in a water depth of 4,344 ft (1,324 m). At the time, it was the world's largest truss spar platform, with a diameter of 150 ft (45.7 m) and a height of 750 ft (229 m). The Mad Dog host is also a truss spar (128D × 550H ft – 39 × 168 m) that was installed in 2004 in a water depth of 4,420 ft (1,347 m). The Atlantis host is a semi-submersible FPS installed in 2006 in a water depth of 7,100 ft (2,164 m). The Na Kika host is a semi-submersible FPS installed in 2003 in 6,350 ft (1,936 m) of water (see Section 2.1.2(c)). The Thunder Horse is also a semi-submersible FPS in 6,300 ft (1,920 m) of water. Details on the size of the decks, the number of wells each serves and their projected production can be found in [2.92].

The transportation system of the Mardi Gras project is complicated by its extent, the water depth, the number of pipelines involved, the uneven relief of parts of the seabed, and other challenges. Five pipelines named after New Orleans Mardi Gras parades constitute the Mardi Gras Transportation System [2.93–2.96], namely the *Okeanos*, *Caesar*, *Proteus*, *Cleopatra* and *Endymion* pipelines (see Figure 3 and Table 1 of [2.93] and Table 1.1). The system's 485 mi (780 km) of pipelines were constructed at a total cost of $1 BIL [2.94]. The Caesar oil pipeline system connects the Holstein–Mad Dog–Atlantis hosts to SS332A in shallow water. It has a total length of about 112 mi (180 km), and consists of a 28 in main line joined by 24 in laterals from each platform (see Figure 2.29). The Cleopatra gas gathering system connects the same three hosts to SS332B. It has a total length of about 115 mi (185 km), with a 20 in main line and 16 in laterals (Figure 2.29). The Okeanos gas gathering system serves the Na Kika and Thunder Horse facilities, as shown in Figure 2.12. It has a 70 mi (112 km), 24 in main line that ends at MP260 and 20 in laterals about 30 mi (48 km) long. The Proteus oil pipeline connects the Thunder Horse platform to SP89E. It consists of a 28 in main line with a 24 in lateral, for a total length of about 72 mi (116 km). The 30 in, 89 mi (143 km) long Endymion oil pipeline connects SP89E to the Louisiana Offshore Oil Port (LOOP) at Clovelly.

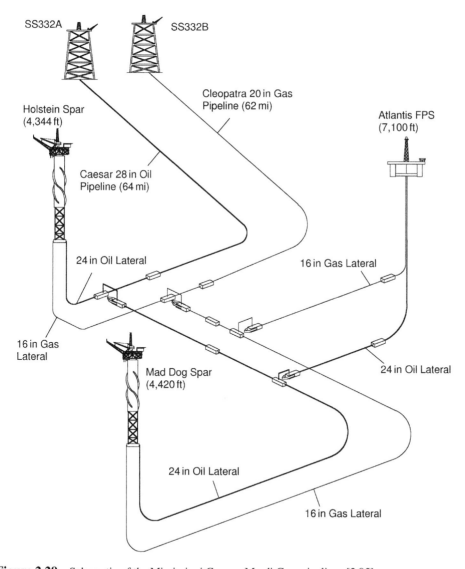

Figure 2.29   Schematic of the Mississippi Canyon Mardi Gras pipelines [2.95].

The 16 in laterals are seamless pipes and the 20, 24 and 28 in diameter pipes are UOE. The wall thicknesses of the various lines vary with depth, resulting in the 13 different values listed in Table 1 of Ref. [2.93], with the D/t s of the pipes varying from about 26.7 to 19.5. The shallower parts of the oil and gas export lines down to depths of about 3,400 ft (1,135 m), were installed by S-lay using Technip's Solitaire. Integral buckle arrestors were installed in these pipelines at intervals of 2,880 ft (878 m) [2.23, 2.24]. All deeper sections, including the steel catenary risers that connect the pipelines to the platforms, were installed by J-lay using Herema's Balder DCV. These had collars every 240 ft (73 m), which were needed for holding the pipeline during installation but also serve as buckle arrestors. The collars were typically 6–7 in (152–178 mm) wide and extend about 4 in (100 mm) beyond the outer surface of the pipe.

The Na Kika project was developed independently by Shell according to the outline in Section 2.1.2(c). On completion in 2003, it was turned over to BP, which became its operator. The other four projects were developed collectively [2.92]. The deepwater parts of the pipelines mentioned have diameters that range from 16 in to 28 in. The 16 in lines are seamless, whereas the 20, 24 and 28 are UOE pipes. Each line is connected to the host by steel catenary risers [2.97, 2.98]. The shallower parts of the pipelines were installed by S-lay using Allseas' Solitaire vessel. The deeper parts of the lines and the risers were installed by J-lay using Herema's DCV Balder [2.62]. The vessel combines a large crane with a J-lay tower, enabling its use in platform as well as pipeline installation. Its large load capacity and special features of its J-lay tower allowed installation of pipelines with attached sleds and piggable wyes at their ends [2.62, 2.93]. All deepwater pipelines have integral buckle arrestors. In the sections installed by J-lay, the arrestors also served as collars for holding the line during installation. Consequently, their spacing is 240 ft (73 m).

At the outset of the project, the only other pipelines installed in comparable water depths were the twin lines of the Blue Stream project in the Black Sea (see Table 1.1). For this reason, an extensive research, development and evaluation collapse testing program was undertaken in support of the project. The main issues of concern were determination of the wall thickness required to ensure that the pipe does not collapse under external pressure and external pressure combined with bending experienced during installation and in the sagbend of catenary risers. Of particular interest in this testing program was the degradation in collapse pressure induced to UOE pipe by the cold forming steps through which it is manufactured. The testing program followed the general guidelines outlined in Chapter 5 and in Section 9.6. The project also considered the collapse pressure improvement gained from heat-treating UOE pipes under conditions that simulate the coating process ([2.95] and Section 5.3). The tests confirmed this increase in collapse pressure, and as a result the UOE pipes used in the deeper section underwent coating under controlled conditions that ensured that they received this beneficial heat treatment. The testing program enabled the selection of safe wall thicknesses for the various Mardi Gras pipelines that are given in Table 1 of [2.93].

REFERENCES

2.1. Leffler, W.L., Pattarozzi, R. and Sterling, G. (2003). *Deepwater Exploration and Production: A Nontechnical Guide*. PennWell. Tulsa, Oklahoma.

2.2. Regg, J.B., Atkins, S., Hauser, B., Hennessey, J., Kruse, B.J., Lowenhaupt, J., Smith, B. and White, A. (2000). *Deepwater Development: A Reference Document for Deepwater Environmental Assessment Gulf of Mexico OCS (1998 through 2007)*. US Department of the Interior, Minerals Management Service, Gulf of Mexico OCS Regional Office.

2.3. Chakrabarti, S.K. Ed. (2005). *Handbook of Offshore Engineering*, Vols. I and II. Elsevier. Oxford, UK.

2.4. API RP 2A-WSD Recommended Practice for Planning, Designing and Constructing Fixed Offshore Platforms-Working Stress Design. 21 Ed. Dec. 2000, Supplement 1, Dec. 2002, Supplement 2, Oct. 2005.

2.5. Tjelta, T.I., Aas, P.M., Hermstad, J. and Andenaes, E. (1990). The skirt piled Gullfaks C platform installation. *Proc. Offshore Technology Conference* **4**, OTC6473, 453–462.

2.6. Svensvik, B. and Kepp, B. (1989). The concrete gravity base structure of the Gullfaks C platform: examples of applied technical development. *Proc. Offshore Technology Conference* **1**, OTC5908, 263–272.

2.7. Sterling, G.H., Krebs, J.E. and Dunn, F.P. (1989). The Bullwinkle project: an overview. *Proc. Offshore Technology Conference* **3**, OTC6049, 53–62.

2.8. Digre, K.A., Brasted, L.K. and Marshall, P.W. (1989). The design of the Bullwinkle platform. *Proc. Offshore Technology Conference* **3**, OTC6050, 63–80.

2.9. Mayfield, J.G., Arnold, P., Eekman, M.M. and Wellink, J. (1989). Installation of the Bullwinkle platform. *Proc. Offshore Technology Conference* **3**, OTC6054, 107–120.

2.10. Langner, C.G. and Wilkinson, H.M. (1980). Installation of Cognac 12-inch pipeline. *Proc. Offshore Technology Conference* **2**, OTC3740, 45–52.

2.11. Walker, A.C. and Davies, P. (1983). A design basis for the J-tube method of riser installation. *ASME J. Energy Resour. Technol.* **105**, 263–270.

2.12. Kyriakides, S., Corona, E., Madhavan, R. and Babcock, C.D. (1989). Pipe collapse under combined pressure, bending, and tension loads. *Proc. Offshore Technology Conference* **1**, OTC6104, 541–550.

2.13. Simon, J.V., Edel, J.C. and Melancon, C. (1999). An overview of the Baldpate project. *Proc. Offshore Technology Conference* **2.2**, OTC10914, 111–119.

2.14. Will, S.A., Edel, J.C., Kallaby, J. and des Deserts, L.D. (1999). Design of the Baldpate compliant tower. *Proc. Offshore Technology Conference* **2.2**, OTC10915, 121–143.

2.15. Clauss, G.F. and Lee, J.-Y. (2003). Dynamic behavior of compliant towers in deep sea. *Proc. 22nd International Conference on Offshore Mechanics and Arctic Engineering*, OMAE2003-37173, June 8–13, 2003, Cancun, Mexico.

2.16. Babin, J.L., Litzelfelner, J.G., Ambrose, M., Edel, J.C. and Will, S. (1999). Design and installation of catenary risers for the Baldpate compliant tower. *Proc. Offshore Technology Conference* **3**, OTC10920, 411–417.

2.17. Enze, C.R., Brasted, L.K., Arnold, P., Smith, J.S., Breaux, J.N. and Luyties, W.H. (1994). Auger TLP design, fabrication and installation overview. *Proc. Offshore Technology Conference* **3**, OTC7615, 379–387.

2.18. Schott III, W.E., Rodenbusch, G., Mercier, R.S. and Webb III, C.M. (1994). Global design and analysis of the Auger tension leg platform. *Proc. Offshore Technology Conference* **2**, OTC7621, 541–552.

2.19. Lohr, C.J., Bowen, K.G., Calkins, D.E. and Kipp, R.M. (1994). Design, fabrication and installation of Auger LMS and tendons. *Proc. Offshore Technology Conference* **2**, OTC7622, 553–563.

2.20. Kyriakides, S., Corona, E. and Fischer, F.J. (1991). On the effect of the UOE manufacturing process on the collapse pressure of long tubes. *Proc. Offshore Technology Conference* **4**, OTC6758, 531–543. Also, *ASME J. Eng. Ind.* **116**, 93–100, 1994.

2.21. Petersen, W.H., Patterson, R.W., Smith, J.D., Denison, E.B., Allen, D.W., Ekvall, A.G., Phifer, E.H. and Li, Y.-S. (1994). Auger TLP well systems. *Proc. Offshore Technology Conference* **2**, OTC7617, 531–540.

2.22. Kopp, F. and Barry, D.W. (1994). Design and installation of Auger pipelines. *Proc. Offshore Technology Conference* **4**, OTC7619, 913–920.

2.23. Kyriakides, S., Park, T.-D. and Netto, T.A. (1998). On the design of integral buckle arrestors for offshore pipelines. *Applied Ocean Research* **20**, 95–114. Also, *Proc. BOSS'97*, Vol. 1, Ed. J.H. Vugts. Pergamon (Elsevier Sciences) 1997, pp. 277–289.

2.24. Langner, C.G. (1999). Buckle arrestors for deepwater pipelines. *Proc. Offshore Technology Conference* **3**, OTC10711, 17–28.

2.25. Springmann, S.P. and Hebert, C.L. (1994). Deepwater pipelaying operations and techniques utilizing J-lay methods. *Proc. Offshore Technology Conference* **4**, OTC7559, 439–448.

2.26. Phifer, E.H., Kopp, F., Swanson, R.C., Allen, D.W. and Langner, C.G. (1994). Design and installation of Auger steel catenary risers. *Proc. Offshore Technology Conference* **3**, OTC7620, 399–408.

2.27. Arthur, T.T. and Meier, J.K. (2001). Diana/Hoover installation: overview. *Proc. Offshore Technology Conference*, OTC13085.

2.28. Lamey, M., Hawley, P. and Maher, J. (2005). Red Hawk project: overview and project management. *Proc. Offshore Technology Conference*, OTC17213.

2.29. Gupta, H., Finn, L. and Halkyard, J. (2002). Spar riser alternatives for 10,000 ft water depth. *Proc. Offshore Technology Conference*, OTC14298.

2.30. Colliat, J.-L. (2002). Anchors for deepwater to ultradeepwater moorings. *Proc. Offshore Technology Conference*, OTC14306.

2.31. Moyer, M.C., Barry, M.D. and Tears, N.C. (2001). Hoover–Diana deepwater drilling and completions. *Proc. Offshore Technology Conference*, OTC13081.

2.32. Gist, G.N. (2001). Diana subsea production system: an overview. *Proc. Offshore Technology Conference*, OTC13082.

2.33. Luyties, W.H. and Freckelton, T.P. (2004). Na Kika – Novel development in record water depth. *Proc. Offshore Technology Conference*, OTC16698.

2.34. Rajasingam, D.T. and Freckelton, T.P. (2004). Subsurface development challenges in the ultra deepwater Na Kika development. *Proc. Offshore Technology Conference*, OTC16699.

2.35. Kenney, J.J., Harris, K.H., Hodges, S.B., Mercier, R.S. and Sarwono, B.A. (2004). Na Kika hull design interface management challenges and successes. *Proc. Offshore Technology Conference*, OTC16700.

2.36. Paton, A.K., Smith, J.D., Newlin, J.A., Wong, L.S., Piter, E.S. and van Beek, C. (2004). Na Kika – Deepwater mooring and host installation. *Proc. Offshore Technology Conference*, OTC16702.

2.37. Kopp, F., Light, B.D., Preli, T.A., Rao, V.S. and Stingl, K.H. (2004). Design and installation of the Na Kika export pipelines, flowlines and risers. *Proc. Offshore Technology Conference*, OTC16703.

2.38. Bastard, A.H. and Bell, M. (2001). Evaluation of buckle arrestor concepts for reeled pipe-in-pipe. *Proc. 20th International Conference on Offshore Mechanics and Arctic Engineering*, June 3–8, 2001, Rio de Janeiro, Brazil, OMAE2001/PIPE4123.

2.39. Clausing, K.M., Williams, V.T. and MacFarlane, J.C. (2004). Na Kika umbilical transport and installation challenges. *Proc. Offshore Technology Conference*, OTC16704.

2.40. Solitaire (2006). *Description of D.P. Pipelay Vessel Solitaire*. Allseas Brochure.

2.41. Kopp, F. (2006). Personal communication.

2.42. Yun, H.D., Peek, R.R., Paslay, P.R. and Kopp, F.F. (2004). Loading history effects for deep-water S-lay pipelines. *ASME J. Offshore Mech. Arctic Eng.* **126**, 156–163.

2.43. Lee, L.-H. and Kyriakides, S. (2004). On the arresting efficiency of slip-on buckle arrestors for offshore pipelines. *Int. J. Mech. Sci.* **46**, 1035–1055.

2.44. Lever, G.V., Dunsmore, B. and Kean, J.R. (2001). Terra Nova development: challenges and lessons learned. *Proc. Offshore Technology Conference*, OTC13025.

2.45. Howell, G.B., Duggal, A.S. and Lever, G.V. (2001). The Terra Nova FPSO turret mooring system. *Proc. Offshore Technology Conference*, OTC13020.

2.46. Boles, B.D. and Mayhall, G.E. (2006). Kizomba A and B: projects overview. *Proc. Offshore Technology Conference*, OTC17915.

2.47. *Xikomba Oil Field*. www.offshore-technology.com/projects/xikomba

2.48. FPSO Mondo. www.singlebuoy.com/HTML/LeaseOperations/Systems

2.49. Serceau, A. and Pelleau, R. (2002). The Girassol development: project challenges. *Proc. Offshore Technology Conference*, OTC14166.

2.50. Bang, P. (2002). Girassol: the FPSO presentation and challenges. *Proc. Offshore Technology Conference*, OTC14172.

2.51. Dendani, H. and Colliat, J.-L. (2002). Girassol: design analysis and installation of suction anchors. *Proc. Offshore Technology Conference*, OTC14209.

2.52. Rouillon, J. (2002). Girassol – The umbilicals and flowlines – presentation and challenges. *Proc. Offshore Technology Conference*, OTC14171.

2.53. Alliot, V. and Carre, O. (2002). Riser tower installation. *Proc. Offshore Technology Conference*, OTC14211.

2.54. Tricard, P., Leijnse, R., Hattet, D. and Seng, D. (2002). Girassol export lines. *Proc. Offshore Technology Conference*, OTC14208.

2.55. Langner, C.G. and Ayers, R.R. (1985). The feasibility of laying pipelines in deep waters. *Proc. 4th International Offshore Mechanics and Arctic Engineering Symposium*, Vol. I, 478–489.

2.56. Castoro Sei (2003). Saipem Brochure (www.saipem.eni.it).

2.57. Heerema, E.P. (2005). Recent achievements and present trends in deepwater pipe-lay systems. *Proc. Offshore Technology Conference*, OTC17627.

2.58. Faldini, R. (1999). S7000: a new horizon. *Proc. Offshore Technology Conference* **3**, OTC10712.

2.59. Palmer, A.C. and King, R.A. (2004). *Subsea Pipeline Engineering*. PennWell. Tulsa, Oklahoma.

2.60. Saipem 7000. Saipem Brochure (www.saipem.eni.it).

2.61. Cavicchi, M. and Ardavanis, K. (2003). J-lay installation lessons learned. *Proc. Offshore Technology Conference*, OTC15333.

2.62. Wolbers, D. and Hovinga, R. (2003). Installation of deepwater pipelines with sled assemblies using the new J-lay system of the DVC Balder. *Proc. Offshore Technology Conference*, OTC15336.

2.63. van der Graaf, J., Wolbers, D. and Boerkamp, P. (2005). Field experience with construction of large diameter steel catenary risers in deep water. *Proc. Offshore Technology Conference*, OTC17524.

2.64. Pulici, M., Trifon, M. and Dumitrescu, A. (2003). Deep water sealines installation by using the J-lay method – The Blue Stream experience. *Proc. International Offshore and Polar Engineering Conference*, 585–590.

2.65. Kunzi, R.E. and Uyeda, S. (1979). Apache: first dynamically positioned vertical reel pipelay ship. *Ocean Industry*, April to May Issues.

2.66. Chickasaw: Reel Pipelay Barge (2005). Global Industries Brochure (www.globalind.com).

2.67. Malahy, R.C. (1995). Installation of DP system and adaptation of the reel barge Chickasaw for deep water pipelay. *Proc. Offshore Technology Conference* **3**, OTC7815, 113–121.

2.68. Palmer, M., McClure, L., Baker, B., Eckert, D. and Malahy, B. (2000). Reel-barge saves time in GOM gathering-line pipelay. *Oil Gas J.* May 1 Issue, 88–93.

2.69. Friman, K.R., Uyeda, S.T. and Birdstrup, H. (1978). First reel pipelay ship under construction – Applications up to 16 inch diameter pipe 3000 feet of water. *Proc. Offshore Technology Conference* **1**, OTC3069, 193–198.

2.70. The fleet: Apache (2006). Technip Brochure (www.technip.com).

2.71. Venkataraman, G. (2001). Reeled risers: deepwater and dynamic considerations. *Proc. Offshore Technology Conference*, OTC13016.

2.72. Netto, T.A., and Lourenco, M.I. and Botto, A. (2007). Fatigue performance of pre-strained pipes with girth weld Full-scale experiments and analyses. *Int. J. Fatigue* (submitted, 2007).

2.73. Tollin, B.I., Wilson, G.J. and Martin, R.G. (1994). Deepwater pipelay: extending the limits. *Proc. Offshore Technology Conference* **4**, OTC7542, 301–311.

2.74. Crome, T. (1999). Reeling of pipelines with thick insulation coating, finite element analysis and local buckling. *Proc. Offshore Technology Conference* **3**, OTC10715, 47–55.

2.75. Tough, G., Denniel, S., Al Sharif, M. and Hutchinson, J. (2001). Nile – design and qualification of reeled pipe in pipe for deepwater. *Proc. Offshore Technology Conference*, OTC13257.

2.76. The fleet: Deep Blue (2005). Technip Brochure (www.technip.com).

2.77. Lecomte, H., Hogben, S., Smith, J., Bendar, J., Day, J. and Palmer, M. (2002). BP Marlin: first flexible pipelay with newbuild deepwater pipelay vessel. *Proc. Offshore Technology Conference*, OTC14185.

2.78. Menier, P. (2003). Devils Tower export lines: 160 miles of 18" OD reeled rigid pipe. *Proc. Offshore Technology Conference*, OTC15338.
2.79. Kavanagh, W.K., Harte, G., Farnsworth, K.R., Griffin, P.G., Hsu, T.M., Jefferies, A. and Desalos, A.P. (2004). Matterhorn steel catenary risers: critical issues and lessons learned for reel-layed SCRs to a TLP. *Proc. Offshore Technology Conference*, OTC16612.
2.80. Cambell, B.R., Jewett, J.B., Gunnion, N.S. and Burton, C.C. (2006). K2 flowlines and risers: from design to precommissioning. *Offshore Technology Conference*, OTC18304.
2.81. Hercules: Pipelay/Derick Barge (2005). Global Industries Brochure (www.globalind.com).
2.82. McLure, L. and Panikkar, A. (2006). Personal communication.
2.83. Subsea 7 Vessel Specifications (http://www.subsea7.com/v_specs.php).
2.84. Hansen, V., Sodahl, N., Aamild, O. and Jenkins, P. (2001). Reeling and J-lay installation of SCRs on Roncador field. *Proc. Offshore Technology Conference*, OTC13245.
2.85. Braestrup, M.W. Ed. (2005). *Design and Installation of Marine Pipelines*. Blackwell Science. Fairfield, New Jersey.
2.86. Brown, R.J. (2006). Past present and future towing of pipelines and risers. *Proc. Offshore Technology Conference*, OTC18047.
2.87. Ley, T. and Reynolds, D. (2006). Pulling and towing of pipelines and bundles. *Proc. Offshore Technology Conference*, OTC18233.
2.88. Alliot, V., Zhang, H., Perinet, D. and Sinha, S. (2006). Development of towing techniques for deepwater flowlines and risers. *Proc. Offshore Technology Conference*, OTC17826.
2.89. Tatsuma, M. and Kimura, H. (1986). Offshore pipeline construction by near-surface tow. *Proc. Offshore Technology Conference* **4**, OTC5337, 425–432.
2.90. Binns, J.R., Heyden, M., Hinwood, J. and Doctors, L.J. (1995). Dynamics of near-surface pipeline tow. *Proc. Offshore Technology Conference* **4**, OTC7818, 139–153.
2.91. da Cruz, I.C.P. and Davidson, J.D. (2006). Deepwater installation of pipelines and risers by towing. *Proc. Offshore Technology Conference*, OTC18196.
2.92. Thurmond, B.F., Walker, D.B.L., Banon, H.H., Luberski, A.B., Jones, M.W. and Peters, R.R. (2004). Challenges and decisions in developing multiple deepwater fields. *Proc. Offshore Technology Conference*, OTC16573.
2.93. Brooks, J., Cook, E.L. and Hoose, J. (2004). Installation of the Mardi Gras pipeline transportation system. *Proc. Offshore Technology Conference*, OTC16638.
2.94. Marshall, R. and McDonald, W. (2004). Mardi Gras transportation system overview. *Proc. Offshore Technology Conference*, OTC16637.
2.95. Karlsen, J.S., McShane, B.M., Rich, S.K. and Vandenbossche, M.P. (2004). Mardi Gras deepwater pipeline design overview. *Proc. Offshore Technology Conference*, OTC16636.
2.96. Chandler, B., Marshall, R. and Vandenbossche, M. (2006). Personal communication.
2.97. Jesudasen, A.S., McShane, B.M., McDonald, W.J., Vandenbossche, M. and Souza, L.F. (2004). Design considerations particular of SCRs supported by spar buoy platform structures. *Proc. Offshore Technology Conference*, OTC16634.
2.98. Kileen, J.P., Dyson, K.C., Reeves, K.D. and Woblers, D. (2005). Large diameter SCR delivery challenges. *Proc. Offshore Technology Conference*, OTC17286.

3
Pipe and Tube Manufacturing Processes

Carbon steel and to a much lesser extent stainless steel are the materials of choice in the great majority of pipelines, flow lines, risers, casings and other onshore and offshore applications of tubulars. The needs are such that during the last 30 years, pipe and tube manufacture has become a technological discipline of its own accord. Industrial production of linepipe and oil well tubulars amounts to several billion dollars per year. While pipe manufacturing companies are spread around the globe, the main centers are in Japan (Nippon Steel, NKK, Sumitomo, etc.), Europe (Corus, Europipe, Tenaris, Vallourec & Mannesmann Tubes, etc.), South America (Tenaris), and North America (US Steel, Lone Star Steel, Oregon Steel Mills, etc.).

The mechanical property requirements of tubulars vary with the application. Pipeline steel requirements are usually the most stringent, as they include high strength, high ductility, high toughness, high corrosion resistance and good weldability [3.1]. For offshore pipelines, these must be met along with the heavy wall thickness required to resist collapse due to external pressure [3.2]. These difficult-to-meet requirements have made pipeline steel the "prince" of steels, and its production a truly high-tech endeavor. The required properties are achieved by carefully controlled alloying and thermo-mechanical processing during the production. Furthermore, each length of pipe is monitored at every manufacturing step, from the casting of steel rounds and slabs, through the rolling steps, to dispatch. A significant amount of non-destructive testing is performed, including ultrasonic, magnetic particle, and X-ray for through-thickness and weld defects. In addition, mechanical testing of the finished pipes includes tensile, Charpy and other tests (see Appendix A). The amount of such testing is often increased by the customer to meet the needs of specific projects. Each pipe is numbered individually and is issued a certificate that includes its exact history through each manufacturing step. With the exception of the aerospace industry, in very few other applications do manufactured metal parts undergo such extensive scrutiny and record keeping.

Oil and gas tubulars are either seamless or seam-welded. Although they can be made to larger diameters, seamless tubes and pipes typically are made with outside diameters of up to 16 inches. Seam-welded pipes, cold formed from individual plates through the UOE (or the JCO) process, are available from 16 to 64 inches (diameter and wall thickness capacities vary among manufacturers). *Electric Resistance Welded* (ERW) pipe, formed through a continuous process from a rolled strip of plate, is available from about 2.375 to

24 inches. Between them, these four major processes produce the great majority of the oil and gas tubulars. A fifth method, in which the plate strip is spirally welded, is also available. Such pipe is often used for water mains in land applications, but it is less common in offshore applications.

This chapter first briefly describes steel making for line pipe. This is followed by outlines of the main steps that comprise the five major manufacturing processes. The effect of the processes on the mechanical properties and the geometric characteristics of the finished pipe will be given particular emphasis, as they affect their mechanical behavior.

3.1 STEELMAKING FOR LINE PIPE

Steel production starts with the melting of iron ore, coke, limestone and manganese ore in a blast furnace. Hot metal is delivered from the blast furnaces to the steel mill in "torpedo" ladles. It then follows the processes shown schematically in Figure 3.1 [3.2–3.4]. Line pipe steel must end up with low amounts of C, S, P, O, H and N. These are removed at different stages of the manufacture. The steel is first desulfurized in the ladle. It is then transferred to a basic oxygen converter with bottom stirring using N_2 and Ar, where C and P levels are adjusted to the required values [3.3]. The molten steel is then transferred to a vacuum degassing converter bottom-stirred with Ar. Here, under vacuum, it is possible to remove remaining unwanted impurities (N, O, H, S). For example, final levels of S, N and O as low as 0.001, 0.005 and 0.002 wt%, respectively, are achievable. Alloying elements are added next, and the mix is checked for content. The process takes about 35 min and finishes with the steel at 1,700°C (3,100°F). The steel goes next to casting in a 180–250 t ladle. This amount of steel constitutes *one heat* from which several slabs, rounds or ingots will be cast. Slabs and rounds are usually cast in continuous casters as shown schematically in Figure 3.1. The advantages of continuous casting will be described in Section 3.2.

3.1.1 Strengthening of Steel

Steel is a polycrystalline material. The small crystals, or grains, contain lattice defects, known as dislocations, which limit their strength, and consequently, that of the material. Motion of dislocations causes plastic deformation of the crystal. Macroscopic plastic strain is the cumulative effect of grain-level plastic deformation. Additional deformation can take place by interaction of the grains at their boundaries. Steel is strengthened in several ways that increase the resistance to dislocation mobility. Some of the main hardening processes are briefly outlined below [3.5].

a. Solid Solution Hardening

Iron exists in two crystal forms: α-iron, or *ferrite*, which has a body-centered cubic lattice (bcc), and γ-iron, or *austenite*, with a face-centered cubic lattice (fcc). Ferrite is stable from low temperatures up to 912°C (1,674°F, see phase diagram in Figure 3.2 [3.6]). Beyond 912°C ferrite transforms to austenite which remains stable up to 1,394°C (2,541°F). Austenite is soft and ductile and well suited for forging and rolling. Thus, most such processes are performed at or above 1,100°C (2,012°F). The oldest and simplest

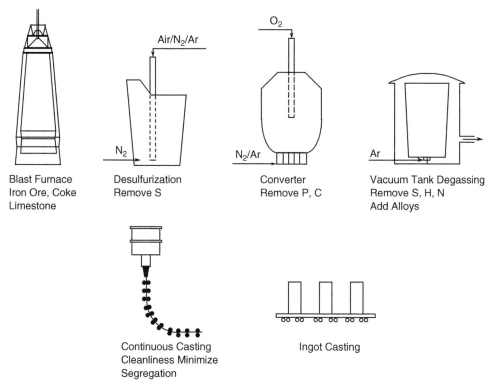

Figure 3.1 Production steps for steel for line pipe.

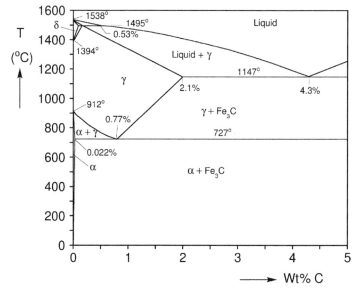

Figure 3.2 Iron-carbon phase diagram [3.6].

form of hardening any metal is to add impurities to form an *alloy*. The impurities form
obstacles to the motion of dislocations and increase the stress required for them to move.
C and N are the oldest impurity choices for making steel. They form *interstitial* solid
solutions (atoms take empty spaces in the two lattices) in γ- and α-iron. The solubility
of the two elements in austenite (up to 2.1 wt% for C) is higher than in ferrite (up to
0.022 wt% for C) because of the larger interstices (interatomic spacings) precipitates as
an iron carbide phase (*cementite*, Fe_3C). Small carbide particles have a strengthening
effect of their own. Alternatively, austenite quenched rapidly transforms to a tetragonal
structure referred to as *martensite*. A supersaturated solid solution of C is trapped in the
tetragonal lattice, resulting in a very significant increase in strength. Martensite has a plate-
like microstructure and can be very hard, brittle and prone to cracking. These unwanted
characteristics can be alleviated while retaining strength by *tempering* (reheating in the
range of 150–700°C––300–1,290°F).

Many other elements (Si, Al, Cu, Cr, Ca, Mn, Mo, Ni, V, Ti) form *substitutional*
solutions (atoms replace iron atoms in lattices, and in the process deform them) in α- and
γ-iron. In practice, several of these alloying elements are also required for other purposes
such as deoxidation (Si), promotion of hardenability (Mo), promotion of dispersion, or
to combine with sulfur (Mn). Therefore, the solid solution hardening contribution can be
viewed as a bonus.

b. Grain Size

Grain refinement by heat treatment and by mechanical work is one of the most important
strengthening methods for steel. This is because the yield stress of the polycrystal (σ_o)
has been shown to be inversely proportional to the square root of grain diameter (d) as
follows:

$$\sigma_o = \sigma_1 + \frac{k}{\sqrt{d}} \tag{3.1}$$

where σ_1 (*friction stress*) and k are material constants. The friction stress is temperature-
dependent, and is regarded as the yield stress of single crystals. Grain sizes down to
5–10 μm are common in modern steels.

Grain refinement is achieved by the addition of elements such as Ni, Ti, V and Al, while
keeping C in the range of 0.03–0.08% and Mn up to 1.5%. The grain refinement of such
micro-alloyed steels takes place by recrystallization of austenite during hot rolling. This
is influenced by both the temperature and the degree of deformation applied during each
pass. The alloying elements enhance the formation of carbide and nitride precipitates,
which tend to pin the grains and prevent their growth. Controlled rolling at specific
elevated temperature levels is called *thermomechanical controlled processing* (TMCP).

c. Dispersion Strengthening

Steels normally will have more than one phase present; indeed, several phases can coexist.
The ferritic or austenitic matrix, strengthened by solid solution and by grain refinement, is
further strengthened by controlling the dispersions of the other phases in the microstructure. The most common of these phases are carbides, nitrides and other intermetallic
compounds. A familiar structure is that of the eutectoid *pearlite*, which is usually a
lamellar mixture of ferrite and cementite.

In summary, the strength and other properties of steels arise from the combined effect of several phenomena. The heat treatment is aimed at adjusting these contributions to achieve the required properties. The $\gamma - \alpha$ phase changes allow a large number of variations in microstructure, resulting in a wide range of properties. The metallic elements, primarily through their influence on the transformation, provide a greater control over microstructure and therefore on the properties [3.5].

Line pipes are by necessity welded to each other, while large-diameter pipes are also seam welded. Thus, good weldability is another required characteristic of line pipe steels. The main hazard in welding is the formation of martensite in the heat-affected zone, which can lead to microcracks. This can be avoided by controlling the hardenability and by preheating the weld area to ensure slower cooling after welding. Another concern is the absorption of hydrogen during welding, which can lead to embrittlement. This can be avoided by using very low hydrogen electrodes, which are dried before use to ensure minimum exposure to moisture. As indicated above, lowering the C content coupled with microalloying (Mo, Ti, V, Nb) increases the strength and the fracture toughness of steel. Moreover, lowering the C and Mn content has the added benefit of improved weldability. Two main formulas are used to judge weldability. The first is the International Institute of Welding (IIW) *carbon equivalent* formula given by:

$$CE_{IIW} = C + \frac{Mn}{6} + \frac{(Cr + Mo + V)}{5} + \frac{(Ni + Cu)}{15} \tag{3.2}$$

where the elements are in wt%. This formula applies to steels with C content of more than 0.12%. It is aimed at the hardenability of the heat-affected zone. In general, the lower the CE, the less chance of weld cracking. A CE of less than 0.32% is a typical value.

The second carbon equivalent formula (Ito-Bessyo or parameter for crack measurement formula) is used strictly for steel with C < 0.12%. It is aimed at cold cracking and is given by:

$$P_{CM} = C + \frac{Si}{30} + \frac{(Cu + Cr + Mn)}{20} + \frac{Ni}{60} + \frac{Mo}{15} + \frac{V}{10} + 5B \tag{3.3}$$

A P_{CM} specified value in the range of 0.18–0.20% is typical [3.1, 3.7].

3.2 PLATE PRODUCTION

Line pipe with diameter larger than 16 in is usually manufactured by the UOE process for both land as well as offshore applications. An essential step for this manufacturing method is the production of high quality plate. We note that ERW pipe, available up to diameters of 24 in and wall thickness of about 18 mm (0.709 in), has similarly high requirements for the rolled plate strip from which it is made. The combined requirements of high strength, high toughness, high ductility, good corrosion resistance and good weldability of line pipe have resulted in special processes for the manufacture of slabs and for the subsequent rolling of the mother plates. The increasing demand for thicker wall pipe, to be installed in ever increasing water depths, has also tightened the requirements for casting and hot rolling. For thicker and/or larger diameter pipes, the first step is the production of high quality slabs. The slab quality is governed by the cleanliness of the steel and by the way it is cast.

Modern steel making is capable of producing highly clean steel to tight specifications on alloy content. Continuous casting coupled with *soft reduction* (see Section 3.2.2) during solidification has been shown to be the best method of producing slabs as thick as 400 mm with reduced segregation, internal cracking and center porosity [3.2, 3.8, 3.12]. The required plate properties can then be achieved during rolling by following tightly controlled *thermomechanical control processing* (*TMCP*). Although these three major manufacturing steps are common to several of the steel and pipe manufacturers, differences between them exist that can reflect on the final plate properties. Following is an outline of these three production steps as they were applied at the Dillinger Hütte plate mill in Dillingen, Germany in 2003.[†]

3.2.1   Steelmaking

Steel production follows the general steps outlined in Section 3.1 and shown schematically in Figure 3.1. Alloys are added to the cleansed steel in the vacuum tank and the mix is checked for content. The steel is then transferred to the continuous caster in a 185 t ladle at 1,700°C (3,090°F). This amount of steel will constitute one *heat*, from which several slabs will be made, each later producing several daughter plates.

3.2.2   Vertical Continuous Casting of Slabs

A unique vertical continuous caster, shown schematically in Figure 3.3, was constructed by Dillinger Hütte in 1998 [3.9, 3.10]. It differs from the more conventional circular arc continuous casters in that it has a 15.6 m (51 ft) vertical section in which the solidification of the cast strand takes place. This compares with a vertical section of 2 to 3 m vertical section for most advanced circular arc continuous casters. The first advantage of the long vertical section is that it allows nonmetallic inclusions and bubbles to rise to the top. The steel is poured gradually into a 50 t mold at the top of the caster, which ensures continuity of the casting when the ladles are switched. The top of the caster has hydraulically set sides to accommodate the onset of solidification. The width of the casting ranges from 1,400 to 2,200 mm (55–87 in) and the thickness from 230 mm (9 in) to a maximum of 400 mm (15.7 in). This section is followed by hydraulically set rollers. The top sections of rollers are kept at a fixed position, while the following segment performs the soft reduction. The strand moves downwards at a chosen speed which ranges between 0.2 and 0.5 m/min (7.9–19.7 in/min). At the top, an outer shell solidifies while the center is still red hot. As the strand moves down the thickness of the solid shell increases, going from about 20% to 100% solid at the end of the vertical section of the caster.

Soft reduction is a technique for reducing macro-segregation toward the center of the strand [3.2, 3.11, 3.12]. This segregation causes the steel composition to deviate at the center of the strand and later the center of the plate. This in turn affects local toughness and resistance to sour-gas service, and can result in cracking of the center. The principle of the technique is to compact the strand by an amount equal to the shrinkage associated with the transformation from liquid to solid phase. The compaction is applied hydraulically in the lower section of movable rollers. Soft reduction is most effective if it is applied

---

[†]We thank Volker Schwinn of Dillinger Hütte for his help in gathering information on modern steel plate manufacturing.

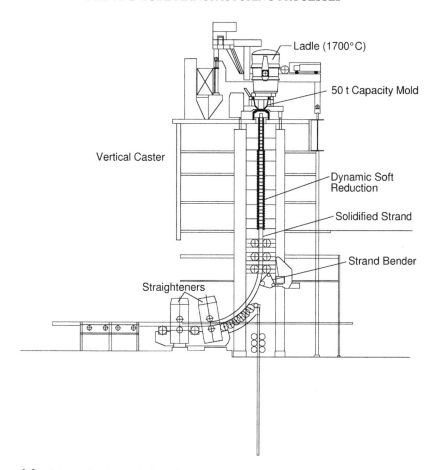

Figure 3.3 Schematic of a vertical continuous slab caster [3.14].

closer to the crater tip (the end of the molten core). The system monitors the position of the crater tip and positions the soft reduction zone optimally. In the system shown in Figure 3.3, the maximum rate of transverse deformation of the strand is 2 mm/m.

The vertical part of the caster terminates into a bending drive roll (Figure 3.3). The now fully solid strand is bent to a radius of 8 m and straightened at the end of the 90° section in a three-point bending straightener. The temperature during bending is controlled by a computerized water spray system. The straight strand is flame-cut into slabs of chosen lengths determined by the desired plate dimensions. The slabs are then stored for later processing in the plate mill. Up to 13 heats can be cast continuously, after which the process is reset.

3.2.3 Plate Rolling

In the plate mill, the slabs are first reheated to about 1,100–1,200°C (2,012–2,192°F) in pusher-type hearth furnaces. The reheating, which can take as long as 3 h, is an

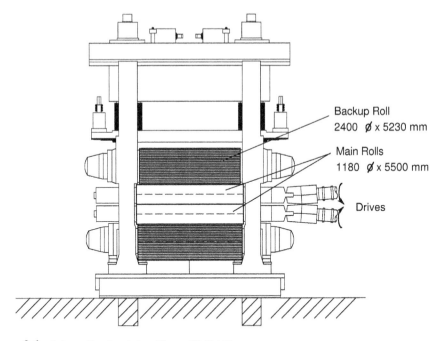

Figure 3.4   Schematic of a plate rolling mill [3.14].

important step in that temperature uniformity is a requirement for more uniform plate properties [3.13]. The scale is first removed and the red-hot slab goes to the main roller stand for primary rolling (or roughing). Dillinger's main roller (Figure 3.4) is a quarto-stand with a rolling width of 5,500 mm (216 in, [3.14]). The two main rolls have diameters of 1,180 mm (46.5 in). Each is supported by a backup roll of 2,400 mm diameter (94.5 in). The stand has a load capacity of 108 MN, a torque capacity of 2 × 4.5 MNm and a thickness capacity of 400 mm (15.75 in). The rolling is reversible and is conducted with water as a lubricant with a typical rolling speed of 7 m/s (276 in/s). Typically the slab undergoes two axial rollings; it is then turned 90° for two rollings in the transverse direction in order to set the plate width. It is subsequently rolled 4–6 more times axially. Each rolling reduces the thickness by a prescribed amount while the length in the rolling direction increases. (The length changes only in the direction of rolling.)

A requirement of TMCP is that recrystallization be facilitated. Thus, the temperature of roughing must be maintained above (usually close to) the recrystallization stop temperature. The power and speed of the Dillinger Hütte roughing stand is such that roughing can be completed before the temperature drops below this critical level. In less powerful systems, the primary rolling may have to be interrupted for the plate to be reheated. This can lead to through-thickness temperature gradients with some effect on through-thickness properties [3.13].

In summary, the temperature of the slab, its uniformity through the thickness and toward the edges of the slab, the amount of thickness reduction at each rolling step, the temperature at which it takes place, the total thickness reduction and the total time of the rolling can all affect the mechanical properties. They are thus variables chosen for each specific plate size. Streisselberger *et al.* [3.13] show data that support the premise that increasing the amount of total reduction, as well as the amount of reduction per pass through the rolls increases the ductility of the material in the middle of the plate (usually the weakest zone). The high load and torque capacities of this quarto-stand allow larger thickness reduction per pass on thicker slabs.

Secondary rolling takes place in the next quarto-stand, which has a rolling width of 4,800 mm (189 in), roller diameters of 1,120 mm (44 in), a load capacity of 90 MN, and a rolling speed of 6 m/s (236 in/s). Here the plate, which has cooled down some, is rolled to essentially the final thickness. Dividing the rolling between two stands makes the process more efficient.

The plate goes next to *accelerated cooling unit* (ACC) which has a width of 4,700 mm (185 in) and a length of 30 m (98 ft). The plate is cooled quickly down to 450°C (842°F) by spraying water at a pressure of a few bar both from the top and bottom. Uneven cooling at the edges and ends can be controlled by masking. The speed of cooling is a variable which can produce required microstructures. Alternatively, the plate can be air cooled, or it can be placed into a thermal box for controlled, slow cooling. In some cases, ACC can also be applied at an intermediate step before final rolling. The objective of TMCP is to reduce austenite grain size and consequently the final ferrite grain size. With ACC, very fine ferrite and bainite can be developed.

The plate moves next to the *hot leveler*, where it is rolled for flatness and straightness. Because of temperature falloff and possibly different cooling rates seen by the edges, their properties vary from those of the main body of the plate. To reduce this effect, all edges are trimmed off. In some cases the plates may undergo one last heat treatment before being cut to size. A final cold leveling may also be applied for nicer surface finish and for improved flatness. The finishing steps involve full surface ultrasonic inspection, marking and weighing. As with seamless pipes, plates are identified by heat and plate numbers. The stress–strain responses of the plates invariably exhibit Lüders banding behavior as shown in Figure 13.3 (see also Section 13.1.2). However, the Lüders bands are usually erased during the cold forming steps through which the plate is formed into a pipe.

In summary, plate making is an advanced technology with significant existing understanding of the role of key parameters of the process in deciding the plate properties. The demanding and tight specifications on properties mentioned earlier require the following:

(a) Clean steel free of impurities and low in C and S.
(b) Precise microalloying.
(c) Rolling at a tightly controlled temperature range, which is influenced by alloy content and cleanliness.
(d) Uniformity in temperature during rolling.
(e) Accelerated cooling applied uniformly to the plate.

Despite this high level of understanding about the process effects on properties, variations within a production schedule are unavoidable. This is because all processes listed above operate within bounds. As a result, some variation in impurities and in alloy content is unavoidable. In addition, the rolling temperature varies within the chosen bounds, which has an effect on properties. Although rolling is computer-controlled, some variations in the rolling parameters from plate to plate can take place, which also have an effect. Despite best efforts, some temperature variation through the slab and plate thickness, as well as between their middle and edges, is always present. Finally, some unevenness in

ACC is also unavoidable. The net result is that in a given order, in a given heat, in a given plate and in a given daughter plate property differences always exist. A variation of 100 MPa (14.5 ksi) in yield strength and tensile strength for plates belonging to the same heat is quite normal (e.g., see [3.14, 3.15]).

Because of the more significant deformation induced by rolling along the length of the plate, the material develops some texture (preferential alignment of grains in the rolling direction). This results in the axial yield strength being systematically lower than the yield strength in the transverse direction. For typical X65 steels, differences of 20–40 MPa (3–6 ksi) in the mean yield strength values in the two directions are typical. There are also some differences in yield strength through the plate wall thickness because of variations in the rate of cooling. Typically, the yield stress is highest at the top surface of the plate, which is cooled most effectively. The lowest yield stress is at mid-thickness while the bottom surface has about the thickness averaged value. Differences of the order of 14–28 MPa (2–4 ksi) between the maximum and minimum have been measured in 30 mm plates. Table 3.1 lists yield stresses at the top (T), center (C) and bottom (B) of a thicker plate (41.3 mm) in the transverse (T) and axial (A) directions. First, the average yield stress in the transverse direction was 70.73 ksi (488 MPa) while that in the axial direction was 65.16 ksi (449 MPa). Second, the center of the plate is seen to be weaker in both directions while the top is seen to be stronger in the transverse direction. In this case the differences are more pronounced because of the thicker wall of the plate.

The effect of alloy content and of thermomechanical processing on the mechanical properties is well understood by manufacturers. However, perhaps because of the competitive nature of the steel industry, such information is rarely published. A representative

Table 3.1 Through-thickness yield stresses in transverse and axial directions in a 1.625 in (41.3 mm) X65 plate.

Spec.	σ_o ksi (MPa)
T-B	71.91 (496)
T-C	66.54 (418)
T-T	73.73 (508)
A-B	67.80 (468)
A-C	60.15 (415)
A-T	67.54 (466)

Table 3.2 Chemical composition in wt% of UOE plate [3.2].

Steel	Grade	C	Si	Mn	*P	*S	Mo	Nb	V	Ti	Al	*B	Others	CE_{IIW}[†]	P_{CM}[†]
1	X80	0.058	0.25	1.61	130	16	0.17	0.051	–	0.015	0.027	–	Ni	0.378	0.163
2	X80	0.081	0.25	1.86	120	15	0.09	0.045	–	0.016	0.025	–	Ni	0.422	0.192
3	X80	0.062	0.23	1.86	100	27	–	0.045	–	0.014	0.003	–	Ni, Cu	0.399	0.183
4	X80	0.080	0.28	1.79	150	20	–	–	–	0.077	0.028	–	–	0.378	0.179
5	X80	0.039	0.20	1.55	170	20	–	0.042	–	0.015	0.024	13	Ni	0.311	0.133
6	X65	0.052	0.22	1.59	180	46	0.27	–	0.069	0.015	0.021	–	Ni	0.415	0.171
7	X70	0.055	0.33	1.54	240	36	0.27	0.044	0.068	0.014	0.025	–	Ni	0.407	0.175

*Values in ppm.
[†] wt%

sample of seven plate steel chemical compositions along with the basic processing followed, published by Nippon [3.2], appear in Table 3.2. The corresponding main mechanical properties are listed in Table 3.3. They are all low carbon steels with somewhat different alloy content produced approximately as described in Section 3.1. Five of the plates are X80 produced by accelerated cooling. One X65 and one X70 plate were produced by controlled rolling. The plates vary in thickness from 15 to 32 mm and the pipe diameters vary between 30 and 42 in. The plates were formed into pipes using the UOE process. The mechanical properties quoted in Table 3.3 were measured in the circumferential direction of the pipes. The following observations can be made from the data:

(a) The C content of the steels is less than 0.1 wt% for all cases while the S is very low. The CE ranges from 0.31 to 0.42 wt% and the P_{CM} ranges between 0.13 and 0.19 wt%.
(b) All yield strength values (API) exceed the grade designation by more than 5–7 ksi (34–48 MPa). This is a common practice followed in order to protect against the variability in strengths in rolled plates mentioned above.
(c) The yield-to-tensile (ultimate) strength ratio varies between 0.83 to 0.91.
(d) The elongation varies between 31% and 37% for X80, is 43% for the X65 and 44% for the X70 materials (based on a 2-inch gage length).
(e) The transition temperatures of all X80 are lower than −80°C and even lower for X65 and X70.
(f) The low temperature Charpy V-notch impact test energy levels are all sufficiently high, indicating adequate toughness for the base material. The Drop Weight Tear Test shear area at the low test temperature specified is 100%, indicating again sufficient toughness to prevent running cracks.

Table 3.3 Major mechanical properties of UOE plate in the transverse direction [3.2].

								CVN			BDWTT	
Steel	Process	D in (mm)	t in (mm)	YS ksi (MPa)	TS ksi (MPa)	YS/TS	Elon. %	T °C	C_V J	vTrs °C	Shear area %	85% Shear °C
1	ACC	40 (1016.0)	0.591 (15.0)	85.3 (588)	93.7 (646)	0.91	37	−20	252	<−80	100	−35
2	ACC	36 (914.4)	0.630 (16.0)	87.7 (605)	98.2 (677)	0.89	33	−20	168	<−80	100	−52
3	ACC	30 (762.0)	0.591 (15.0)	83.8 (578)	97.0 (669)	0.86	35	−46	163	<−80	100	−60
4	ACC	30 (762.0)	0.591 (15.0)	85.1 (587)	100.6 (694)	0.85	31	−20	115	<−80	100	−53
5	ACC	36 (914.4)	0.591 (15.0)	85.0 (586)	100.8 (695)	0.84	33	−20	334	<−80	100	−35
6	CR	36 (914.4)	1.063 (27.0)	70.0 (483)	84.4 (582)	0.83	43	−45	164	<−100	100	−73
7	CR	42 (1066.8)	1.260 (32.0)	75.1 (518)	90.0 (621)	0.83	44	−60	159	<−120	100	<−80

ACC: accelerated cooling; CR: controlled rolling; CVN: Charpy V-notch test; BDWTT: Battelle Drop Weight Tear Test.

The thermomechanical processing, as well as the exact chemistry of the steels, varies to some degree between manufacturers, as their casters, rolling mills, and cooling systems differ. The processing and alloy content are also varied by manufacturers by design depending on the thickness of the required plate, as well as on other customer specified requirements such as corrosion resistance or low operating temperature of the pipeline. In view of this, the data given in Tables 3.2 and 3.3 should be viewed just as representative of modern practice.

3.3 SEAMLESS PIPE

Most seamless tubulars start as round billets produced by continuous casting [3.12] for smaller diameters, or are cast conventionally as ingots for larger diameter pipes (this varies among manufacturers). The billets are pierced at elevated temperature through the Mannesmann process. How they are subsequently processed depends on the desired pipe diameter and wall thickness. In this section, continuous casting of rounds is first outlined, followed by brief descriptions of three seamless pipe manufacturing processes. The tube sizes produced by each process, as well as the finishing mechanical steps and heat treatment, differ to some degree between manufacturers. The latter are also altered to achieve project-specific requirements. The ones described here are mainly based on the basic processes of Vallourec & Mannesmann Tubes as they were used in 2004 [3.19].[†]

3.3.1 Continuous Casting of Round Billets

Round billets used in seamless pipe manufacturing are most commonly produced by continuous casting. Although some vertical casters do exist, today most steel mills produce round billets by circular arc continuous casters, such as the one of HKM of Germany shown schematically in Figure 3.5 [3.16, 3.17]. The benefits of continuous casting are similar to those outlined for the slab continuous caster in Section 3.2b: reduced segregation, porosity and cracking in the center of the round and more uniform properties in a given heat. These are achieved while simultaneously maintaining high productivity [3.12, 3.18]. A necessary condition for such results is clean, low carbon content steel. Clean and microalloyed liquid steel, produced more or less in the manner described in Section 3.1, comes to the top of the caster in 225 t ladles at a temperature of about 1,700°C (3,090°F, [3.16]). It is gradually poured into a tundish that distributes the liquid steel in a controlled manner to several strands being cast simultaneously (Figure 3.5 shows just one circular arc caster). The tundish, with a capacity of 25 t, maintains a relatively constant liquid head and acts as a buffer and as a remixing vessel. In addition, inclusions tend to float to the top, facilitating casting of a cleaner product. An important aspect of the transfer of steel from the ladle to the tundish and then to the caster is prevention of oxidation [3.2, 3.12, 3.16]. This is achieved by argon sealing between the ladle and tundish and by the use of long liquid submerged nozzles to transfer the steel to the molds.

The liquid steel goes from the tundish to a circular mold, which is the major component of the caster [3.12]. The mold is about 18 mm thick and 700 mm long and is made out

---

[†]We thank Dr. Markus Ring of Vallourec & Mannesmann Tubes for his help in gathering technical information on seamless pipe manufacture.

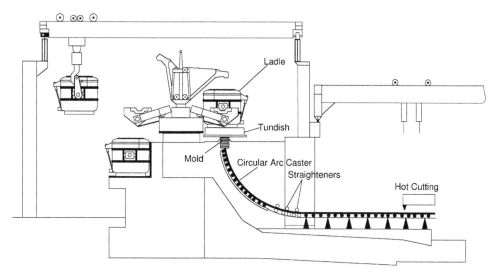

Figure 3.5 Schematic of a circular arc continuous caster for rounds [3.17].

of copper. It is concentrically placed into a steel jacket with pressurized cooling water being circulated through the annulus between the two. Thus, solidification begins at the top of the mold and continues as the strand moves slowly downwards. At the exit of the mold, a thick enough shell must be formed in order to contain the liquid core of the strand. A defining feature is a slight internal taper in the mold introduced in order to accommodate shrinkage during solidification. The taper prevents separation of the solidifying shell from the mold, which in turn improves heat flow and reduces porosity formation in the center of the round. The taper depends on the alloy composition and on the diameter of the round and is determined for optimal heat flow [3.18]. The mold is lubricated, and the assembly is usually vibrated to prevent the steel from sticking to the copper. The round shape of the mold and the taper result in uniform and efficient solidification and therefore more uniform properties.

The strand exits the mold with a solid shell outer layer 11–19 mm thick depending on the casting speed. It moves gradually downwards, supported by v-shaped grooved rollers. It conforms to the circular shape of the caster while it is simultaneously cooled by water spraying. The water spraying is intensive just below the mold and more moderate away from it. The radius of the caster varies with the diameter of the strand. For example, HKM's smaller casters have radii of 10.5 or 19.5 m, a metallurgical length of 30 m and produce rounds with diameters of 177 mm (6.97 mm) and 220 mm (8.66 in), respectively. The casting speed can be up to 3.5 m/min but 2–2.5 m/min is more common. The larger casters have radii of 10.5, 13.5, 18 or 30.5 m, and produce rounds of 117, 220, 270, 310 and 340 mm (4.61, 8.66, 10.6, 12.2, 13.4 in), respectively. The metallurgical length is 36.5 m and the casting speed is up to 2.5 m/min. At the bottom of the 90° bend, the strand goes through a pair of powered three-point benders that straighten it. It continues to solidify and cool down as it is conveyed on a horizontal bed of rollers. The strand is finally flame cut into billets ranging from 9 to 14 m. The billets continue to cool in a cooling bed and can be cut once more to lengths required by the rolling mill.

3.3.2 Plug Mill

The *plug* manufacturing process is named after the main forming step where a pierced hot round is rolled over a plug inserted in the annulus with a long rod [3.19–3.21]. The plug mill is typically used for tube diameters between 7 and 16 inches. All but the largest diameter pipes are made from continuously cast rounds. The larger diameter pipes are made from ingots. The wall thickness capacity varies to some degree with the diameter. For example, for 7 in tubes it ranges from 5.6 to 25 mm, while for 16 in tubes from 11 to 36 mm. The main steps of the process are shown schematically in Figure 3.6. The process starts with continuously cast rounds with diameters of 180, 220, 270 or 310 mm (these numbers are mill-specific). The round diameter and length are chosen based on the required dimensions of the final product. The rounds are preheated in a furnace and then heated to the forming temperature of about 1,280°C in a gas rotary-hearth furnace. The hot round goes to the pierce rolling mill, which is a modern version of the original Mannesmann piercing process (invented by the Mannesmann brothers in 1886). The round is engaged by two large double-conical work rollers, connected with long shafts to large electrical motors. The rollers rotate the round and simultaneously force it forward. The forward motion forces the round onto a conical insert that is held in place by a long water-cooled rod as shown in the figure. The round gets simultaneously pierced and elongated. By the time the whole length is pierced, the conical insert gets hot. It is replaced by a new one while the hot one is sent to a water bath to cool down. Although the process is computer-operated with precision, some eccentricity is introduced to the hollow during piercing. This eccentricity will remain in some magnitude in the final product (see specimens 1 and 3 in Figure 4.31). In addition, the rotary motion of the rollers introduces some spiral imperfection to the OD. The wavelength of the spiral depends on the feed speed and the dimensions of the round. Although most of this imperfection will be removed in subsequent steps, some will remain in the finished pipe as seen, for example, in Figure 4.30.

Next, the hollow round moves to the plug mill, which constitutes the main forming step of this process. It engages two rollers which work over a plug inserted into it and held in place by a long rod, as shown in the figure. The rolling reduces the diameter and wall thickness and simultaneously moves the part forward.

The tube moves next to the reeling mill where both the OD and ID are smoothened. Large diameter cylindrical rollers at a small inclination to the axis of the tube work over a shaped insert, causing some dimensional change. The rotation of the rollers turns the part and forces it forward. By the time the tube exits the reeling mill, it has cooled down and goes to a furnace, where it is reheated to about 1,200°C.

The next step is the sizing mill for a final OD reduction. This mill consists of several sets of three rollers, each arranged 60° out-of-phase with its neighbors. For larger diameter tubes, four roller sets are used arranged to be 45° out-of-phase with their neighbors. The roller sets are set to increasingly smaller diameters, gradually reducing the OD of the tube to the required final dimension. In this mill, the tube does not rotate; the rotary action of the rollers move it forward. Since in this mill there is no internal support, the alternating angular arrangement of the roller sets can introduce a star-shaped imperfection to the wall thickness of the finished tube (e.g., specimens 2 and 4 in Figure 4.31). The thickness imperfection has six corners for the three roller sets, and eight for the four roller sets. Such imperfections are superimposed on the eccentricity introduced at the piecing mill.

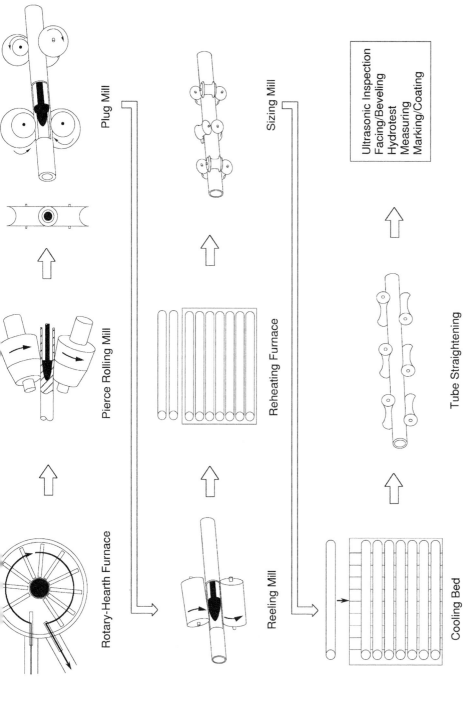

Figure 3.6   Schematic representation of major manufacturing steps of a plug mill for seamless pipe [3.19].

On exiting the sizing mill, the tubes cool off on a cooling bed in order to achieve the required microstructure. The cold tubes finally pass through the straightening mill. The mill consists of relatively flat rollers arranged at an angle to the axis of the tubes as shown in Figure 3.6. They are set to close tolerance and simultaneously rotate, push forward, and straighten the pipe. Although changes in OD introduced at this step are generally small, they can be sufficient to remove the Lüders bands, which are usually inherent in low carbon linepipe steels. Some mills routinely quench and temper the tubes after straightening, which results in the reappearance of Lüders bands. Thus, if the tubes are required to have a monotonic stress–strain response, this must be specified to the pipe mill. Cold-finished tubes may exhibit some yield anisotropy, which can be modeled as mentioned in Appendix B. Quenched and tempered pipes usually exhibit isotropic yielding.

The tubes move next to the finishing lines. Here they can be quenched and tempered; the final wall thickness can be measured ultrasonically; ultrasound, stray flux testing, and magnetic particle testing are available for defect detection. The ends are usually beveled to accommodate welding, or in the case of casing and some risers, the ends can be threaded to specified connector threads (standard or custom). The tubes are usually hydrotested to a specified pressure. They are finally weighed and marked with their heat and individual ID numbers and packaged for shipping.

3.3.3 Mandrel Mill

In the *mandrel* manufacturing process, the main forming step involves rolling over a solid mandrel insert [3.22]. This process is used for the continuous manufacture of tubes with diameters ranging from about 1 to 7 inches. The range of wall thickness available varies with the diameter, from 2 mm, at the lower end, to 32 mm for larger sizes. The raw material is again in the form of continuously cast steel rounds up to 5 m long and with diameter of 180 mm. The main steps of the process are shown in Figure 3.7. The rounds are again heated in a gas rotary-hearth furnace to about 1,280°C. The hot rounds are then pierced using a Mannesmann piercer, which operates in a manner similar to the one described in the previous section.

The pierced tube blank next passes through six three-roller stands that comprise the reducing mill. The mill reduces the OD to a desired size in a step-by-step manner. This reduction allows the use of just one single diameter round for the whole range of finished diameter tubes. On exiting the reducing mill, the wall thickness is measured ultrasonically and displayed in the operating room. The continuous dimensional check allows direct intervention and correction should the need arise.

The hot tube blank moves next to the mandrel mill, which is the major forming step of this process. A solid mandrel is inserted in the annulus as shown in the figure. The assembly passes through eight pairs of stands each arranged at 90° to its neighbors. Each stand reduces the OD a certain amount working against the mandrel. At the exit of the mill, the tube diameter is either 119, 152 or 189 mm and the tube can be as long as 30 m. An important aspect of mandrel forming is speed. Speed is necessary in order to keep the part hot enough to allow removal of the mandrel. The extraction is accomplished using a special mandrel extractor.

The tube blanks exit the mandrel mill at a temperature of about 700°C. They are reheated in a furnace to about 1,000°C and then go to the stretch-reducing mill. Here up to 28 three-roller stands are used to progressively reduce the tube diameter and wall

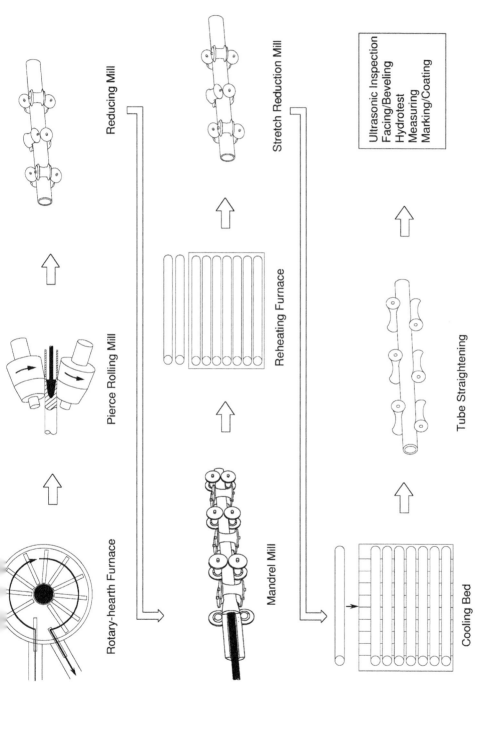

Figure 3.7 Schematic representation of major manufacturing steps of a mandrel mill for seamless pipe [3.19].

thickness to the desired dimensions. The three roller stands are again 60° out-of-phase with adjacent stands. As a result, a six-corner, star-shaped ID superimposed on the eccentricity introduced at the piercing mill is again possible (e.g., see specimen 4 in Figure 4.31). The combined effect of the four forming steps is that the original 5 m round has been formed into a tube measuring as much as 160 m in length.

The hot tubes are left to cool down in a large cooling bed. Subsequently they are cut to desired lengths and are cold straightened as shown in the figure. The testing and finishing steps, including heat treatment, are similar to those described above for the plug mill.

3.3.4 Pilger Mill

The *pilger* mill process is another invention of the Mannesmann brothers [3.19]. The name "pilger" (pilgrim) is associated with a unique forming step at the start of the process described below. Presently, pilger mills are used to manufacture thicker wall tubes. Typical diameter capacities are 9.625 to 26 in, while the thickness varies from 25 to 60 mm for the smaller diameter tubes and from 20 to 110 mm for the 26 in ones. Pilger forming has the advantage that it can manufacture tubes to specified IDs, a capability very advantageous for the manufacture of buckle and crack arrestors. Today the number of such mills around the world is relatively small, being used primarily to manufacture oil field tubulars, boiler tubing and tubes for other applications.

The main steps of the process are shown schematically in Figure 3.8. The starting material is conventionally cast, relatively short and thick round ingots, weighing up to 7 t. They are heated in a rotary-hearth furnace to about 1,280°C. Each hot ingot is center-punched using a high load capacity press. As in other piercing processes, some eccentricity is introduced to the part in this step, which will remain to some extent in the finished pipe. The punch leaves a plug at the bottom, which is removed by piercing in the next step.

The pierced ingot goes next to the pierce rolling mill for the first diameter and wall thickness reduction. The part is rolled by double conical rolls over a conical die held in the annulus with a long rod, as shown in the figure. The rollers operate in a manner similar to that of the pierce rolling mill of the plug mill, rotating the part while simultaneously forcing it to move forward.

Next, the hollow ingot moves to the pilger mill, the main forming step of the process. A long mandrel is inserted in the annulus, and the part is engaged by two kidney-shaped rollers (see Figure 3.8) running eccentrically. The process is reciprocating, where in the first phase the part moves forward while it engages the thick part of the rollers. The second phase commences when the thin part of the rollers comes into play. The part disengages from the rollers and is pulled backwards half the forward distance. In the next forward phase, the part is rotated by 90° and the cycle is repeated. This "two steps forward, one backwards" cycle is reminiscent of a certain pilgrim procession, and thus the process was given the name "pilger." The reciprocating rolling leaves small hills and valleys in the OD, which have some reflection on the finished tube. Once the whole length is formed in this manner, the mandrel is removed and is placed in a water bath for cooling. A new mandrel is used to roll the next part. At the exit of the pilger mill the back end of the tube, which was engaged to the head of the mandrel and distorted, is removed by hot sawing.

The tube goes next to a cooling bed, where it is allowed to cool for grain refinement. It is subsequently reheated in a furnace to about 1,200°C and is moved to the sizing mill. The sizing mill is similar to the one described in the plug mill section. It consists of

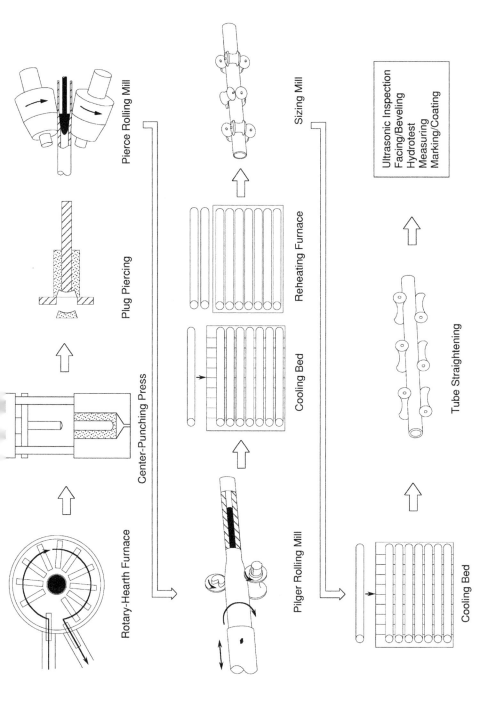

Rotary-Hearth Furnace

Pierce Rolling Mill

Plug Piercing

Center-Punching Press

Sizing Mill

Reheating Furnace

Cooling Bed

Pilger Rolling Mill

Ultrasonic Inspection
Facing/Beveling
Hydrotest
Measuring
Marking/Coating

Tube Straightening

Cooling Bed

Figure 3.8 Schematic representation of major manufacturing steps of a pilger mill for seamless pipe [3.20].

several sets of three or four rollers, each of which reduces the tube diameter and its wall thickness. Because adjoining roller sets are at a 60° or 45° angle to their neighbors, the ID of the finished tube can be star-shaped, superimposed on the eccentricity introduced at the punching press.

On exiting the sizing mill, the tubes cool off on a cooling bed in order to achieve the required microstructure. The cold tubes finally pass through the straightening mill. The mill consists of relatively flat rollers arranged at an angle to the axis of the tubes, as shown in Figure 3.8.

The tubes move next to the finishing lines, where they can undergo heat treatment, end machining, and the inspections and markings described in Section 3.3.2.

The stress–strain response of seamless steel pipe produced by any of the three methods described above will exhibit Lüders banding behavior (see Figures 13.3 and 13.4) if it is either hot finished or quenched and tempered at the end of the manufacturing process. This has implications in some mechanical performance criteria. The Lüders bands are most often erased by cold finishing the pipe. They can also be avoided by special alloying, but this is less common. Some variations in casting as well as in the rolling and heat treatment are unavoidable. As a result, within an order or a heat, variations in yield and ultimate strength usually exist. Variations in the range of 100–120 MPa (14–17 ksi) are typical. In order to meet the minimum specified yield strength requirement for all pipes, the average value of yield strength in an order usually exceeds the specified value.

Seamless pipe usually has excellent circularity, with ovality values (Δ_o) ranging between 0.1% and 0.2%. Wall eccentricity values (Ξ_o) are typically below 7%. Both are much lower than the minimum specified in API Spec. 5L [3.7].

3.4 ELECTRIC RESISTANCE WELDED PIPE

In the *electric resistance welded* (*ERW*) pipe manufacturing process, the pipe starts as a long strip of plate. The strip is formed into a closed circular shape, and the seam is welded by a heat induction process. ERW pipe is more economical than both UOE and seamless pipes because the process is continuous and has fewer steps, and as a result is more efficient. It is available in diameters which range from 2.375 to 24 in and wall thicknesses from 2 mm for the small diameter pipes to 18 mm for larger diameters [3.23, 3.24].[†]

The plate strip is rolled out of either an ingot or slab produced by a continuous vertical casting process similar to that described in Section 3.2. Degassing, desulfurizing and accurate alloying enable production of the fine-grained steel required to meet the multiple demands of linepipe. The long strip is hot-rolled into coils which can be as long as 1,000 ft (300 m, note that this varies with wall thickness and diameter) and sent to the ERW mill. The pipe manufacturing process is shown schematically in Figure 3.9. The plate is uncoiled and leveled by passing it through a series of rollers. The edges of the flat plate strip are then trimmed by milling, preparing them for welding. The plate is inspected for defects ultrasonically and is formed into a closed circular shape. A series of shaped rollers form it first into an open "C" shape. The C is forced closed by a pair of rollers with semicircular shape. The two edges are further pressed together by a four-roll stand.

[†]We thank Beau Urech of Lone Star Steel for providing information on the ERW process.

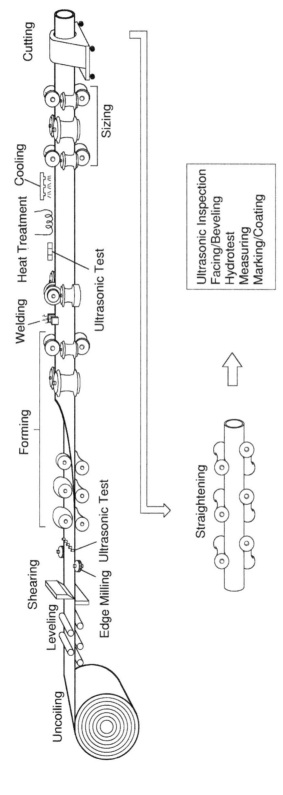

Figure 3.9 Schematic representation of major manufacturing steps of an ERW pipe mill [3.23, 3.24].

A high frequency (HF) induction heater locally melts the mating surfaces, and the pressure causes them to weld together (i.e., no welding electrode is used). This is the most important step in the process, as inadequate or excessive heat input can make the weld defective. The heat input is kept at the optimum level by closed-loop computer control of the heaters. The resulting weld bead is then removed both from the inside and outside and the weld is inspected ultrasonically for defects. Electric resistance welding results in grain coarsening in the weld itself and in the heat affected zone. This is corrected by local heat treatment of the weld. Carefully calibrated heat treatment can produce nearly homogeneous properties in the main body and the welds. The weld is water-cooled, and the pipe then passes through a series of sizing roller stands where it is cold-formed to the final diameter. The result is a high circularity cross section. Following the sizing, the pipe is cut into sections of 40 to 60 ft (12–18 m) but even longer if required. The efficiency of ERW manufacturing lies in the fact that all the steps, from uncoiling of the plate to cutting the finished pipe into the desired lengths, take place along one linear continuous process. Depending on the diameter and thickness, mills can produce 60 to 240 pipes per hour.

Subsequently, the pipe sections go through a straightening mill usually involving rollers such as those shown schematically in Figure 3.9. The rollers turn the pipe while simultaneously forcing it forward and straightening it. Instead of the local heat treatment of the weld shown in the figure, some mills will quench and temper the pipe following straightening, whereas others heat treat the whole pipe after welding. This breaks the continuity of the process to some degree and reduces the throughput, but may improve corrosion resistance. The final finishing and testing steps for ERW pipes are similar to those of seamless pipe, although here a last inspection of the weld is often performed (usually ultrasonically).

In addition to its more efficient production, ERW pipe can be produced to very uniform thickness held to tight tolerance. For example, eccentricity is negligibly small and can be assumed to be zero in collapse calculations. This reduces the net weight of the pipe, resulting in a cost benefit. In addition, the pipe can be produced with relatively low ovalization. The uniform thickness and good circularity help speed up welding either in the field or on a lay-barge, which provides an additional cost reduction.

Despite these benefits, today's ERW pipe is mainly used for land pipelines and for less demanding offshore applications. ERW pipe produced in the early stages of the development of the process had a poor track record. The pipe welds developed problems, and the heat-affected zone had a tendency to corrode. Although these problems have largely been overcome by technological developments such as HF induction heating and more careful heat treatment, there remains some lingering resistance to placing ERW pipe in demanding offshore applications, despite the cost savings it can provide. By contrast, ERW tubes are more widely used in casing applications.

3.5 SPIRAL WELD PIPE

Another continuous process of producing pipe of large diameter (typically diameters 20–100 inches) is the *spiral weld* process. A continuous plate strip is rolled to the required width and wall thickness and hot-rolled into a coil similar to the ERW plate coils. At the pipe mill, pipe is formed as shown schematically in Figure 3.10. The plate is uncoiled and leveled by passing it through a series of rollers. The edges are then trimmed and

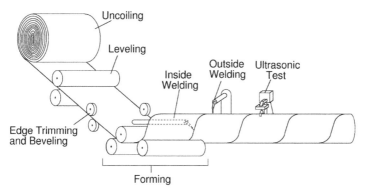

Figure 3.10 Schematic representation of a spiral weld pipe mill.

beveled. The strip, appropriately angled, goes next to a precision forming stand where it is rolled into a circular shape as shown in the figure. The pipe diameter is decided by the spiral angle and by the plate width. The seam is welded usually through the *submerged arc welding (SAW)* method (see Section 3.6). The pipe is first welded on the inside and then on the outside. The weld is first inspected ultrasonically in line. The welded pipe then passes through a series of rollers (not shown in the figure) in order to enhance its circularity. The pipe is cut into the required lengths, which can be ordered longer than standard lengths for UOE pipe (12 or 18 m). Each length goes though radiographic inspection, hydrotesting, edge beveling and marking, as outlined in the preceding manufacturing processes.

Spiral weld pipe can be made to larger diameters than customarily produced by other processes. The continuous process through which it is made is one of the most efficient, a fact that has a corresponding impact on its price. Because the pipe is rolled from plate, wall thickness uniformity and tolerances are very good. On the other hand, out-of-roundness can be more significant than in UOE pipe. The long spiral weld and the out-of-roundness have made spiral weld pipe less common in offshore applications involving high internal or external pressure. Offshore, it is used as conductors, as large diameter piles, and in some low pressure applications. On land it is commonly used in water mains, tunnel liners, power plants and in other low pressure applications.

3.6 UOE PIPE MANUFACTURE

Pipe larger than 16 inches is usually cold-formed from plate through the *UOE* process. The plate is typically produced in the manner described in Section 3.2. It is formed into a circular shape through the four mechanical steps shown schematically in Figure 3.11.[†] The plate edges are first crimped into circular arcs. The plate is then formed into a U-shape in the "U-press" (see also Figure 3.12(a)). It is then pressed into a circular shape in the "O-press" (see Figure 3.12(b) and (c)). The seam is subsequently welded using submerged arc welding. The final step involves mechanical expansion of the welded pipe in order to improve its circularity (Figure 3.12(d)). The name UOE stems from the initials

[†]We are grateful to David Brereton, Peter Tait and Mark Fryer of Corus Tubes, Inc. for many helpful exchanges on UOE pipe forming offered over a period of several years.

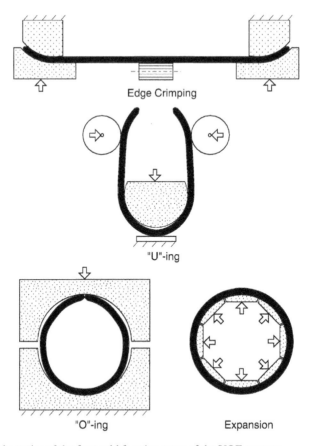

Figure 3.11   Schematics of the four cold forming steps of the UOE process.

of three of these mechanical steps. Such pipe is sometimes also called *SAW* because of the way it is welded. UOE pipe is available up to a diameter of 64 in and lengths of 40 or 60 ft (12.2 or 18.3 m). Wall thickness capacities vary with steel grade and diameter. Presently, for X65 grade steel, 1.75 in (45 mm) is the maximum wall thickness capacity available industry-wide.

The process was initially developed by Kaiser Steel Co. of California. Today, modern versions of the process exist in Japan, Europe, South America and the USA. Figure 3.13 shows a more detailed flow chart of the steps followed in the manufacture of UOE pipe [3.25]. In most mills, the steps up to the expansion are arranged linearly. This makes the UOE mill up to 1 km long. The mills are highly automated via computer control, and can run three shifts a day for increased productivity. Despite the thorough inspection plates undergo at the plate mill, the process starts with ultrasonic inspection of the plate for through-thickness flaw detection. Tabs are then welded at the four corners. The tabs are later used to start welding outside the pipe in order to ensure continuity of the weld in the pipe body. The tabs are removed after welding is completed. The longitudinal edges of the plate are trimmed by milling, bringing the width to the exact required value. Simultaneously, the ends are beveled to later form v-grooves in the circular skelp to accommodate the welding.

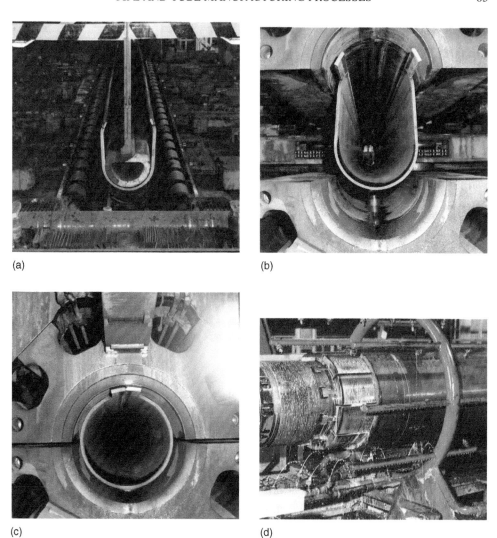

(a)   (b)

(c)   (d)

Figure 3.12   Photographs of (a) U-press, (b) and (c) O-press, and (d) pipe expansion (courtesy, Corus Tubes, UK).

The first mechanical step involves crimping of the edges of the plate into circular arcs over a width of about one radius on each side. This is achieved by pressing the ends between two shaped dies as shown in Figure 3.11. Crimping is done in a step fashion, involving lengths of one to four pipe diameters at a time (depending on the wall thickness).

The plate moves next to the U-press, where it initially rests centered between a pair of side rollers that run along its entire length. The U-punch moves down and bends the entire plate through three-point bending. The U-punch stops when the plate contacts a series of anvils set at a predetermined height. The U-punch is then held in place, and the side rollers are moved inwards as shown in Figure 3.11. The vertical position and inward travel of the rollers are selected such that the final position of the straight arms of the U-shaped skelp are nearly vertical.

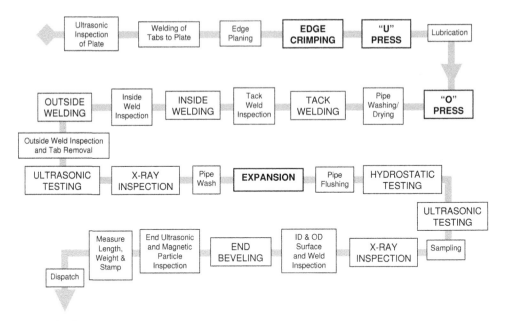

Figure 3.13   Flow chart of the major steps of the UOE pipe manufacturing process.

The skelp is then conveyed to the O-press, which consists of two semi-circular stiff dies as shown in Figure 3.12(b). A lubricant is applied to the skelp to reduce friction during the O-ing, and the top die is actuated downwards. This forces the skelp into a nearly circular shape (Figure 3.12(c)). Once the pipe is thus formed, the dies are pressed further together, producing a net compressive strain of 0.1–0.2%. Corus' 50,000 t O-press is presently the most powerful in the industry. It can produce thicker pipes at 40 ft (12.2 m) lengths.

After leaving the O-press, the pipe is washed and dried in preparation for welding. The pipe is passed through a series of pressure rolls to ensure the two edges are aligned, and is tack welded along its entire length. After the tack welds are inspected, the pipe is passed onto submerged arc welding machines that weld it first on the inside and then on the outside. This involves several electric arcs struck between bare metal electrodes (typically 4) and the pipe. A blanket of fusible material covers the arcs and the molten metal for protection. Figure 3.14 shows a macrograph of such a weld. The inside and outside welds are seen with their characteristic "bowler hat" shapes, forming small peaks at their outer surfaces. Heat affected zones (HAZs) can be seen at the interface between the base material of the welds. Extensive ultrasonic examination is performed on the weld before the pipe is expanded. Defects located either at the interface between the two welds or between the welds and the base material are detected, verified by X-ray radiography and repaired. The welds and HAZs are also evaluated by hardness and Charpy V-notch impact tests (see Appendix A). For the particular case shown in Figure 3.14, at mid-thickness the pipe steel had Hv10 (Vickers10) hardness of 212, the HAZs an average of 192 and the weld 235 (see Appendix A). This is indicative of the overmatching of yield strength of welds to that of the pipe that is a usual practice.

Expansion is accomplished by an internal mandrel shown in Figure 3.12(d). The mandrel consists of 8, 10, or 12 segments. Segments are chosen so that their radii are near

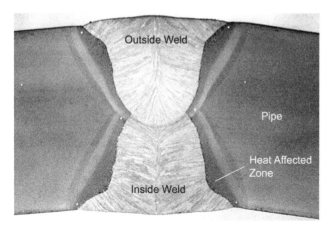

Figure 3.14 Macrograph showing the weld of a 30 in diameter, 1.575 in (40 mm) wall UOE pipe. (Courtesy Corus Tubes)

that of the inside of the pipe. One of the mandrel segments has a groove to accommodate the pipe weld. The mandrel is hydraulically actuated, and in one step it typically expands a length of one half to one diameter (depending on the wall thickness). The mandrel is inserted in the pipe and expanded radially while a lubricant is simultaneously applied. It is then contracted and moved down the pipe, maintaining some overlap between the expanded and unexpanded pipe sections, and the process is repeated. Expansion improves the roundness and straightness of the pipe and brings it to its desired final size. To achieve low ovality, the pipe is typically expanded 0.8–1.3% from its diameter after the O-step.

The pipe is washed once more and hydrotested to an internal pressure specified by the customer. The seam is ultrasonically inspected once more and X-rayed. The final inspection of both the inside and outside of the pipe is manual. The ends are then beveled, the pipe is weighed, marked with its individual pipe number, etc., and sent for dispatch.

UOE pipe has been widely used for land pipelines, including the Trans-Alaska and Trans-Siberia pipelines. In the last 15 years, it has also been increasingly used in offshore applications where collapse under external pressure is a primary design consideration and as a result high circularity is required. Figure 3.15 shows the cross section of a 24 in UOE pipe. The shape was measured both on the inside and outside surfaces as described in Sections 4.5.1 and 4.5.2. Radial displacement variations are amplified 20 times. UOE pipe usually has nearly uniform thickness. The nearly circular shape has some inperfections, which depend on each of the four mechanical steps. Ovality on the range of 0.1–0.3% is typical, but can be slightly higher for thicker wall pipe. Straightness is usually within the API specified value of 0.2% of the length. The deviation can be larger for underexpanded thicker pipes.

An important consideration for collapse-sensitive pipes is the compressive response of the pipe material. In particular, the cold work through which UOE pipe is formed introduces changes in the compressive stress–strain response of the pipe in the circumferential direction. The plate stress–strain response is usually linear up to yield. It is followed by a stress plateau associated with Lüders banding, which typically extends to a strain of 1%–3%. This is followed by hardening behavior. The cold work, and in particular the

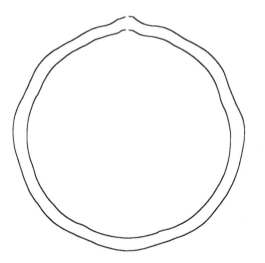

Figure 3.15 Cross section of a 24 in diameter, 1.250 in (31.75 mm) UOE pipe ($w \times 20$).

expansion step, result in rounding of the response and lowering of the yield stress in compression [3.26, 3.27]. As a consequence, the collapse pressure of UOE pipe is lower than that of seamless pipe of the same grade. The extent of this reduction in collapse pressure and remedies that can alleviate it are discussed in Chapter 5.

3.7 JCO FORMING

An alternative to the UOE large-diameter pipe forming process is the so-called *JCO* process (e.g., see [3.28]). In this process the U and O steps are replaced by the JCO forming steps. The plate preparation up to, and including, crimping follows the same steps as those in the flow chart in Figure 3.13. The crimped plate is fed into a press where the whole length is shaped simultaneously by local pressing of part of the circumference by a shaped tool, as shown in Figure 3.16. Pressing starts at one edge of the plate and progresses towards the mid-width. The first N strokes result in a J-shaped cross section as shown in Figure 3.16(a). The process is then repeated, this time starting from the other edge of the plate. After $2N$ stokes, the cross section has been formed into a C-shape as shown in Figure 3.16(b). The last stroke, centered at the mid-width, closes the C into an O shape as shown in Figure 3.16(c). The pipe seam is first tack-welded and then welded using submerged arc welding. The inside is welded first, followed by the outside, as described in Section 3.6. The welded pipe is finished by circumferential expansion of about 1%, using expanders similar to those employed in the UOE process. Figure 3.17 shows the cross section of a 30 in X60 pipe with a wall thickness of 1.383 in (35.13 mm) formed by JCO. With an ovality of $\Delta_o = 0.144\%$ the pipe circularity is excellent. The inner and outer surfaces were scanned individually as described in Sections 4.5.1 and 4.5.2. The radial deviation from the corresponding mean circles is exaggerated $20\times$ and consequently the locations where the indention tool contacted the inner surface are visible.

Overall JCO pipe is very similar in both shape and mechanical properties to UOE pipe. Consequently it also exhibits a lower collapse pressure than corresponding seamless pipe of the same D/t and grade. An advantage of JCO is that thicker walls can be accommodated

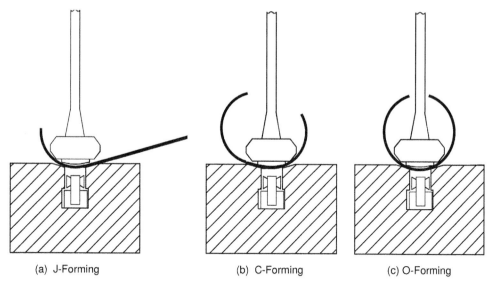

(a) J-Forming   (b) C-Forming   (c) O-Forming

Figure 3.16 Forming of plate by incremental pressing by a shaped tool in the JCO process

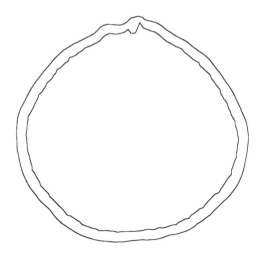

Figure 3.17 Cross section of a 30 in diameter, 1.383 in (35.13 mm) JCO pipe ($w \times 20$).

more easily, as the step-by-step forming does not require as large a force as that required in the O-press. A disadvantage is that in this process, expansion at the level of 1% is essential for good circularity. As a result, the potential benefits that can be achieved by increasing the O-strain and simultaneously decreasing the expansion strain, recommended in Chapter 5, are not feasible here.

REFERENCES

3.1. Craig, B.D. (1993). *Practical Oilfield Metallurgy and Corrosion*, 2nd Edition. Pennwell Publishing Co., Tulsa, Oklahoma.

3.2. Tamehiro, H. and Chino, H. (1991). The progress in pipeline material properties. Nippon Steel Corporation. *Proc. AIB-Vicotte Leerstoel*, April 1991, National Fonds Voor Wetenschappelijk Onderzoek Laboratorium Soete, Rijksuniversiteit Gent.

3.3. Fruehan, R.J. Ed. (1998). Association for Iron & Steel Engineers Steel Foundation. *The Making, Shaping and Treating of Steel, 11th Edition: Steelmaking and Refining Volume*. Pittsburgh, Pennsylvania.

3.4. Wakelin, D.H. Ed. (1999). Association for Iron & Steel Engineers Steel Foundation. *The Making, Shaping and Treating of Steel, 11th Edition: Ironmaking Volume*. Pittsburgh, Pennsylvania.

3.5. Honeycombe, R.W.K. (1981). *Steels: Microstructure and Properties*. Edward Arnold, London.

3.6. Hawkins, D.T. and Hultgren, R. (1972). Constitution of binary alloys. In *Metals Handbook. Vol. 8, Metallography, Structures and Phase Diagrams*, 8th Edition, T. Lyman Ed. American Society for Metals, pp. 251–338. Metals Park, Ohio.

3.7. API Specifications 5L: Specifications for Line Pipe, American Petroleum Institute, 43rd Edition, 2004.

3.8. Harste, K., Klingbeil, J., Schwinn, V., Bannenberg, N. and Bergmann, B. (2000). The new continuous caster at Dillinger Hütte to produce semi products for high quality heavy plates. *Stahl und Eisen* **120**(2), 53–59.

3.9. Bannenberg, N., Bruckhaus, R., Harste, K. and Lachmund, H. (2000). Steelmaking: production of steel for high-grade plate. *Millenium Steel*, 140–144.

3.10. *Steel Plant–Technical Information*. Dillinger Hütte Brochure.

3.11. Tsuchida, Y., Nakada, M., Sugawara, I., Miyahara, S., Murakami. K. and Tokushige, S. (1984). Behavior of semi-macroscopic segregation in continuously cast slabs and technique for reducing the segregation. *Trans. Iron Steel Instit. Jpn* **24**, 899–906.

3.12. Cramb, A. Ed. (2003). Association for Iron & Steel Engineers Steel Foundation. *The Making, Shaping and Treating of Steel, 11th Edition: Casting Volume*. Pittsburgh, Pennsylvania.

3.13. Streisselberger, A., Flüss, P. and Kolling, A. (1998). Controlling homogeneity and reproducibility in the production of heavy plates. *Steel Res.* **69**(4,5), 136–142.

3.14. Streisselberger, A., Flüss, P., Bauer, J. and Bennett, C.J. (1999). Modern line pipe steels designed for sophisticated subsea projects for sweet and sour gas. *Proc. International Society on Offshore and Polar Engineering*, Vol. II, Brest, France, May 30 to June 4, 1999, pp. 125–131.

3.15. Bauer, J., Schwinn, V. and Tacke, K.-H. (2002). Recent quality achievements on steel plate for line pipe. *Proc. 4th Int'l Pipeline Conference*, Calgary, Canada, September, 2002.

3.16. Liestmann, W.D., Gruner, H., Sardemann, J. and Schrewe, H.F. (1983). The six-strand bowcaster for rounds of Mannesmannrohren-werke AG. *Iron and Steelmaker*, December 1983, pp. 26–33.

3.17. Wünnenberg, K. and Jacobi, H. (1988). Casting of round sections for tubemaking. *Steel Technology International*, pp. 225–230.

3.18. Jacobi, H. and Wünnenberg, K. (1997). Final solidification, centre segregation and precipitation phenomena in continuously cast round billets. *Steel Res.* **68**(6), 258–265.

3.19. *The Dusseldorf-Rath Works*. Vallourec & Mannesmann Tubes, Brochure V&M-112e.

3.20. *Nippon Steel Pipe & Tube*. Nippon Steel Corporation, Brochure Cat. No. EXE 501 1992.8.

3.21. *United States Steel Tubular Products*. ADUSS 44-8698-03, July 2001.

3.22. *The Mülheim Mandrel Mill*. Vallourec & Mannesmann Tubes, Brochure S 28/29.

3.23. *HF-T ERW Line Pipe*. Sumitomo Metals, Brochure No. 15421, 1984.

3.24. *Nippon Steel ERW Pipe*. Nippon Steel, Brochure EXE520, 2000.6.

3.25. *42″ Pipe Mill*. British Steel Brochure No. TD 376/3E/96R, 1996. Also, *42″ SAW Pipe Mill*, Corus Tubes, Brochure CT13:3000:UK:08/2001.

3.26. Kyriakides, S., Corona, E. and Fischer, F.J. (1991). On the effect of the U-O-E manufacturing process on the collapse pressure of long tubes. *Proc. Offshore Technology Conference*, **4**, OTC6758, 531–543, 1991. Also, *ASME J. Engg Indus.* **116**, 93–100, 1994.

3.27. Herynk, M.D., Kyriakides, S., Onoufriou, A. and Yun, H.D. (2007). Effects of the UOE/UOC manufacturing process on pipe collapse pressure. *Intl J. Mech. Sci.* **49**, 533–553.

3.28. "Longitudinal SAW Pipes." Process. Welspun Gujarat Stahl Rohren, Ltd. 11 March 2006. http://www.welspunpipes.com/longisawpipe.asp

4
Buckling and Collapse Under External Pressure

Offshore pipelines are commonly installed empty in order to reduce the installation tension due to the weight of the suspended section (see Chapter 2). In addition, during operation they are periodically depressurized for maintenance. Thus, external pressure is an important load parameter in design; indeed, it is often the prime parameter. This chapter deals with the mechanics of buckling and collapse of long pipes under external pressure. Thinner pipes used in shallower waters buckle elastically, but collapse due to postbuckling inelastic action. The classical elastic buckling pressure is derived in Section 4.1, followed by the derivation of Timoshenko's design formula for the onset of collapse of an initially ovalized pipe. Thicker pipes used in deeper waters buckle and collapse in the plastic range. In Section 4.2, plastic buckling equations are derived. Their relative simplicity makes them useful tools in design. Practical factors that affect collapse include initial imperfections such as ovality and wall thickness variations. Other factors include residual stresses, yield anisotropy, etc. The influence of such factors is best treated numerically. Section 4.3 presents the external pressure part of the formulation and solution procedure of the custom computer code BEPTICO [4.1], capable of including these aspects of the problem. Section 4.4 presents a parametric study of buckling and collapse of elastic–plastic pipe under external pressure. Section 4.5 outlines how actual pipe imperfections can be measured. Examples of typical shape and wall thickness imperfections in seamless pipes are presented and used in collapse calculations.

4.1 ELASTIC BUCKLING

We consider a long, circular cylindrical shell of mean radius R and wall thickness t under external pressure. Such a structure buckles into axially uniform modes. Thus, problem variables depend only on the polar coordinate θ defined in Figure 4.1(a). The nonlinear, small strain, moderate rotation kinematics appropriate for establishing the buckling pressure are as follows [4.2–4.4]:

$$\varepsilon_{\theta\theta} = \varepsilon_{\theta\theta}^o + z\kappa_{\theta\theta}, \tag{4.1a}$$

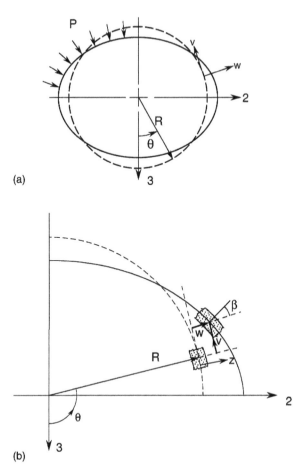

Figure 4.1 (a) Pipe deformed and undeformed cross section and (b) definition of mid-surface displacements.

where

$$\varepsilon^o_{\theta\theta} = \frac{v' + w}{R} + \frac{1}{2}\left(\frac{v - w'}{R}\right)^2 \quad \text{and} \quad \kappa_{\theta\theta} = \frac{v' - w''}{R^2}. \tag{4.1b}$$

Here v and w are respectively the circumferential and radial displacements of the mid-surface, as shown in Figure 4.1(b) and $(\bullet)' \equiv (\bullet)_{,\theta}$. (Observe that Eqs. (4.1) are consistent with Sanders' equations in Appendix D.) The corresponding equilibrium equations can be derived from the potential energy given by

$$V = \int_0^{2\pi} \frac{1}{2}\left[N_{\theta\theta}\varepsilon^o_{\theta\theta} + M_{\theta\theta}\kappa_{\theta\theta}\right]R\,d\theta + PR\int_0^{2\pi}\left[w + \frac{1}{2R}(v^2 + w^2 - vw' + v'w)\right]d\theta, \tag{4.2}$$

where the external pressure P is positive [4.5]. The force and moment intensities in (4.2) are defined in the customary manner as follows

$$N_{\theta\theta} = \int_{-t/2}^{t/2} \sigma_{\theta\theta}\, dz = \frac{Et}{(1-v^2)}\varepsilon_{\theta\theta}^o \equiv C\varepsilon_{\theta\theta}^o,$$

(a)

$$M_{\theta\theta} = \int_{-t/2}^{t/2} \sigma_{\theta\theta} z\, dz = \frac{Et^3}{12(1-v^2)}\kappa_{\theta\theta} \equiv D\kappa_{\theta\theta}.$$

Using variational calculus, the following equilibrium equations are derived from (4.2)

$$RN_{\theta\theta}' + M_{\theta\theta}' - RN_{\theta\theta}\beta - PR^2\beta = 0,$$

(4.3)

$$M_{\theta\theta}'' - RN_{\theta\theta} - R(N_{\theta\theta}\beta)' - PR(v'+w) = PR^2,$$

where $\beta = (v - w')/R$ is the rotation of the normal to the mid-surface shown in Figure 4.1(b). The prebuckling solution is the uniform membrane state $N_{\theta\theta o} = -PR$ and $\varepsilon_{\theta\theta o}^o = w_o/R$ related through Eq. (a). Bifurcation buckling equations are established by perturbing the prebuckling state with the buckling mode $(\tilde{v}, \tilde{w})$ and linearizing (4.3), which results in the following:

$$RN_{\theta\theta}' + M_{\theta\theta}' = 0,$$

(4.4a)

$$M_{\theta\theta}'' - RN_{\theta\theta} + PR^2\left(\frac{\tilde{v}' - \tilde{w}''}{R}\right) - PR(\tilde{v}' + \tilde{w}) = 0,$$

along with the following linearized kinematical relationships:

$$\varepsilon_{\theta\theta}^o = \frac{\tilde{v}' + \tilde{w}}{R} \quad\text{and}\quad \kappa_{\theta\theta} = \frac{\tilde{v}' - \tilde{w}''}{R^2}.$$

(4.4b)

The following buckling mode satisfies (4.4) and has the required periodicity

$$\tilde{w} = a\cos n\theta \quad\text{and}\quad \tilde{v} = b\sin n\theta.$$

(4.5)

Substituting (4.5) into (4.4) results in

$$\left[\begin{matrix} n(1+\rho n^2) & n^2(1+\rho) \\ (1+\rho n^4) - \gamma(n^2-1) & n(1+\rho n^2) \end{matrix}\right]\left\{\begin{matrix} a \\ b \end{matrix}\right\} = \mathbf{0},$$

(b)

where

$$\rho = \frac{D}{CR^2} = \frac{1}{12}\left(\frac{t}{R}\right)^2 \quad\text{and}\quad \gamma = \frac{PR(1-v^2)}{Et}.$$

For nontrivial solutions the determinant of matrix (b) must be zero, which results in the following sequence of eigenvalues:

$$P_n = \frac{(n^2-1)}{12(1+\rho)}\frac{E}{(1-v^2)}\left(\frac{t}{R}\right)^3, \quad n = 2, 3, \ldots$$

(c)

Since for the relatively high R/t tubes for which elastic buckling is applicable $\rho \ll 1$, it is neglected by comparison to 1. The lowest eigenvalue corresponds to $n = 2$, which is an ovalization-type mode. The critical buckling pressure is then given by the following classical result (updated version of ring buckling pressure of Levi [4.6, 4.7])

$$P_C = \frac{E}{4(1 - v^2)} \left(\frac{t}{R}\right)^3 = \frac{2E}{(1 - v^2)} \left(\frac{t}{D_o}\right)^3 \quad (D_o = 2R). \quad (4.6)$$

4.1.1 Imperfect Pipe

We now consider a pipe with an axially uniform geometric imperfection defined by $(\bar{v}, \bar{w})$. The linearized buckling Eqs. (4.4) become

$$RN'_{\theta\theta} + M'_{\theta\theta} = 0,$$

$$M''_{\theta\theta} - RN_{\theta\theta} + PR^2 \left(\frac{v' - w''}{R} - \frac{\bar{v}' - \bar{w}''}{R}\right) - PR(v' + w - \bar{v}' - \bar{w}) = 0, \quad (4.7a)$$

with

$$\varepsilon^o_{\theta\theta} = \frac{v' + w}{R} \quad \text{and} \quad \kappa_{\theta\theta} = \frac{v' - w''}{R^2}, \quad (4.7b)$$

where v and w are measured from the circular reference geometry. We specialize the imperfection to the first buckling mode (uniform ovality) given by

$$\bar{w} = -a \cos 2\theta \quad \text{and} \quad \bar{v} = \frac{a}{2} \sin 2\theta. \quad (4.8)$$

By inspection, the solution is

$$w = A \cos 2\theta \quad \text{and} \quad v = B \sin 2\theta, \quad (d)$$

which results in the following equations for A and B

$$\begin{bmatrix} 2(1 + 4\rho) & 4(1 + \rho) \\ (1 + 16\rho) - 3\gamma & 2(1 + 4\rho) \end{bmatrix} \begin{Bmatrix} A \\ B \end{Bmatrix} = \begin{Bmatrix} 0 \\ -3a\gamma \end{Bmatrix}. \quad (e)$$

The solution of (e) is as follows:

$$w = \frac{-aP}{P_C - P} \cos 2\theta \quad \text{and} \quad v = \frac{aP}{2(P_C - P)} \sin 2\theta, \quad (4.9)$$

in which again $\rho \ll 1$ and was neglected by comparison to 1. Figure 4.2 shows a plot of the pressure–maximum displacement response for several values of imperfection amplitude (the initial ovality variable $\Delta_o = a/R$). The tube ovality grows with pressure and becomes singular as the critical pressure of the perfect structure is approached. Along with the ovality, the bending stresses grow, and the material eventually yields.

Timoshenko [4.8] suggested that in design, the onset of yielding should be considered as a conservative upper bound for the collapse pressure. The tube yields when the membrane

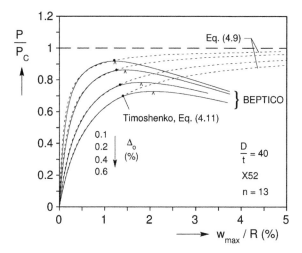

Figure 4.2 Pressure–maximum displacement responses for an imperfect tube. Results from linear elastic analysis and from BEPTICO for elastic–plastic material.

and bending stresses add up to the yield stress (σ_o). The moment and force intensities are

$$M_{\theta\theta} = \frac{D}{R^2}(v' - w'') = \frac{-E}{4(1-v^2)}\left(\frac{t}{R}\right)^3 \frac{PRa}{(P_C - P)}\cos 2\theta \quad \text{and} \quad N_{\theta\theta} \cong -PR.$$

Thus, the condition for first yielding is

$$\sigma_o = \left|\frac{N_{\theta\theta}}{t}\right| + \left|\frac{6M_{\theta\theta\,\text{max}}}{t^2}\right| = \frac{PR}{t} + \frac{6aRPP_C}{(P_C - P)t^2}. \tag{f}$$

If we associate collapse with the onset of yielding, then (f) is equivalent to

$$P_{CO}^2 - (P_o + \psi P_C)P_{CO} + P_o P_C = 0, \tag{g}$$

where P_o is the yield pressure defined as

$$P_o = \frac{\sigma_o t}{R} = \frac{2\sigma_o t}{D_o}, \tag{4.10}$$

and

$$\psi = \left(1 + 3\Delta_o \frac{D_o}{t}\right).$$

The solution of (g) is

$$P_{CO} = \frac{1}{2}\left\{(P_o + \psi P_C) - \left[(P_o + \psi P_C)^2 - 4P_o P_C\right]^{1/2}\right\}, \tag{4.11}$$

which is the long tube/pipe equivalent to Timoshenko's result derived for inextensional ring kinematics. For example, for an X52 pipe with $D/t = 40$ this formula yields the collapse pressures marked with solid dots on the $P - w_{\max}$ results in Figure 4.2. Clearly, according to this criterion the collapse pressure is imperfection sensitive. The validity of this formula and its limitations are discussed in Section 4.4.3.

4.2 PLASTIC BUCKLING

The linearized buckling equations in terms of the force and moment intensities are similar to those of the elastic case, but now are expressed in incremental form represented here by $(\dot{\bullet})$ as follows [4.9,4.10] $((\dot{\bullet})$ represents the increment of $(\bullet))$:

$$R\dot{N}'_{\theta\theta} + \dot{M}'_{\theta\theta} = 0,$$

$$\dot{M}''_{\theta\theta} - R\dot{N}_{\theta\theta} + PR^2 \left(\frac{\dot{v}' - \dot{w}''}{R} \right) - PR(\dot{v}' + \dot{w}) = 0. \tag{4.12}$$

The appropriate linearized kinematic relations are

$$\dot{\varepsilon} = \dot{\varepsilon}^o + z\dot{k} = \left\{ \begin{matrix} \dot{\varepsilon}^o_{xx} \\ \dot{\varepsilon}^o_{\theta\theta} \end{matrix} \right\} + z \left\{ \begin{matrix} \dot{k}_{xx} \\ \dot{k}_{\theta\theta} \end{matrix} \right\},$$

where

$$\dot{k}_{xx} = 0, \tag{4.13}$$

$$\dot{\varepsilon}^o_{\theta\theta} = \frac{\dot{v}' + \dot{w}}{R} \quad \text{and} \quad \dot{k}_{\theta\theta} = \frac{\dot{v}' - \dot{w}''}{R^2}.$$

Constitutive Equations

When calculating prebuckling equilibrium states involving nonproportional stress histories, the J_2 flow theory with isotropic hardening is the most appropriate (see Section 13.2). For the present problem (13.20) reduce to

$$\left\{ \begin{matrix} \dot{\varepsilon}_x \\ \dot{\varepsilon}_\theta \end{matrix} \right\} = \frac{1}{E} \begin{bmatrix} 1 + Q(2\sigma_x - \sigma_\theta)^2 & -v + Q(2\sigma_x - \sigma_\theta)(2\sigma_\theta - \sigma_x) \\ -v + Q(2\sigma_x - \sigma_\theta)(2\sigma_\theta - \sigma_x) & 1 + Q(2\sigma_\theta - \sigma_x)^2 \end{bmatrix} \left\{ \begin{matrix} \dot{\sigma}_x \\ \dot{\sigma}_\theta \end{matrix} \right\}, \tag{4.14}$$

$$Q = \frac{1}{4\sigma_e^2} \left(\frac{E}{E_t} - 1 \right).$$

For bifurcation checks, the J_2 deformation theory of plasticity is preferred (see Section 13.3). For the present problem (13.26) reduce to

$$\left\{ \begin{matrix} \dot{\varepsilon}_x \\ \dot{\varepsilon}_\theta \end{matrix} \right\} = \frac{1}{E_s} \begin{bmatrix} 1 + q(2\sigma_x - \sigma_\theta)^2 & -v_s + q(2\sigma_x - \sigma_\theta)(2\sigma_\theta - \sigma_x) \\ -v_s + q(2\sigma_x - \sigma_\theta)(2\sigma_\theta - \sigma_x) & 1 + q(2\sigma_\theta - \sigma_x)^2 \end{bmatrix} \left\{ \begin{matrix} \dot{\sigma}_x \\ \dot{\sigma}_\theta \end{matrix} \right\}, \tag{4.15}$$

$$q = \frac{1}{4\sigma_e^2}\left(\frac{E_s}{E_t} - 1\right) \quad v_s = \frac{1}{2} + \frac{E_s}{E}\left(v - \frac{1}{2}\right).$$

If the material exhibits anisotropic yielding, (4.14) and (4.15) are replaced by their anisotropic versions (13.21b) and (13.28) suitably reduced. Inverting (4.14) and (4.15) yields

$$\dot{\boldsymbol{\sigma}} = C_d \dot{\boldsymbol{\varepsilon}}$$

or

$$\begin{Bmatrix} \dot{\sigma}_x \\ \dot{\sigma}_\theta \end{Bmatrix} = \begin{bmatrix} C_{11} & C_{12} \\ C_{12} & C_{22} \end{bmatrix} \begin{Bmatrix} \dot{\varepsilon}_x \\ \dot{\varepsilon}_\theta \end{Bmatrix}. \tag{a}$$

The incremental force and moment intensities become

$$\dot{N} = \int_{-t/2}^{t/2} \dot{\boldsymbol{\sigma}}\, dz = t\, C \dot{\boldsymbol{\varepsilon}}^o \quad \text{and} \quad \dot{M} = \int_{-t/2}^{t/2} \dot{\boldsymbol{\sigma}} z\, dz = \frac{t^3}{12} C \dot{\boldsymbol{\kappa}}. \tag{b}$$

The buckling Eqs. (4.2) can be expressed in terms of the displacements using (4.13) and (b). It can then be easily recognized that the following buckling mode satisfies the resulting differential equations and the appropriate periodicity conditions

$$\tilde{w} = a \cos n\theta \quad \text{and} \quad \tilde{v} = b \sin n\theta. \tag{c}$$

We now consider long pipes with the three end conditions shown in Figure 4.3.

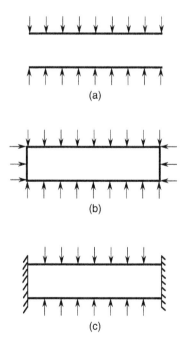

Figure 4.3 Pipe under external pressure and three types of axial loading: (a) lateral pressure, (b) hydrostatic pressure, and (c) plane strain pressurization.

4.2.1 Lateral Pressure

If the pipe is free axially (*lateral pressure*, see Figure 4.3(a)), the prebuckling stress state is given by

$$N_{xxo} = 0, \quad N_{\theta\theta o} = -PR, \quad \text{and} \quad \sigma_e = \frac{PR}{t}. \tag{d}$$

For this proportional loading case, at bifurcation the constitutive matrix (4.15) reduces to

$$D_d = \frac{1}{E_t} \begin{bmatrix} \dfrac{1}{4}\left(1 + 3\dfrac{E_t}{E_s}\right) & -v_t \\ -v_t & 1 \end{bmatrix}, \quad v_t = \frac{1}{2} + \frac{E_t}{E}\left(v - \frac{1}{2}\right). \tag{e}$$

By requiring $\dot{N}_{xx} = 0$ at bifurcation

$$\dot{N}_{\theta\theta} = tC_{22}\Omega\dot{\varepsilon}_{\theta\theta}^o \quad \text{and} \quad \dot{M}_{\theta\theta} = \frac{t^3}{12}C_{22}\dot{k}_{\theta\theta},$$

where

$$\Omega = \left(1 - \frac{C_{12}^2}{C_{11}C_{22}}\right). \tag{f}$$

Using (c) and (f) in Eq. (4.2) results in the following eigenvalue problem

$$\begin{bmatrix} n(\Omega + \rho n^2) & n^2(\Omega + \rho) \\ (\Omega + \rho n^4) - \gamma(n^2 - 1) & n(\Omega + \rho n^2) \end{bmatrix} \begin{Bmatrix} a \\ b \end{Bmatrix} = 0, \tag{g}$$

where

$$\rho = \frac{1}{12}\left(\frac{t}{R}\right)^2 \quad \text{and} \quad \gamma = \frac{PR}{C_{22}t}.$$

For nontrivial solutions of (g), the following buckling pressures are obtained

$$P_n = \frac{(n^2 - 1)}{12} \frac{C_{22}}{\left[1 + \dfrac{1}{12\Omega}\left(\dfrac{t}{R}\right)^2\right]} \left(\frac{t}{R}\right)^3, \quad n = 2, 3, \ldots \tag{h}$$

The critical buckling pressure occurs for $n = 2$, thus,

$$P_C = \frac{1}{4} \frac{C_{22}}{\left[1 + \dfrac{1}{12\Omega}\left(\dfrac{t}{R}\right)^2\right]} \left(\frac{t}{R}\right)^3. \tag{4.16a}$$

Using (d) and (e),

$$C_{22} = \frac{E_t}{\Omega}, \quad \Omega = 1 - \frac{4v_t^2}{\left(1 + 3\dfrac{E_t}{E_s}\right)}. \tag{4.16b}$$

The second term in the square bracket in the denominator of (4.16a) is $\ll 1$ and can be eliminated (see also [4.11]). For a linearly elastic material, the resultant equation reduces to the classical result in Eq. (4.6). The constitutive constants on the RHS of (4.16a) depend on the pressure, and thus P_C is evaluated iteratively. A simple procedure is to progressively increase P, evaluate the constitutive constants from (4.15), obtain an estimate of P_C from (4.16), and compare it with the value of P used. The critical buckling pressure is established when $P = P_C$. A suitable numerical technique, such as the bisection method, can be used to obtain a solution to a desired accuracy.

4.2.2 Hydrostatic Pressure

Hydrostatic pressure loading represents a tube that is closed with end-caps and pressurized uniformly (Figure 4.3(b)). The prebuckling stress state is given by

$$N_{xxo} = -\frac{PR}{2}, \quad N_{\theta\theta o} = -PR \quad \text{and} \quad \sigma_e = \frac{\sqrt{3}}{2}\left(\frac{PR}{t}\right). \tag{i}$$

This is also a proportional stress state and the constitutive matrix in (4.15) takes the form

$$D_d = \frac{1}{E_s}\begin{bmatrix} 1 & -v_s \\ -v_s & \frac{1}{4}\left(1 + 3\frac{E_s}{E_t}\right) \end{bmatrix}. \tag{j}$$

The critical buckling pressure is calculated in a similar fashion as in the case above, and takes the same form as (4.16a) with

$$C_{22} = \frac{E_s}{\frac{1}{4}\left(1 + 3\frac{E_s}{E_t}\right)\Omega} \quad \text{and} \quad \Omega = 1 - \frac{4v_s^2}{\left(1 + 3\frac{E_s}{E_t}\right)}. \tag{k}$$

4.2.3 Pressure with Zero Axial Strain

This is plane strain pressurization (Figure 4.3c), for which the prebuckling solution is

$$N_{\theta\theta o} = -PR \quad \text{and} \quad \varepsilon_x = 0. \tag{l}$$

The stress state induced by this loading is not proportional. The prebuckling stress–strain relationships can be established incrementally from (4.14) using (l). The critical buckling pressure is given by

$$P_C = \frac{1}{4}\frac{C_{22}}{1 + \frac{1}{12}\left(\frac{t}{R}\right)^2}\left(\frac{t}{R}\right)^3 \tag{4.17}$$

in which C_{22} can be evaluated from (4.15) by substituting the prebuckling stress combinations calculated from the flow theory. In the linearly elastic case, $C_{22} = E/(1 - v^2)$.

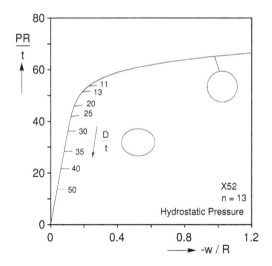

Figure 4.4   Bifurcation pressures of pipes of various D/t values marked on an X52 stress–strain response.

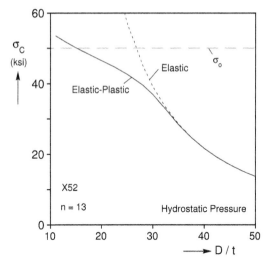

Figure 4.5   Bifurcation buckling stress as a function of D/t for elastic and elastic–plastic (X52) materials.

Hydrostatic pressure loading is considered the most representative of a long pipeline installed on the sea floor, and is generally preferred in design. It is also commonly used in collapse tests. Figure 4.4 shows a sample of bifurcation pressures for X52 pipes of various D/t values. The critical pressures converted to critical stresses are marked on the stress–strain response with the symbol ⌐ ($n = 13$, see Appendix C). Note that even for rather low D/t values, bifurcation occurs at relatively small strain levels where the tangent modulus is still relatively high.

The bifurcation stress ($\sigma_C = P_C R/t$) for the same material is plotted against the pipe D/t in Figure 4.5. Included for comparison is the corresponding elastic plot. The two sets

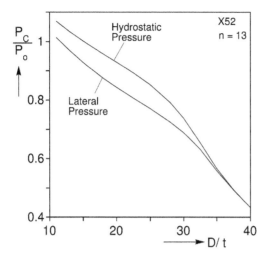

Figure 4.6 Bifurcation pressure as a function of D/t for hydrostatic and lateral pressure loadings for an X52 steel.

of results agree down to a $D/t \sim 35$. For lower D/t values, the plastic results are seen to be increasingly below the elastic ones. Note that for low D/t pipes, the plastic critical stress is higher than the nominal yield stress of this material.

The critical bifurcation pressure yielded by hydrostatic loading is the highest of the three and the one for lateral pressure is the lowest. This difference is illustrated in Figure 4.6 for an X52 material. P_C corresponding to the two loading conditions is plotted against D/t. The difference, which depends to some degree on the shape of the stress–strain response, is of the order of 10% for lower D/t pipes, but decreases for higher D/ts (for an elastic material both formulas reduce to Eq. (4.6)). Plane strain loading is less common in practice. The bifurcation pressures corresponding to this loading lie between the two shown in Figure 4.6.

4.3 NONLINEAR FORMULATION

A more complete analysis of plastic buckling and collapse of long pipes under external pressure requires numerical treatment. Because both elastic and plastic buckling exhibit initially stable postbuckling behavior, it suffices to consider a long cylinder that deforms uniformly along the length. Yeh and Kyriakides [4.12] developed such a formulation based on large deflection, small strain kinematics, and elasto–plastic material behavior. The formulation was used to study collapse and the major factors that affect it. Concurrently, the custom program BEPTICO [4.1] was developed, which is capable of analyzing long pipes under combined external pressure, bending, and tension. In what follows, the aspects of BEPTICO's formulation relevant to the external pressure problem will be described. In subsequent chapters, extensions to combined loadings will be added.

4.3.1 Kinematics

Consider a long, nearly circular pipe with radius R and wall thickness t loaded by uniform external pressure. The pipe cross section is allowed to have a small, arbitrary initial geometric imperfection. The imperfection is uniform along the length of the pipe, and is expressed in terms of the radial displacement through

$$\bar{w}(\theta) = R \sum_{n=1}^{N_R} (A_n \cos n\theta + B_n \sin n\theta). \tag{4.18}$$

The wall thickness is also allowed to vary around the circumference, in a way that is also represented by a series expansion as follows

$$t(\theta) = t \left[1 + \sum_{m=1}^{M_t} (C_m \cos m\theta + D_m \sin m\theta) \right]. \tag{4.19}$$

Making the usual thin-walled section assumptions, the following small strain/finite rotations strain–displacement equations are adopted. The circumferential strain is given by

$$\varepsilon_\theta = \left(\varepsilon_{\theta\theta}^o + z\kappa_{\theta\theta} \right) - \bar{\varepsilon}_\theta, \quad -\frac{t}{2} \le z \le \frac{t}{2}, \tag{4.20a}$$

where $\bar{\varepsilon}_\theta$ is the apparent strain of the initial imperfection, which is assumed not to result in stresses. The membrane component of (4.20a) is related to the displacements (v, w) (see Figure 4.1 and [4.4]) via

$$\varepsilon_{\theta\theta}^o = \left(\frac{v' + w}{R} \right) + \frac{1}{2} \left(\frac{v' + w}{R} \right)^2 + \frac{1}{2} \left(\frac{v - w'}{R} \right)^2, \tag{4.20b}$$

[4.4] while the local curvature is given by

$$\kappa_{\theta\theta} = \left(\frac{v' - w''}{R^2} \right) \Big/ \sqrt{1 - \left(\frac{v - w'}{R} \right)^2}. \tag{4.20c}$$

Thus, here we include additional nonlinearities which allow for larger rotations of the normals. The axial strain of the pipe is ε_x.

4.3.2 Constitutive Behavior

The pipe material is modeled as an elastic–plastic solid using the J_2 flow theory of plasticity with isotropic hardening (see Section 13.2). Hill's anisotropic yield function, as specialized in Appendix B, is adopted. In the examples presented in this chapter $S_r = S_\theta$ is often assumed, in which case (13.21a) reduces to

$$f = \left[\sigma_x^2 - \sigma_x \sigma_\theta + \frac{1}{S^2} \sigma_\theta^2 \right]^{1/2} = \sigma_{e\,\text{max}}, \quad S = \frac{\sigma_{o\theta}}{\sigma_{ox}}. \tag{4.21}$$

In (4.14) and (13.21b) E_t is evaluated either from a piecewise linear fit or from a continuous fit of a measured stress–strain response (e.g. Ramberg–Osgood – see Appendix C).

4.3.3 Principle of Virtual Work

Equilibrium is satisfied through the Principle of Virtual Work (PVW), which for this problem can be expressed as

$$R \int_0^{2\pi} \int_{-t/2}^{t/2} (\hat{\sigma}_x \, \delta \dot{\varepsilon}_x + \hat{\sigma}_\theta \, \delta \dot{\varepsilon}_\theta) \left(1 + \frac{z}{R}\right) dz \, d\theta = \delta \dot{W}_e, \qquad (4.22a)$$

where $(\dot{\bullet})$ denotes a variable increment and $(\hat{\bullet}) \equiv (\bullet + \dot{\bullet})$. The RHS of (4.22a) is the virtual work increment due to the pressure load, given by

$$\delta \dot{W}_e = -\hat{P} R \int_0^{2\pi} \left[\delta \dot{w} + \frac{1}{2R} (2 \hat{w} \, \delta \dot{w} + 2 \hat{v} \, \delta \dot{v} + \hat{w} \, \delta \dot{v}' + \hat{v}' \, \delta \dot{w} - \hat{v} \, \delta \dot{w}' - \hat{w}' \, \delta \dot{v}) \right] d\theta.$$

$$(4.22b)$$

In pipe collapse tests, the ends of the specimen are often capped, so the pressure loading is hydrostatic. In that case, the term $(-\hat{P} \pi R^2 \, \delta \dot{\varepsilon}_x)$ must be added to (4.22b) (ε_x is the uniform axial strain).

The problem domain is discretized by using the following series expansions for the two displacement components

$$w \cong R \left[a_o + \sum_{n=1}^{N} (a_n \cos n\theta + b_n \sin n\theta) \right] \quad \text{and} \quad v \cong R \sum_{n=2}^{N} (c_n \cos n\theta + d_n \sin n\theta).$$

$$(4.23)$$

Terms containing the coefficients c_0, c_1, and d_1 are omitted in order to preclude rigid body motion.

When (4.23) are substituted into Eqs. (4.20) and then into Eq. (4.22), a system of $4N$ nonlinear algebraic equations results. These equations are solved for the unknowns $\{\dot{a}_o, \dot{a}_1, \ldots, \dot{a}_N, \dot{b}_1, \ldots, \dot{b}_N, \dot{c}_2, \ldots, \dot{c}_N, \dot{d}_2, \ldots, \dot{d}_N, \dot{\varepsilon}_x\}$ by prescribing increments in pressure. To properly identify limit pressure instabilities, pressure loading can also be accomplished by prescribing increments of the volume enclosed by a unit length of pipe. In this case, $\dot{P}$ is an additional unknown and the following equation is added to the system outlined above

$$\hat{v} - v^* = \pi R^2 \left\{ 1 + \frac{1}{\pi R} \int_0^{2\pi} \left[\hat{w} + \frac{1}{2R} \left(\hat{v}^2 - \hat{v}\hat{w}' + \hat{v}'\hat{w} + \hat{w}^2 \right) \right] d\theta \right\} - v^* = 0, \quad (4.24)$$

where $\hat{v}$ is the volume calculated from the current value of the unknowns and v^* is the prescribed volume [4.13].

The system of nonlinear equations is solved using the Newton–Raphson method. The integrals in (4.22) are evaluated via a Gauss integration scheme. The number of

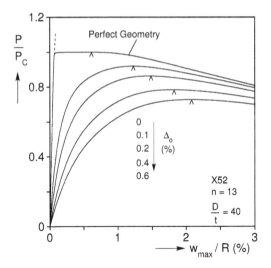

Figure 4.7   Pressure–maximum displacement responses for perfect and imperfect tubes with $D/t = 40$. Perfect case buckles elastically and collapses due to inelastic action.

terms in the series (4.23) and the number of integration points around the circumference depend on the features of the initial imperfection that need to be modeled (see next section).

For the most common imperfection case of simple initial ovality ($\bar{w}/R = -\Delta_o \cos 2\theta$), the deformation becomes symmetric about planes $x_1 - x_2$ and $x_1 - x_3$. For compatibility with other loading cases, such as a combined bending and external pressure, only symmetry about plane $x_1 - x_3$ is taken advantage of in BEPTICO. In this case, $b_n = c_n = 0$, $\forall n$ and the number of equations reduces to $2N + 1$. Five to seven integration points through the thickness of the pipe and twelve around the half circumference are usually sufficient [4.14] (note that the number of integration points usually must be increased when the stress–strain response exhibits low hardening).

4.3.4   Examples

Illustrative examples of nonlinear responses obtained with this type of formulation are shown in Figures 4.7 and 4.8. Figure 4.7 shows pressure–maximum deflection responses for an X52 ($n = 13$), $D/t = 40$ pipe with several initial ovality values (amplitude Δ_o), including the perfect geometry. This set of results characterize pipes of higher D/t that buckle elastically but collapse plastically. The perfect case buckles in the $n = 2$ mode at $P_C = 1{,}111$ psi (76.6 bar, Eq. (4.6)). The pipe initially ovalizes uniformly with a nearly unchanged pressure. The growing bending stress combined with the membrane stress eventually cause the material to yield. The resultant reduction in rigidity causes the response to start a downward trajectory. The maximum pressure achieved represents the collapse pressure (P_{CO}).

In the presence of initial ovality in the form of Eq. (4.8), the response becomes more compliant and the collapse pressure is seen to decrease. This significant imperfection sensitivity of P_{CO} points to the importance of initial ovality as a design parameter. The

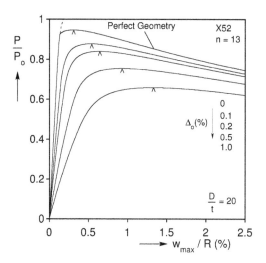

Figure 4.8 Pressure-maximum displacement responses for perfect and imperfect tubes with $D/t = 20$. Perfect case bifurcates plastically.

results for the imperfect cases are also included in Figure 4.2, along with the corresponding elastic results based on Eq. (4.9). The initial parts of the $P - w$ responses are essentially coincident with the elastic results, but the two deviate as plastic action develops. It is, however, interesting to compare the first yielding collapse pressure yielded by Timoshenko's formula (4.11) with the numerical collapse pressures represented by the pressure maxima. The two are very close indeed, which illustrates the dependability of Eq. (4.11) for pipes that buckle elastically.

Results characteristic of pipes which buckle in the plastic range are shown in Figure 4.8 for an X52 pipe with $D/t = 20$. For such cases, the pressure will be normalized by the yield pressure P_o defined in (4.10). The perfect case bifurcates at $P_C = 4900$ psi (275.9 bar, hydrostatic pressure loading (4.16a) and (k)). The pipe initially ovalizes uniformly with increasing pressure as is characteristic of plastic buckling problems. Inelastic action leads to a pressure of maximum 1.84% higher than the bifurcation pressure. Thus, for the perfect case P_C is a dependable lower bound for the collapse pressure. As in the previous case, in the presence of an imperfection the response becomes increasingly more compliant and the collapse pressure decreases significantly. The effect of the type of loading and of the shape of the stress–strain response on the collapse pressure will be discussed in the next section.

In these calculations, the collapse pressure has been associated with the pressure maximum of a uniformly deforming long pipe. In experiments in a stiff pressure vessel, collapse is always localized as shown in Figures 1.3 and 4.9. Reproducing such localized collapse requires a 3-D formulation [4.15]. However, because localization takes place after the pressure maximum, the collapse pressure yielded by the 2-D analysis suffices for most cases. Exceptions are pipes with local imperfections, in the form of dents (see Chapter 6). The adequacy of this important conclusion will be demonstrated in the next section, where such predictions are compared to experiments.

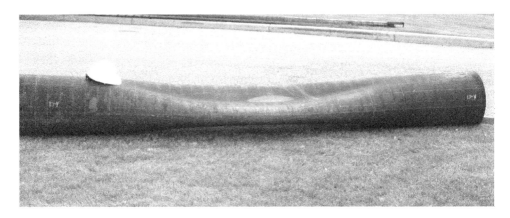

Figure 4.9   Photograph of a 20-inch, X80 pipe collapsed under external pressure demonstrating localized collapse (courtesy C-FER Technologies).

4.4   FACTORS AFFECTING PIPE COLLAPSE

4.4.1   Collapse Pressure Experiments

The most common method of establishing the collapse pressure of pipes is by enclosing an end-capped specimen in a pressure vessel, as shown in Figure 4.10 (hydrostatic pressure loading). An extensive experimental study of collapse conducted in this manner is reported in Kyriakides and Yeh [4.16]. It involves seamless SS-304 tubes in the D/t range of 12.8 to 51. The data set discussed here includes results from this study as well as results from additional tests performed later. The diameters of test specimens ranged from 1.0 to 2.0 in (25–51 mm), and their lengths were $20D$ or more. The following testing practices have been found to produce dependable results and are recommended. The length of the test section of the specimen must be at least $10D$. The tube initial imperfections are measured prior to each test as follows. The diameter variation around the circumference is measured at intervals of $(2–3)D$. The mean value of all measurements is designated as the diameter (D) of the tube. An ovality parameter Δ_o is established at every axial station using

$$\Delta_o = \frac{D_{\max} - D_{\min}}{D_{\max} + D_{\min}}, \qquad (4.25)$$

where $D_{\max}$ and $D_{\min}$ are the maximum and minimum diameters measured (Figure 4.11(a)). The maximum value of Δ_o along the tube length is adopted as the "initial ovality" of the specimen [4.12, 4.17, 4.18]. The wall thickness is measured at approximately 12 circumferential positions at each end of the tube. The mean value of all measurements is designated as the "wall thickness" (t) of the tube. For seamless tubes and pipes, a common thickness imperfection is wall thickness eccentricity (Figure 4.11(b)). An eccentricity parameter is calculated for each end as follows [4.12, 4.17]

$$\Xi_o = \frac{t_{\max} - t_{\min}}{t_{\max} + t_{\min}}. \qquad (4.26)$$

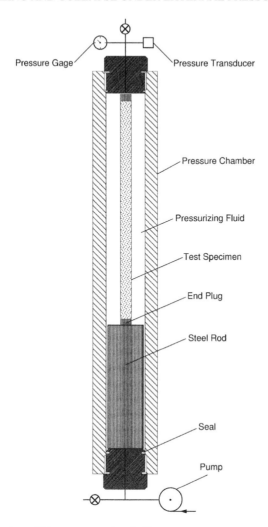

Figure 4.10   Relatively "stiff" pressure testing facility.

Ξ_o is evaluated for the two ends, and the mean of the two values is designated as the wall eccentricity of the tube.

Uniaxial tests are performed on axial test coupons cut from each tube tested, as described in Appendix A. The stress–strain response is either fitted with a Ramberg–Osgood fit (Appendix C) or represented by a piecewise linear fit. An additional test is commonly conducted in which the yield stress in the circumferential direction is measured, as described in Appendices A and B.

To conduct the test, the ends of the tube are sealed and the specimen is placed in a high-pressure capacity vessel in the arrangement shown in Figure 4.10. The cavity is filled with water and the system is pressurized at a slow rate using a positive displacement water pump. This scheme approximates "volume-controlled" pressurization. Collapse is sudden and localized as shown in Figures 1.3 and 4.9. It is accompanied by a drop in pressure. The extent of the local collapse depends on the stiffness of the pressurizing

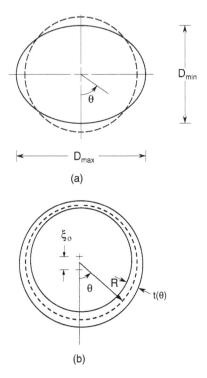

Figure 4.11   Idealized pipe cross-sectional imperfections: (a) initial ovality and (b) wall thickness eccentricity.

system, so that in a more compliant system the local collapse will start to spread. The maximum pressure recorded is the collapse pressure (P_{CO}).

A number of such results are listed in Table 4.1. Results from 29 additional tests can be found in Table 1a in [4.16]. The initial ovality ranged from 0.1% to 0.25% (one tube had a value of 1% due to damage). The wall eccentricity ranged from approximately 1.5% to 9% while 3% to 6% was typical. The yield stress varied from approximately 38 to 51 ksi (262–352 MPa). The anisotropy variable S in Eq. (4.21) varied from 0.82 to 1.0.

The measured collapse pressures that are plotted versus D/t in Figure 4.12. The collapse pressures vary from 356 psi (24.6 bar) for $D/t = 50.76$ to 7200 psi (496.5 bar) for $D/t = 12.78$, illustrating the strong influence of this variable. The observed scatter in the data is due to variations in initial ovality and material properties not captured by the adopted normalization of P_{CO}.

4.4.2   Prediction of Collapse Pressures

The experiments reported above were simulated using the individual measured geometric and material characteristics in BEPTICO [4.1]. The maximum value of measured ovality was assumed to be uniformly distributed, so that (4.18) takes the form

$$\frac{\bar{w}}{R} = -\Delta_o \cos 2\theta, \tag{4.27}$$

Table 4.1 Tube geometric and material parameters and measured (P_{CO}) and calculated ($\hat{P}_{CO}$) collapse pressures.

Material	D in (mm)	D/t	Ξ_o (%)	Δ_o (%)	E Msi (GPa)	σ_o ksi (MPa)	n	S	P_{CO} psi (bar)	$\hat{P}_{CO}$ psi (bar)	$\dfrac{\hat{P}_{CO} - P_{CO}}{P_{CO}}$ (%)
SS-304	1.250 (31.75)	18.66	3.0	0.08	28.5 (197)	38.7 (267)	13.0	0.82	3427 (236.3)	3254 (224.4)	−5.1
SS-304	1.9997 (50.79)	21.07	2.8	0.06	28.1 (193)	41.24 (284)	9.0	0.97	3154 (217.5)	3318 (228.8)	5.2
SS-304	1.9996 (50.79)	24.09	5.9	0.19	30.8 (212)	40.24 (277)	11.5	0.917	2484 (171.3)	2474 (170.6)	−0.4
SS-304	1.2545 (31.86)	25.34	5.1	0.12	27.6 (191)	49.8 (344)	14.5	0.88	2563 (176.8)	2664 (183.7)	4.7
SS-304	1.3750 (34.93)	28.65	2.1	0.15	28.7 (198)	50.1 (346)	14.0	0.88	2092 (144.3)	2145 (147.9)	2.5
SS-304	2.0032 (50.88)	29.89	6.9	0.19	29.4 (202)	37.12 (256)	9.5	1	1606 (110.8)	1640 (113.1)	2.1
SS-304	1.248 (31.70)	34.67	2.8	0.05	26.6 (184)	43.5 (300)	12.0	0.85	1300 (89.7)	1269 (87.5)	−2.4
SS-304	1.7579 (44.65)	37.24	2.5	0.06	28.0 (193)	40.31 (278)	7.0	1	1173 (80.9)	1134 (78.2)	3.4
SS-304	1.9993 (50.78)	42.09	2.7	0.05	29.0 (200)	40.21 (277)	17.0	1	803 (55.4)	864 (59.6)	7.1
X42-1	4.000 (101.6)	27.87	3.8	0.05	29.7 (205)	48.0 (331)	–	0.93	2720 (187.6)	2660 (183.4)	−2.2
X42-2	4.010 (101.9)	23.80	5.0	0.15	29.6 (204)	54.3 (374)	–	0.93	3307 (228.1)	3408 (235.0)	3.1
X65-3	4.012 (101.9)	31.97	3.6	0.27	29.7 (205)	75.0 (517)	6.5	1	1764 (121.7)	1824 (125.7)	3.4
X65-4	4.003 (101.7)	24.44	2.9	0.16	30.2 (208)	104 (717)	–	0.95	4187 (288.8)	4345 (299.7)	3.8

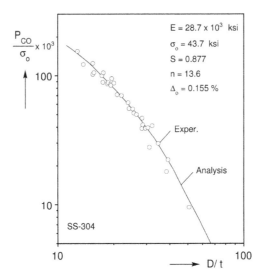

Figure 4.12   Measured collapse pressures versus D/t and predictions based on average geometric and material parameters.

where Δ_o is the maximum value measured in the tube. Collapse pressure predictions $(\hat{P}_{CO})$ for nine representative cases are listed in Table 4.1. The absolute difference between predicted and measured values ranges from 0.4% to 7.1%, while the mean difference is 3.6%. This demonstrates the adequacy of the 2-D formulation developed above for the problem. The predictions can be improved somewhat by including wall eccentricity and residual stresses, both neglected in these simulations. Additional experience, garnered from many simulations conducted for full scale pipes, has shown that the type of measurements outlined above used in BEPTICO can be expected to yield collapse pressure that are within 5% of the test values.

Individual simulations were also conducted for the remaining experiments using a value of $S = 0.88$. The predicted collapse pressures are compared to the measured values directly in Figure 4.13 (includes the nine cases in Table 4.1; see also Table 1a in [4.16]). The comparison is once more favorable; the absolute difference ranges from 0.15% to 13.2% while the mean difference is 5.5%. The wider disparity can be reduced by measuring the actual value of S for each case.

The data in [4.16] was used to develop mean values of the material properties and of the ovality, which are listed in Figure 4.12. The collapse pressure of tubes with these properties were calculated, and the results are plotted as a solid line in Figure 4.12. The trajectory of the predictions is seen to represent well the trend of the experimental results. For $D/t > 35$, the log–log plot is essentially linear, with a slope approaching -3. For lower D/t values, the slope decreases due to plasticity.

4.4.3   Effect of Initial Ovality

The results in Figures 4.7 and 4.8 showed that the collapse pressure is strongly sensitive to initial ovality for all D/ts of interest to pipelines. Experimental verification of this was

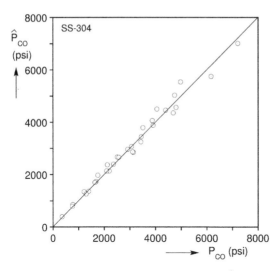

Figure 4.13 Direct comparison of measured (P_{CO}) and calculated ($\hat{P}_{CO}$) collapse pressures [4.21].

presented in [4.12]. Tubes of five D/t values in the range of 14 to 39 were permanently ovalized by crushing them to various degrees between two rigid plates. Axially uniform ovality values ranging from 0.5% to 5% were produced over tube lengths of 12D. The tubes were subsequently collapsed under external pressure in the manner described in Section 4.4.1. The measured collapse pressures normalized by P_o ($= 2\sigma_o(t/D_o)$) are plotted against Δ_o in Figure 4.14 (data in [4.16]). Each set of experiments was simulated numerically (neglecting residual stresses). The predictions, drawn in solid lines in the figure, are seen to reproduce the experimental results very well for all D/ts. The results clearly demonstrate the strong sensitivity of the collapse pressure to initial ovality. Ovality of 1% causes a reduction in P_{CO} of 30–40%. For 5% ovality, the reduction is more than 50%.

For pipes which buckle elastically, Timoshenko's design formula (4.11) can be used to predict the effect of ovality on P_{CO}. Figure 4.15 shows the $P_{CO} - \Delta_o$ experimental results for $D/t = 39.12$, along with the predictions from BEPTICO and Eq. (4.11). Both predictions closely follow the trend of the experimental results. By contrast, for the lower D/t cases in Figure 4.14 (e.g. 13.74 and 19.26), Eq. (4.11) does not perform well. A rule of thumb for when to stop using (4.11) is as follows. The D/t at which the yield pressure in Eq. (4.10) equals the elastic buckling pressure (4.6) is

$$\left.\frac{D_o}{t}\right|_{tr} = \sqrt{\frac{E}{(1 - v^2)\sigma_o}}. \tag{4.28}$$

This can be viewed as the transitional value which depends on the ratio of E/σ_o (for materials with monotonic stress–strain responses, use stress at elastic limit in place of σ_o). For pipes with higher D/t values, the Timoshenko's formula and Eq. (4.6) can be used. For lower D/t values, the collapse pressure should be calculated numerically.

Because of its strong influence on P_{CO}, initial ovality must always be considered in design. At the same time, the bifurcation buckling pressures are easier to establish, even

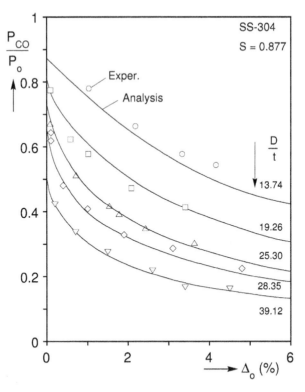

Figure 4.14 Collapse pressure versus initial ovality for five tube D/ts. Comparison of experiments and predictions.

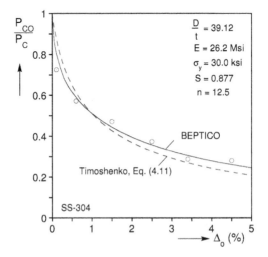

Figure 4.15 Collapse pressure versus initial ovality for $D/t = 39.1$. Predictions based on Eq. (4.11) in good agreement with experimental and numerical results.

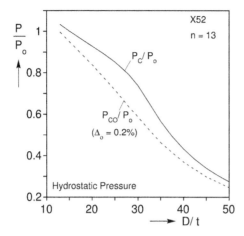

Figure 4.16 Comparison of bifurcation pressures of perfect tubes with collapse pressures of imperfect tubes with for various D/t values.

for plastic buckling. It is thus useful to compare how P_{CO} differs from P_C over a range of D/t values. Figure 4.16 shows such a comparison for an X52 material ($n = 13$) and $\Delta_o = 0.2\%$. P_{CO} is of course lower than P_C, but the difference varies quite significantly with D/t. This difference will increase with Δ_o and also depends on the shape of the stress–strain response. In general, calculations for design should include an imperfection.

More complex imperfections than initial ovality are discussed in Section 4.5. Often, they can be idealized as initial ovality. In some cases, such imperfections can be represented as a Fourier series of the type in Eq. (4.18) and used directly in a model like BEPTICO to produce a more accurate prediction of the collapse pressure.

4.4.4 Type of Pressure Loading

Two alternate collapse testing methods which approximately correspond to lateral pressure loading are shown schematically in Figure 4.17(a). In the first, shown in Figure 4.17(a), a solid rod is placed through the test specimen, which connects to rigid end-caps. Small gaps are left between the end-caps and the pipe to ensure free axial expansion of the pipe. The gaps are sealed with a polymeric seal capable of sustaining the pressure loading but providing no axial restraint to the pipe. When the assembly is pressurized, the axial component of the pressure loading is reacted by the rod while the pipe reacts the radial pressure. An advantage of this scheme is that the specimens can be shorter (as short as $L/D \sim 3$). Shorter test sections will statistically produce somewhat higher collapse pressures, because the probability of having an imperfection with larger amplitude is smaller than in a longer specimen (Weibull-like argument). When using this method, care must be taken to ensure that the pipe is free to expand axially and that the ends do not provide any interference with the process of collapse.

In the second lateral pressure testing technique, shown in Figure 4.17(b), the pipe specimen protrudes through the ends of the pressure chamber. Special seals which contact but allow axial motion of the pipe are used at the exit ports. Such devices are used by some pipe mills [4.19, 4.20] for collapse testing, and also for collapse under combined

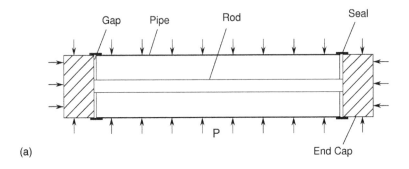

(a)

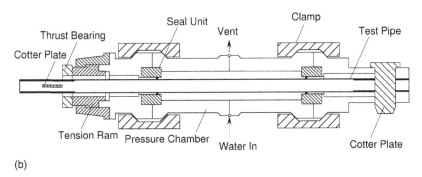

(b)

Figure 4.17 (a) Lateral pressure collapse testing set-up and (b) schematic of industrial combined tension–external pressure testing facility [4.19].

external pressure and axial tension. For combined loading, one of the protruding ends is held fixed while the other is loaded in tension by a special hydraulic ram operated under load control.

In Section 4.2, it was demonstrated that for buckling in the plastic range the bifurcation buckling pressure is influenced by the type of loading. The collapse pressure of imperfect pipes is similarly affected. Figure 4.18 shows a comparison of calculated collapse pressures of X52 pipes for hydrostatic and lateral pressure loading. The pipes have an assumed initial ovality of 0.2%. Lateral pressure loading lowers the collapse pressure also. The difference from the hydrostatic loading values is however somewhat lower than the difference in the corresponding bifurcation pressures in Figure 4.6. Again, for higher D/ts the influence of inelastic effects is reduced and this reduces the difference between the two collapse pressures. Because both types of loading are used in practice for lower D/t pipes, distinction between them is important.

4.4.5 Wall Thickness Variations

Seamless pipes produced by piercing exhibit some variation in wall thickness. Measurements of thickness variations appear in [4.21]. Often, such variations do not change significantly over a length of 10D. Furthermore, they can be idealized as eccentricity

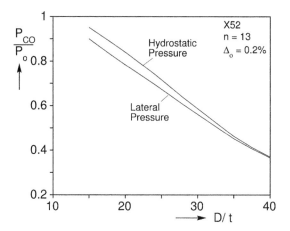

Figure 4.18 Collapse pressure versus D/t for hydrostatic and lateral pressure loadings.

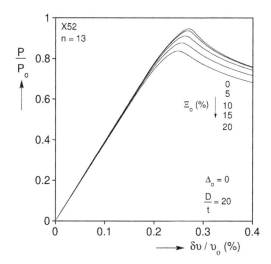

Figure 4.19 Pressure-change in volume responses of tubes with various wall eccentricity values $(D/t = 20)$.

between the inner and outer surfaces of the pipe, as shown in Figure 4.11(b). In this case, the wall thickness can be approximated as

$$t(\theta) \cong t\left(1 + \frac{\xi_o}{t}\cos\theta\right), \qquad (4.29)$$

where ξ_o is the distance between the centers of the two circular surfaces and t is the mean wall thickness. Using (4.29) in (4.26) yields $\Xi_o = \xi_o/t$.

The effect of such eccentricity on the pressure-change in volume response $(P - \delta v)$ of an X52 pipe with $D/t = 20$ is shown in Figure 4.19. Results for the perfect case and for four values of Ξ_o are shown. The effect on the prebuckling response is relatively small. The wall thickness variation causes a nonuniform membrane stress distribution around

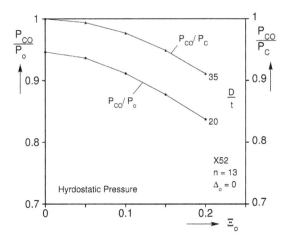

Figure 4.20   Collapse pressure versus wall eccentricity parameter for tubes with D/ts of 20 and 35.

the circumference. This leads to bending deformation and the development of a limit load that is lower than that of the perfect case. For $\Xi_o = 5\%$ the collapse pressure is reduced by 1%, for $\Xi_o = 10\%$ by 3.7%, and $\Xi_o = 20\%$ by 11.5%. The calculated collapse pressure is plotted against Ξ_o in Figure 4.20, along with corresponding results for a pipe of the same material with $D/t = 35$, representative of pipes which buckle elastically. The trend of the results is similar for the two cases. Overall, the effect of wall eccentricity on the collapse pressure is relatively minor, provided $\Xi_o < 10\%$ (see also [4.22]).

In these results, the ovality was neglected. Ovality and thickness variation coexist and interact. Yeh and Kyriakides [4.12, 4.21] examined the effect of the orientation of the two imperfections on the collapse pressure. The effect was found to be small, and as a result the imperfection orientations can be neglected in design calculations.

More complex thickness variations than simple eccentricity are discussed in Section 4.5. If the variation exceeds 5% of t, then a sufficient number of circumferential measurements of the wall thickness can be conducted to produce a Fourier series expansion of $t(\theta)$ (see Eq. (4.19)). The expansion can then be used in a model like BEPTICO to estimate the effect on P_{CO}.

4.4.6   Effect of Material Stress–Strain Response

The yield stress (σ_o) of the pipe steel governs the transitional value of D/t (4.28), which separates elastic and plastic buckling. For plastic buckling, the collapse pressure can be viewed as being proportional to σ_o, which makes it a key variable in the design of pipelines for deeper waters. The shape of the stress–strain response can also affect the collapse pressure. This is demonstrated through collapse calculations performed for the three X52 stress–strain responses in Figure C2. The collapse pressures normalized by the yield pressure, P_o', which is based on the common value of σ_o' (stress at a strain of 0.5%) are plotted against D/t in Figure 4.21. The elastic-perfectly plastic response (EP) yields the highest collapse pressures, and the one with $n = 13$ the lowest. The difference between the three sets of results is more pronounced for lower D/t values where the two extremes differ by as much as 15%. For $D/t > 30$, the difference decreases significantly.

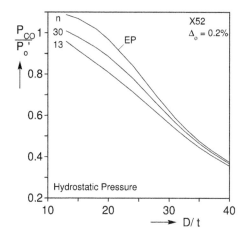

Figure 4.21 Collapse pressure versus D/t for X52 tubes with different material hardening behaviors.

We observe that for relatively small values of initial ovality, such as the 0.2% value used in the results shown in the figure, both the maximum and the mean strain values at collapse tend to be smaller than 0.5%. In this strain regime, the stress is highest for the EP material and lowest for $n = 13$ (see Figure C.2). This in turn is responsible for the order in which the collapse pressures fall in Figure 4.21.

4.4.7 Residual Stresses

In most major manufacturing processes, pipe is finished by cold forming, which often induces residual stress fields. The collapse pressure of pipes is influenced to some degree by circumferential-type bending stresses. The amplitude of this particular residual stress field is determined in a simple pipe ring splitting test as described in Appendix A. Such measurements performed on seamless X42 and X65 pipes produced σ_R/σ_o in the range of 0.1 to 0.45. Similar amplitudes have also been measured in small diameter seamless SS-304 tubes.

The effect of this type of residual stress field on collapse of X52 pipes of several D/t values is illustrated in Figure 4.22. Calculated collapse pressures in the presence of residual stress normalized by the corresponding values for $\sigma_R = 0$ are plotted against σ_R (see also [4.22]). For low D/t pipes, the effect of σ_R on the collapse pressure is minor. The effect grows with D/t, so that for $D/t = 30$ a residual stress with amplitude one-half the yield stress reduces the collapse pressure by about 13.5%. Interestingly, the effect decreases once more for higher D/ts as the influence of plasticity is reduced (see results for $D/t = 40$). The D/t most influenced by σ_R depends on the yield stress, and to a smaller degree on the shape of the stress–strain response.

4.4.8 Anisotropic Yielding

Tubes and pipes often exhibit some initial anisotropy in yielding. For collapse under external pressure, the most relevant form of anisotropy is a difference between the yield stress in the axial and circumferential directions. The extent and type of anisotropy depend on the manufacturing process, as discussed in Chapter 3. The collapse pressure depends

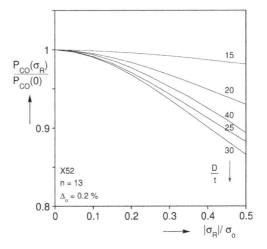

Figure 4.22   Collapse pressure as a function of residual stress amplitude σ_R for various tube D/ts.

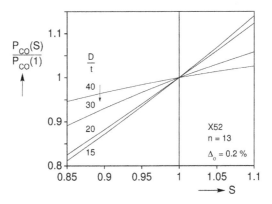

Figure 4.23   Collapse pressure as a function of anisotropy parameter S for various tube D/ts.

mainly on the stress–strain response of the material in the circumferential direction (for UOE must be compressive). For smaller diameter tubes and pipes, this can be obtained using the experimental scheme of Kyriakides and Yeh [4.23] described in Appendix B. For larger diameter pipes, circumferential test specimens extracted as described in Appendix A can be used for this purpose.

For seamless pipe, the yield stress in the circumferential direction can differ from the axial yield stress. The difference can be accentuated when the pipe is cold finished. This type of anisotropy is accounted for in the yield function (4.21) through the parameter S, defined as

$$S = \frac{\sigma_{o\theta}}{\sigma_{ox}}, \tag{4.30}$$

where σ_{ox} and $\sigma_{o\theta}$ are respectively the yield stresses in the axial and circumferential directions. Typical measured values of S can be found in Table 4.1.

The effect of this type of anisotropy on the collapse pressure is illustrated through calculations on X52 pipes of various D/ts, which appear in Figure 4.23. The collapse

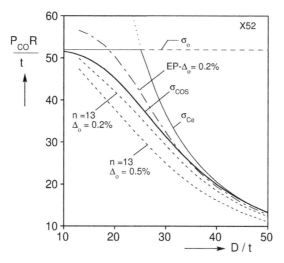

Figure 4.24   Collapse pressure (stress) versus D/t. Comparison of collapse formula [4.32] and numerical predictions for two material stress–strain responses and two initial ovality values.

pressures in the presence of anisotropy, normalized by the corresponding value for $S = 1$, are plotted against S. $S < 1$ reduces the collapse pressure, while $S > 1$ results in an increase. For all D/t values considered, P_{CO} essentially varies proportionately to S. The effect, however, of S on collapse pressure is larger for lower D/t pipes that buckle in the plastic range (e.g. $D/t = 15$ and 20). For this reason, when S is known, the yield pressure defined in (4.10) can be modified to

$$P_o = 2S\sigma_o \left(\frac{t}{D_o} \right). \tag{4.31}$$

For higher D/t pipes which buckle elastically, the dependence of P_{CO} on S is weaker.

UOE pipe has much more complicated yield anisotropy, resulting from the cold forming processes through which it is manufactured. This needs to be modeled in a different manner, which is described in Chapter 5.

4.4.9   An Approximate Estimate of Collapse Pressure

A quick first estimate of P_{CO} can be obtained by the so-called *Shell* collapse pressure formula [4.17, 4.18], given by

$$P_{COS} = \frac{1}{\left[\dfrac{1}{P_C^2} + \dfrac{1}{P_o^2} \right]^{1/2}}. \tag{4.32}$$

The expression is based on the premise that elastic buckling will occur at P_C (4.6) and plastic buckling at P_o (4.10). Its particular form is a way of smoothly connecting these two assumed critical states from below. Its performance is illustrated in Figure 4.24, where the hoop stress (PR/t) corresponding to (4.32) is plotted as a function of D/t for an X52 material. The main weakness of the formula is that it does not account for all the factors mentioned in this chapter that could influence P_{CO}. These include initial ovality, the shape of the stress–strain response, residual stresses, the UOE effect, etc. Thus, in some cases it produces conservative estimates, and in others unconservative estimates. This point is illustrated in Figure 4.24 using three sets of numerical predictions (BEPTICO [4.1]) of P_{CO} for two X52 materials (all were generated for hydrostatic pressure loading). The first has a monotonic stress–strain response with $n = 13$; the second is elastic–perfectly plastic (EP, see Figure C.2). For the $n = 13$ material, results for two initial ovality values are included, $\Delta_o = 0.2\%$ and 0.5%. The results for the EP material are underpredicted by $\sigma_{COS} (= P_{COS}R/t)$, with the difference growing as the pipe D/t decreases (this difference is partly due to the fact that the calculated values of P_{CO} are for hydrostatic pressure loading). By contrast, σ_{COS} overpredicts both sets of the results for the $n = 13$ material, with the difference for $\Delta_o = 0.5\%$ being quite significant for the whole range of D/t values shown. The discrepancy between numerical prediction and Eq. (4.32) can be further aggravated when additional factors such as residual stresses, wall thickness variations, and yield anisotropy are considered in the calculation of P_{CO}.

Despite these deficiencies, the simplicity of the Shell collapse formula makes it an alternative to P_o as a first estimate of the collapse pressure of pipes that buckle in the plastic range (P_C and P_{CO} calculated via Timoshenko's Eq. (4.11) for imperfect pipes are the best options for pipes that buckle elastically). Expression (4.32) can be used in preliminary design to get approximate estimates of the required variables. These results can then be improved using numerical models such as BEPTICO.

4.5 REPRESENTATIVE SEAMLESS PIPE IMPERFECTIONS

In the simulations of collapse experiments presented in Section 4.4, the geometric imperfections of tubes and pipes analyzed were idealized as initial ovality (4.25) and as wall eccentricity (4.26). In this section, actual imperfections measured in seamless pipes are examined in more detail. Subsequently, the effect of such imperfections on the calculated collapse pressure is discussed vis-à-vis corresponding calculations based on the recommended idealized imperfection measures. The discussion is based on four 4-inch seamless pipes analyzed by Yeh and Kyriakides [4.21]. The pipes analyzed were subsequently collapsed and the collapse pressure recorded. The collapse pressure was then calculated in several different ways. Imperfection measurements, collapse experiments, and simulations of similar nature are also reported in [4.24] for 7 and 9.625 in seamless pipes.

4.5.1 Imperfection Scanning System

The four seamless pipes analyzed were manufactured by Sumitomo Metal Industries according to API Spec 5LX. They were provided in 18 ft (5.5 m) lengths. Two were nominally X42 grade and two X65. Nine to ten diameter long specimens were cut out of approximately the middle of the pipe lengths. The dimensions of the specimens were first measured manually, as described in Section 4.4.1. The mean dimensional properties, the maximum ovality and the wall eccentricity are listed in Table 4.1 (specimens numbered as 1–4). Solid end-plugs were then bonded to each specimen, and the assembly was mounted in a lathe. It was held between centers which allowed free rotation about the axis of the

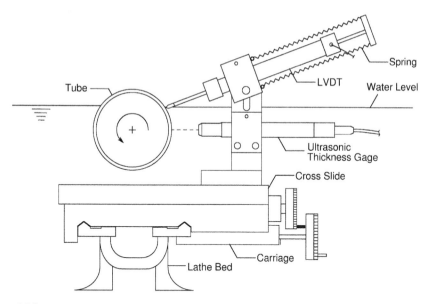

Figure 4.25   Imperfection measuring set-up with pipe rotated in a lathe bed [4.21].

pipe. The angular position of the rotating cylinder was determined by an emitter/receiver photodiode as follows: a graduated black/white tape was placed on one of the end-plugs; the diode generated a short-duration pulse each time a white graduation passed by it, which triggered the data acquisition system and data from the transducers were recorded. The system acquired 90 sets of data per revolution of the pipe.

The deviation of the pipe outer surface from a fictitious, reference circular cylinder was measured with an LVDT mounted on the cross slide of the carriage of the lathe as shown in Figure 4.25. An ultrasonic thickness gage also mounted on the cross slide was used to measure the wall thickness at the same axial position. The lathe carriage was used to move the two transducers along the axis of the pipe. The axial position was monitored by a second LVDT sliding over an inclined beam at the back of the lathe. The ultrasonic thickness gage used required a liquid coupling medium between the sensor and the surface of the pipe. As a result, the whole assembly was placed in a water bath as shown in Figure 4.25 (the water bath can be avoided with modern ultrasonic thickness gages).

Circumferential scans, consisting of 90 sets of shape and wall thickness measurements, were typically taken at axial locations of 1 in (25.4 mm) apart.

4.5.2   Data Reduction

The shape of an imperfect circular cylinder can be established by reducing the measurements into axially- and circumferentially-varying harmonic functions of the deviation from a perfect cylinder ($\bar{w}(x, \theta)$ [4.21, 4.25]). Due to the nature of the problem, correlation between the axial and circumferential directions can be avoided. Thus, we establish only the harmonic components of the measured imperfections for each circumferential scan recorded. The data reduction scheme consists of several steps. First, the radius and position of the "best-fit circle" are calculated at every axial station scanned. The best-fit circle is the least-squares circular fit of the data (see Figure 4.26). To find this circle let

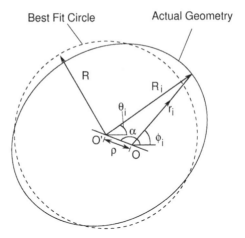

Figure 4.26   Best-fit circle and coordinate system used to measure in-plane shape deviation.

$\{(r_i,\ \phi_i), i = 1, M\}$ be the coordinates of the measured points relative to the axis of rotation of the pipe. Assume the best-fit circle to be centered at point O' with coordinates (ρ, α) and to have a radius R. Now transform the coordinates from the system centered at O to the one centered at O' and let the new coordinates of the measured points be $\{(R_i, \theta_i), i = 1, M\}$. The three unknown variables $\{R, \rho, \alpha\}$ can be evaluated using the following procedure. Let

$$\Pi = \sum_{i=1}^{M} (R - R_i)^2, \tag{a}$$

where

$$R_i = [\rho^2 + r_i^2 - 2\rho r_i \cos(\alpha - \phi_i)]^{1/2}. \tag{b}$$

Require that

$$\frac{\partial \Pi}{\partial R} = \frac{\partial \Pi}{\partial \rho} = \frac{\partial \Pi}{\partial \alpha} = 0. \tag{c}$$

The resultant three nonlinear algebraic equations can be solved numerically for the unknowns (R, ρ, α). The transformation of coordinates is completed by evaluating R_i from Eq. (b) and θ_i from the following

$$\theta_i = \alpha - \sin^{-1}\left[\frac{r_i}{R_i}\sin(\alpha - \phi_i)\right]. \tag{d}$$

The geometry of the imperfect profile is now expressed in terms of a harmonic series relative to the polar coordinate system centered at O' so that

$$R(\theta) = R\left[1 + \sum_{n=1}^{N_R}(A_n \cos n\theta + B_n \sin n\theta)\right], \tag{e}$$

where

$$A_n = \frac{1}{\pi R} \sum_{i=1}^{M} R_i \cos n\theta_i (\Delta\theta)_i \quad \text{and} \quad B_n = \frac{1}{\pi R} \sum_{i=1}^{M} R_i \sin n\theta_i (\Delta\theta)_i. \tag{f}$$

Thickness deviations are also represented through a harmonic series given by

$$t(\theta) = t \left[1 + \sum_{m=1}^{M_t} (C_m \cos m\theta + D_m \sin m\theta) \right], \tag{g}$$

with

$$t = \frac{1}{2\pi} \sum_{i=1}^{M} t_i (\Delta\theta)_i, \quad C_m = \frac{1}{\pi t} \sum_{i=1}^{M} t_i \cos m\theta_i (\Delta\theta)_i, \quad \text{and} \quad D_m = \frac{1}{\pi t} \sum_{i=1}^{M} t_i \sin m\theta_i (\Delta\theta)_i. \tag{h}$$

Typically, $M = 90$, $N_R = 10$ and $10 < M_t < 20$ are sufficient.

4.5.3 Four Examples

Radial and thickness imperfections that have been established with this scheme for the four pipes analyzed are plotted in Figures 4.27–4.30. Results for seven equidistant circumferential scans are presented for specimens 1 to 3. The imperfections of specimen 4 varied along the length. Capturing this axial variation required 31 circumferential scans. The two imperfections are defined as

$$\bar{w}(\theta, x_i) = R(\theta, x_i) - \bar{R}, \quad \bar{R} = \frac{1}{L} \sum_{i=1}^{L} R(x_i) \tag{i}$$

and

$$\Delta t(\theta, x_i) = t(\theta, x_i) - \bar{t}, \quad \bar{t} = \frac{1}{L} \sum_{i=1}^{L} t(x_i). \tag{j}$$

The variables $\bar{w}$ and Δt are plotted in phase for all x_i for comparison purposes. Figure 4.31 shows a set of exaggerated views of the cross sections of specimens at their respective initial positions ($x_i = 0$) approximately 3 in (76 mm) from one end. The thickness is exaggerated by a factor of 2, while $\bar{w}$ and Δt are exaggerated by a factor of 20. These exaggerations introduce distortions to the geometry, and as a result the profiles are strictly for qualitative observations.

The first observation that can be made from the measurements is that the imperfections (see Δ_o and Ξ_o values in Table 4.1) are quite small in comparison to the API Spec. 5L [4.26] accepted values. The specified tolerances are motivated by welding considerations rather than collapse performance. In general, modern manufacturing processes produce pipes with much smaller ovalities and wall thickness variations than those in these specifications.

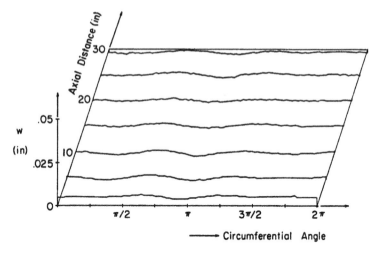

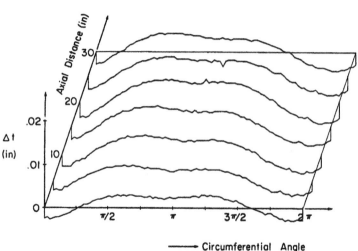

Figure 4.27   Imperfections of test specimen no. 1.

The second observation, made by comparing the four pairs of imperfection maps, is that the radial imperfections retained their shape and orientation along the length of tube scanned for test specimens 1 to 3. Any spiraling of the imperfections introduced during piecing, if present, must have been of a wavelength much longer than $10D$ which should have a negligible effect on the collapse pressure. Test specimen 4 was found to have imperfections whose orientation varied along the tube length. In fact, the main wave of the imperfections can be seen to shift by approximately 2π radians over a length of about $8D$. The relatively short wavelength of the spiral suggests that it may have been introduced during the last stage of the manufacturing process, where a dimensional reduction through a three-roller reducer was applied while the tube was rotating. The thickness imperfections retain their shape and orientation along the length of the specimen for all four cases.

The third point is that thickness imperfections are of two categories. For test specimens 1 and 3, the thickness variation is characterized approximately by one full wave around

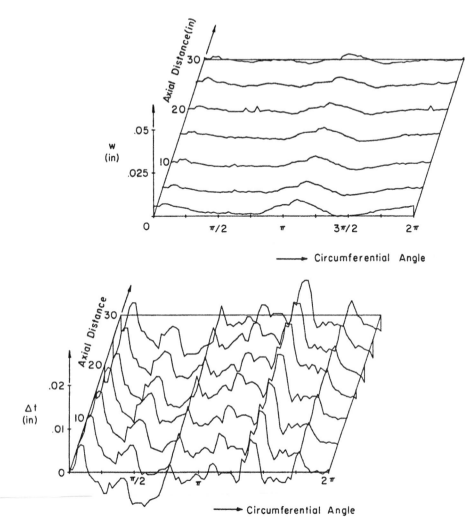

Figure 4.28   Imperfections of test specimen no. 2.

the circumference. This is reminiscent of the idealized eccentricity type of imperfection in Eq. (4.29) and Figure 4.11(b). Such wall eccentricity is usually introduced during piercing. In the cases of test specimens 2 and 4, the thickness varied in more complex fashions producing "star-shaped" inner surfaces. Star-shaped wall thicknesses are introduced by multi-roller reduction stands, used in the sizing and stretch reduction mills discussed in Section 3.3 (see Figures 3.6 and 3.7). Specimen 2 shows a four-wave variation, suggesting that it was reduced in a four-roller stand. Specimen 4 has an even more complex shape, indicating that successive multi-roller reduction stands may be out of phase with their neighbors as discussed in Section 3.3.

The material stress–strain responses were measured on axial coupons cut from each mother pipe. The measured responses appear in Figure 4.32. Specimens 1 and 2 exhibit Lüders bands, and it is thus concluded that they were either hot-finished or heat-treated

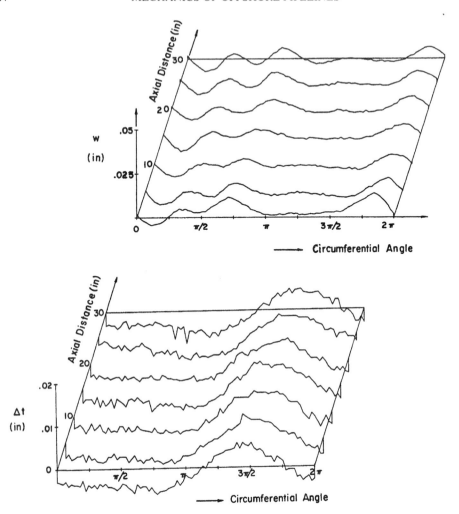

Figure 4.29   Imperfections of test specimen no. 3.

at the end of manufacture. The respective yield stresses are 48 and 54.3 ksi (331 and 374 MPa). Specimen 3 exhibits a rounded hardening response, indicating that it was probably cold finished. It has a yield stress of 73 ksi (503 MPa). Specimen 4 has a yield stress of 104 ksi (717 MPa) and has a relatively flat post-yield response. However, some nonlinearity is also exhibited in the initial, stiff part of the response. In the calculations that follow, these responses were approximated as piecewise linear in order to capture all their features. Yeh and Kyriakides in their original simulations [4.21] fitted three of them as bilinear and the third with a Ramberg–Osgood fit. This contributes to small differences between their predictions and the ones presented here. Three of the pipes exhibited some anisotropy, in the form of lower yield stress in the circumferential direction (see variable S in Table 4.1).

The four specimens were collapsed under external hydrostatic pressure, and the measured collapse pressures appear in Table 4.1. Several collapse calculations were performed

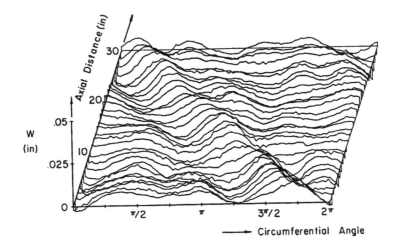

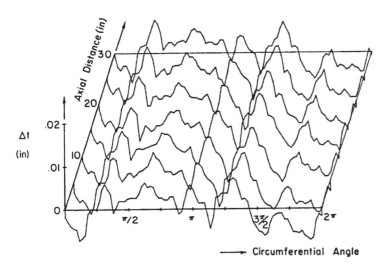

Figure 4.30   Imperfections of test specimen no. 4.

for each of the four pipes, using the custom analysis BEPTICO. The pressure loading is hydrostatic and all variables are assumed not to vary along the length. The mean diameter, wall thickness, initial ovality, eccentricity, and anisotropy used appear in Table 4.1. Table 4.2 shows calculated collapse pressures ($\hat{P}_{CO}$) for six different representations of the imperfections. In the first row, the pipes are assumed to be perfectly circular with constant wall thickness. Consequently, the calculated collapse pressures are higher than the measured values ((P_{CO}) in last row of Table 4.2). In the second row, the shape variation is idealized as initial ovality and the wall thickness variation is neglected. This brings the calculated collapse pressures within 5% of the measured values. This demonstrates once more that: first ovality is the main component of the shape imperfection that influences collapse and second that ovality is an important parameter of the problem. In the third and fourth rows, the wall eccentricity is included but the orientations of eccentricity are

Table 4.2   Comparison between calculated ($\hat{P}_{CO}$) and measured (P_{CO}) collapse.

Spec. No.	1	2	3	4
D in (mm)	4.000 (101.6)	4.010 (101.9)	4.012 (101.9)	4.003 (101.7)
t in (mm)	0.1435 (3.645)	0.1685 (4.280)	0.1255 (3.188)	0.1638 (4.161)
$\hat{P}_{CO}$ psi (bar)	2730 (188.3)	3939 (271.7)	2053 (141.6)	4895 (337.6)
Δ_o (%)	0.05	0.15	0.274	0.157
Ξ_o (%)	0	0	0	0
$\hat{P}_{CO}$ psi (bar)	2563 (176.8)	3441 (237.3)	1731 (119.4)	4353 (300.2)
Δ_o (%)	0.05	0.15	0.274	0.157
Ξ_o (%)	3.8	5.0	3.6	2.9
$\hat{P}_{CO}$ psi (bar)	2551 (175.9)	3414 (235.4)	1727 (119.1)	4345 (299.7)
Δ_o (%)	0.05	0.15	0.274	0.157
Ξ_o (%)	3.8	5.0	3.6	2.9
$\hat{P}_{CO}$ psi (bar)	2551 (175.9)	3414 (235.4)	1727 (119.1)	4345 (299.7)
$R(\theta)$	*	*	*	–
t in (mm)	0.1441 (3.660)	0.171 (4.343)	0.1235 (3.137)	–
$\hat{P}_{CO}$ psi (bar)	2649 (182.7)	3415 (235.5)	1830 (126.2)	–
$R(\theta)$	*	*	*	–
$t(\theta)$	*	*	*	–
$\hat{P}_{CO}$ psi (bar)	2660 (183.4)	3408 (235.0)	1824 (125.8)	–
P_{CO} psi (bar)	2720 (187.6)	3307 (228.1)	1764 (121.7)	4187 (288.8)

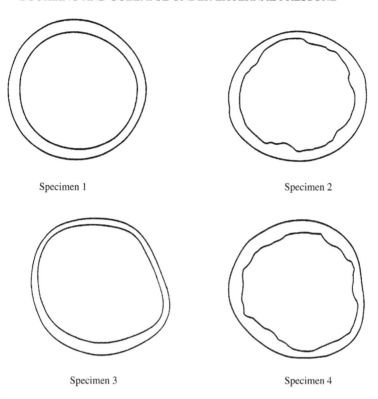

Specimen 1   Specimen 2

Specimen 3   Specimen 4

Figure 4.31   Typical cross sections of four pipes scanned illustrating effect of different manufacturing processes on final shape (imperfections amplified).

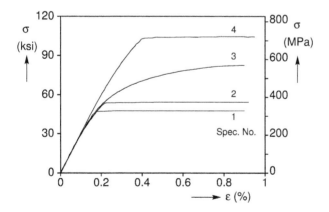

Figure 4.32   Stress–strain responses of four pipes analyzed.

90° out of phase between the two. The wall thickness variations measured were relatively small, and as a result, the effect of eccentricity on the collapse pressure is <1%. The difference in the orientation of the eccentricity had no effect on the calculated collapse pressures.

In row five, the actual shape imperfection of the most deformed cross section (biggest value of $|D_{max} - D_{min}|$) represented by a Fourier series expansion as in (e) is used, but the wall thickness is assumed to be uniform. The coefficients of the Fourier expansions appear in [4.21] and will not be repeated here. The calculated collapse pressures are somewhat higher than in row two for specimens 1 and 3, and quite close for specimen 2. Using just ovality can in general be expected to produce a slightly lower collapse pressure. In the sixth row, the wall thickness variation is also represented by a Fourier series as in (g). Including the actual wall thickness variation changes the collapse pressure only marginally from the value in row five. The wall thickness variations were not large enough to have a significant impact on the collapse pressure. Specimen 4 exhibited a spirally varying shape, and as a result the last two representations could not be performed.

In summary, the imperfections become more representative from top to bottom of Table 4.2, with row six being the most realistic one. The results clearly indicate first that a uniform tube analysis leads to very good predictions of the collapse pressure. Second, idealizing the shape imperfections as initial ovality and wall eccentricity leads to predictions of the collapse pressure which are well within engineering accuracy of the measured values. Both conclusions are in concert with similar ones drawn from the much larger experimental data set and corresponding predictions of smaller diameter tubes in Section 4.4.2 and in [4.12].

4.6 CONCLUSIONS AND DESIGN RECOMMENDATIONS

The main design load of land pipelines and lines installed in shallow waters is internal pressure. By contrast, in the case of pipelines installed in deeper offshore waters, the primary load is external pressure, thereby making the collapse pressure the major design parameter. For material and geometries of interest in pipeline applications, establishing the collapse pressure of a long circular cylinder under external pressure is sufficient and appropriate. The major factors that affect collapse are the pipe D/t, the elastic modulus (E), and yield stress (σ_o) of the material in the circumferential direction, as well as small geometric imperfections, the major one being initial ovality (Δ_o). Higher D/t pipes buckle elastically but experience plastic collapse. Lower D/t pipes buckle and collapse in the plastic range. In both cases, collapse localizes in the manner shown in Figures 1.3 and 4.9. However, the onset of collapse is adequately captured by models that consider a long cylinder with a uniform initial ovality. The elastic buckling pressure is accurately represented by Eq. (4.6). When this is lower than the yield pressure given in Eq. (4.31), the collapse pressure can be accurately predicted by Timoshenko's collapse formula in Eq. (4.11). Plastic buckling pressures can be evaluated iteratively from Eqs. (4.16), which requires the shape of the stress–strain response as input. Plastic collapse pressures can be evaluated via a uniform ovality numerical model (BEPTICO or corresponding FE model). The following additional guidelines should be helpful in designing against collapse:

- Initial ovality can reduce the collapse pressure significantly (e.g. ovality of 0.5% can reduce P_{CO} by 20–30%).
- The pipe yield stress in the circumferential direction can differ (it is often lower) from that in the axial direction. For collapse pressure calculations, the circumferential material properties should be adopted.

- Hydrostatic pressure loading is more representative of the loading seen by a pipeline on the sea floor. Collapse tests performed under lateral pressure loading usually result in lower collapse pressures than tests under hydrostatic pressure loading.
- Wall thickness variations do not significantly influence P_{CO}. Eccentricity (Ξ_o) $<10\%$ is desirable.
- Circumferential bending-type residual stress fields can influence P_{CO}. $\sigma_R < 0.3\sigma_o$ is recommended.
- The shape of the stress–strain response can influence P_{CO}. The effect becomes more prominent as the D/t decreases.
- UOE pipe has lower collapse pressure than seamless pipe of the same D/t and steel grade. The collapse performance of this type of pipe is discussed in Chapter 5.

REFERENCES

4.1. Kyriakides, S., Dyau, J.-Y. and Corona, E. (1994). Pipe collapse *Under Bending, Tension and External Pressure* (BEPTICO). *Computer Program Manual*. University of Texas at Austin, Engineering Mechanics Research Laboratory Report No. 94/4.

4.2. Boresi, A. (1955). A refinement of the theory of buckling of rings under uniform pressure. *ASME J. Appl. Mech.* **22**, 95–102.

4.3. Smith, C.V. and Simitses, G.J. (1969). Effect of shear and load behavior on ring stability. *Proc. ASCE.* **EM3**, 559–569.

4.4. Brush, D.O. and Almroth, B.O. (1975). *Buckling of Bars, Plates and Shells*. McGraw-Hill.

4.5. Pearson, C.E. (1956). General theory of elastic stability. *Quart. Appl. Math.* **14**, 133–144.

4.6. Levy, M. (1884). Memoire sur un nouveau cas integrable du probleme de l'elastique et l'une de ses applications. *J. Math. Pures Appl.* **10**, 5–42.

4.7. Timoshenko, S.P. and Gere, J.M. (1961). *Theory of Elastic Stability*, 2nd Edn. McGraw-Hill.

4.8. Timoshenko, S.P. (1933). Working stresses for columns and thin-walled structures. *ASME J. Appl. Mech.* **1**, 173–183.

4.9. Dubey, R.N. (1969). Instabilities in thin elastic–plastic tubes. *Intl. J. Solids Struct.* **5**, 699–711.

4.10. Ju, G.-T. and Kyriakides, S. (1991). Bifurcation versus limit load instabilities of elastic–plastic tubes under bending and pressure. *ASME J. Offshore Mech. Arc. Eng.* **113**, 43–52.

4.11. Chakrabarty, J. (1973). Plastic buckling of cylindrical shells subjected to external fluid pressure. *J. Appl. Math. Phys. (ZAMP)* **24**, 270–280.

4.12. Yeh, M.-K. and Kyriakides, S. (1986). On the collapse of inelastic thick-walled tubes under external pressure. *ASME J. Energy Resour. Technol.* **108**, 35–47.

4.13. Corona, E. and Kyriakides, S. (2000). Asymmetric collapse modes of pipes under combined bending and external pressure. *ASCE J. Eng. Mech.* **126**, 1232–1239.

4.14. Corona, E. and Kyriakides, S. (1988). On the collapse of inelastic tubes under combined bending and pressure. *Intl J. Solids Struct.* **24**, 505–535.

4.15. Dyau, J.-Y. and Kyriakides, S. (1993). On the localization and collapse in cylindrical shells under external pressure. *Intl J. Solids Struct.* **30**, 463–482.

4.16. Kyriakides, S. and Yeh, M.-K. (1985). Factors affecting pipe collapse. Final report to American Gas Association, Project PR-106-404. Also, University of Texas at Austin, Engineering Mechanics Research Laboratory Report No. 85/1.

4.17. Murphey, C.E. and Langner, C.G. (1985). Ultimate pipe strength under bending, collapse and fatigue. *Proc. of the 4th Int. Conference on Offshore Mechanics and Arctic Engineering*, Vol. 1, 467–477.

4.18. American Petroleum Institute Recommended Practice 1111 (1999). *Design, Construction, Operation and Maintenance of Offshore Pipelines (Limit State Design)*, 3rd Edn, July.

4.19. Kyogoku, T., Tokimasa, K., Nakanishi, H. and Okazawa, T. (1981). Experimental study of axial tension load collapse strength of oil well casing. *Proc. Offshore Technology Conference II*, OTC 4108, Vol. II, 387–395.

4.20. Tamano, T., Mimura, H. and Yanagimoto, S. (1981). Examination of commercial casing collapse strength under axial loading. *Proc. of the 1st Offshore Mechanics, Arctic Eng. Conference, Deepsea Systems Symposium, ASME*, Vol. 1, 113–118.

4.21. Yeh, M.-K. and Kyriakides, S. (1988). Collapse of deepwater pipelines. *ASME J. Energy Resour. Technol.* **110**, 1–11.

4.22. Tokimasa, K. and Tanaka, K. (1986). FEM Analysis of collapse strength of a tube. *ASME J. Pressure Vessel Technol.* **108**, 158–164.

4.23. Kyriakides, S. and Yeh, M.-K. (1988). Plastic anisotropy in drawn metal tubes. *ASME J. Eng. Ind.* **110**, 303–307.

4.24. Assanelli, A.P., Toscano, R.G., Johnson, D.H. and Dvorkin, E.N. (2000). Experimental/numerical analysis of the collapse behavior of steel pipes. *Eng. Comput.* **17**, 459–486.

4.25. Arbocz, J. and Babcock, C.D. (1969). The effect of general imperfections on the buckling of cylindrical shells. *ASME J. Appl. Mech.* **36**, 28–38.

4.26. API Specification 5L (2004). *Specification for Line Pipe*, 44th Edn. American Petroleum Institute, March.

5

Collapse of UOE Pipe Under External Pressure

Pipe larger than approximately 16 inches in diameter is commonly manufactured by cold forming of long plates through the UOE process (or the JCO process). The various steps of the process, and in particular its four cold forming steps, are outlined in Section 3.6 (see Figures 3.11 and 3.12).

UOE pipe has been widely used for land pipelines, including the Trans-Alaska and Trans-Siberia pipelines. In the last 15 years, it has also been increasingly used in offshore applications where collapse under external pressure is a primary design consideration, therefore requiring therefore high circularity (see Chapter 4). Modern UOE pipe mills are capable of delivering this demand for higher circularity pipe. Figure 3.15 shows the inside and outside shapes of a 24 in UOE pipe measured and reproduced as described in Section 4.5. For each surface, the radial deviation from the best-fit circular shape (w) is amplified $\times$ 20. The ovality of this pipe is 0.175%, while typical values range between 0.10% and 0.35%. Despite their low ovalities, experiments have demonstrated that the collapse pressure of UOE pipes can be significantly lower than that of seamless pipes of the same D/t and the same grade steel. For typical pipe geometries and plate material properties, the degradation in collapse pressure of UOE pipe can be higher than 30%. Because of the importance of UOE pipe to offshore applications, this chapter examines the causes of this degradation, and outlines remedies to it.

The degradation in collapse performance is demonstrated in Section 5.1, using results from full-scale collapse experiments. A procedure for predicting the collapse pressure of UOE pipe analytically is described in Section 5.2. Heat treatment procedures that have been shown to improve the collapse pressure of UOE pipe are discussed in Section 5.3. Section 5.4 describes a relatively simple one-dimensional model of the forming process that can be used to establish the induced changes in mechanical properties. Section 5.5 describes a more elaborate two-dimensional model of the process, which can capture the changes in mechanical properties as well as the final geometry of the formed pipe. Results from the two models are used to examine forming changes that can improve the collapse pressure of UOE pipe.

5.1 COLLAPSE PRESSURE OF UOE PIPE

One of the earlier uses of UOE pipe in moderately deep waters offshore was as tendons for TLP platforms [5.1–5.3]. In such applications, collapse under external pressure is one

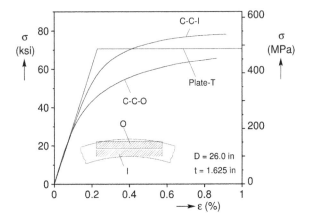

Figure 5.1 Comparison of stress–strain responses from transverse UOE pipe specimens tested in compression with the response measured in the original plate.

of the design concerns. This prompted an experimental program to evaluate the collapse performance of UOE pipe in which its reduced performance by comparison to seamless pipe was first observed. The reasons behind the reduction in collapse pressure were first identified in Kyriakides *et al.* [5.4]. It was demonstrated that the four cold forming steps, and in particular the final expansion, introduce changes to the compressive stress–strain response of the pipe in the circumferential direction. This is illustrated in Figure 5.1 for a 26-inch, X65 (nominally) pipe with wall thickness of 1.625 in (41.3 mm). Shown are compressive (C) stress–strain responses from circumferential (C) specimens extracted from two locations through the thickness (inside (I) and outside (O)) as shown in the inset. Included is the stress–strain response of the original plate in the transverse (T) direction (properties vary to some degree through the thickness, so this is the tensile response that would be measured by a full thickness transverse specimen). As is often the case, the plate exhibited Lüders banding that extended to a strain of approximately 3%, followed by hardening. The yield stress was 70.7 ksi (488 MPa). The mechanical work rounded the pipe's stress–strain responses significantly and lowered the stress in the critical strain range for collapse of 0.3–0.5% (see Figure 5.1). A significant gradient in properties developed through the thickness is illustrated by the much lower stress recorded for the outer specimen. The collapse pressure of this pipe was 7,144 psi (492.7 bar). By contrast, when the same pipe geometry is assigned the plate mechanical properties, the collapse pressure increases to 9,568 psi (659.9 bar). Thus, if we take the latter as the nominal collapse capacity of this pipe geometry and steel grade, the cold forming process degraded the collapse pressure by 25%.

For comparison, Figure 5.2 shows the tensile circumferential stress–strain responses measured at the same through-thickness locations. The forming, and in particular the final expansion, has had a strengthening effect on these responses. The extensive rounding is absent and the average yield stress exceeds that of the plate. Thus the UOE process has a beneficial effect on pipes destined for internal pressure loading. The process is consequently well-suited for land pipelines.

The high level of degradation in collapse pressure alluded to above was demonstrated more systematically in full-scale tests performed in support of the Oman–India pipeline

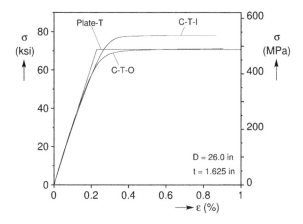

Figure 5.2   Comparison of stress–strain responses from transverse UOE pipe specimens tested in tension with the response measured in the original plate.

[5.5]. Eleven collapse tests were performed on 20-inch pipe with D/t of approximately 18, and two tests were performed on 26-inch pipe with D/t of nearly 16. The 20-inch pipes were tested in a US Navy pressure chamber in Carderock, Maryland, while the 26-inch pipes were tested in a new pressure chamber built for the project by C-FER Technology in Edmonton, Calgary. This facility has internal diameter 48 in (1219 mm), internal length 34 ft (10331 mm) and working pressure 8,000 psi (55 bar). It has since been used in several full-scale collapse testing programs, results from which will be cited below. The pipes were nominally X65 grade and were supplied by four different manufacturers: Europipe, Nippon Steel Corp., Sumitomo Metal Industries and British Steel (presently Corus Tubes). The major geometric and material parameters of the test specimens are listed in Table 5.1. With one exception, the ovalities ranged between 0.1% and 0.2%, which confirms the ability of pipe mills to deliver high-circularity pipes. Included in the table are the yield stresses measured in uniaxial compression tests on round specimens extracted from two locations through the thickness at 180° from the seam (see Figure A.2(b)). The yield stresses are designated as σ_{oO} for the outer specimen and σ_{oI} for the inner one. The stress–strain responses were rounded and followed the trend shown in Figure 5.1. In all cases the outer half of the pipe wall exhibited yield stresses that were much lower than 65 ksi (448 MPa). In several cases this was also true for the inner half.

The pipes were tested either under hydrostatic pressure (HP) or under lateral pressure (LP) (in the latter case using the scheme shown in Figure 4.17(a)). Nine of the 20-inch pipes were tested in the as-received condition (AR). Their collapse pressures (P_{CO}) ranged from 5,270 to 6,671 psi (363.4–460.1 bar). The average collapse pressure of the five pipes tested under LP was 5,544 psi (382.3 bar), while that of the four tested under HP was 6,154 psi (424.4 bar). The trend confirms that hydrostatic pressure loading leads to higher collapse pressures, as pointed out in Section 4.4.4.

Included in Table 5.1 are collapse pressures calculated numerically using the measured values of diameter, wall thickness and ovality, and an assumed elastic-perfectly plastic stress–strain response with a yield stress of 65 ksi (448 MPa); in other words, the nominal properties of the material ($\hat{P}_{CON}$). The predictions are uniformly higher. The average value of the specimens tested under lateral pressure loading is 7,136 psi (492.1 bar), which is

Table 5.1 Pipe parameters and collapse pressures of UOE pipes [5.5].

NO.	EP1	EP2	EP3	EP4	EP5	NI1	NI2	NI3	SU1	SU2	SU3	BS1	BS2
D in	19.969	19.965	19.958	19.956	19.963	20.008	20.018	19.999	20.047	20.026	20.040	25.971	25.970
t in	1.127	1.125	1.114	1.108	1.122	1.1256	1.1346	1.132	1.119	1.126	1.128	1.6232	1.6328
D/t	17.72	17.75	17.92	18.01	17.79	17.78	17.64	17.67	17.92	17.79	17.77	16.00	15.91
L/D	9	9	9	9	9	9	9	9	9	9	12	13.3	14.5
Δ_o % (max)	0.173	0.183	0.170	0.153	0.178	0.127	0.110	0.105	0.172	0.295	0.130	0.2	0.18
Condition	AR	AR	AR	AR	HT	AR	AR	AR	AR	AR	HT	AR	AR
$|\sigma_R|$ ksi	8.882	8.172	9.467	9.173	0.724	6.72	6.84	6.65	6.86	6.73	0	10.0	11.1
σ_{oO} ksi	−57.6	−55.1	−61.8	−63.0	−69.2	−52.3	−51.8	−53.7	*	−60.4	−69.0	−54.5	−54.4
σ_{oI} ksi	−75.2	−65.1	−76.7	−71.1	−72.9	−58.0	−57.2	−58.0	*	−61.9	−69.0	−71.9	−61.6
HP/LP	LP	HP	HP*	LP	HP	LP	HP	LP	LP	HP	HP	HP	LP
P_{CO} psi	**6,113**	**6,321**	**6,671**	**5,713**	**8,839**	**5,289**	**5,634**	**5,270**	**5,337**	**5,990**	**8,429**	**7,144**	**7,117**
$\hat{P}_{CO}$ psi	6,471	6,157	6,857	5,889	8,335	5,440	5,730	5,559	*	5,897	8,450	7,324	6,850
$\hat{P}_{CON}$ psi	7,098	7,754	7,700	7,005	7,746	7,223	8,204	7,361	6,993	7,272	8,009	9,568	8,751

28.7% higher than the average of the measured collapse pressures. The average of the calculated values of the four HP loaded pipes is 7,733 psi (533.2 bar), which is 25.6% higher than the average of the measured values. The actual yield stress of the plate material is usually higher than the nominal value. As a result, these predictions are generally conservative; they do, however, clearly illustrate that the UOE cold forming lowers the collapse pressure significantly.

A more accurate measure of the degradation induced by the UOE process was obtained by heat-treating two of the 20-inch pipes and subsequently collapsing them (HT condition). The heat treatment took the pipes to normalizing temperature, so that the sharp yield point and Lüders banding behavior were recovered and the residual stresses erased. The yield stresses measured at the two locations increased for both pipes to values around 70 ksi (483 MPa) (this value is assumed to be close to the yield stress of the original plate). Furthermore, one of the pipes showed no difference in yield stress at the two thickness locations, while in the second the difference was down to about 3.5 ksi (24 MPa). The shapes of the cross sections of the two pipes were measured before and after heat treatment, and were found to remain the same. The collapse pressures increased to 8,429 psi (581.3 bar), and 8,839 psi (609.6 bar), respectively. Since the cross sectional shape was not altered by heat treatment, the increase is directly attributable to the recovery in yield strength. Comparing the average values of the collapse pressures of the two heat-treated pipes to the average value measured in the nine as-received pipes, one concludes that the forming decreased the collapse pressure by about 33%.

The two pipes with $D/t \approx 16$ were tested in the as-received state in the C-FER pressure chamber. Some perspective on the size of such specimens can be developed from Figure 5.3, which shows a collapsed pipe being removed from this pressure vessel. The collapse pressures of the pipes were 7,144 psi (492.7 bar) and 7,117 psi (490.1 bar). The significantly higher collapse pressure compared to the 20-inch pipes is a direct result of their lower pipe D/t. In this case, plate properties were available and the transverse yield stress was slightly higher than 70 ksi (483 MPa). Using this value and the measured diameter, wall thickness and ovality, the calculated average collapse pressures are

Figure 5.3   24-inch UOE pipe collapsed under external pressure (courtesy C-FER Technologies).

9,568 psi (659.9 bar) and 8,751 psi (603.5 bar), respectively. Both predictions demonstrate once more the degrading effect of the UOE process.

5.2 PREDICTION OF COLLAPSE PRESSURE OF UOE PIPE

Prediction of the collapse pressure of a given UOE pipe requires measurement of several parameters. The pipe specimen must be at least 10 diameters long. The measurements follow the procedure outlined in Section 4.4.1 with some additions. Diameter measurements are performed at axial stations one pipe diameter apart. At every station, the diameter is measured at least at the angular positions shown in Figure 5.4. The average of all diameter measurements is designated as the pipe diameter. Although the shape of the cross section of UOE pipe is rather complex (see Figure 3.15), establishing the ovality is sufficient for calculating the pipe collapse pressure. The ovality is calculated at each station using Eq. (4.25). The maximum value measured along the length is designated as the ovality of the pipe (Δ_o). The wall thickness is measured at least at the two ends at the angular positions shown in Figure 5.4. (More measurements can be performed in the body of the pipe with an ultrasonic thickness gage.) The average value of the measurements is designated as the thickness of the pipe (t). Thickness variations in UOE pipe are generally small, and as a result can be neglected in collapse calculations. A mean value of residual stresses can be obtained using the split-ring test described in Appendix A. The amplitudes (σ_R) of the stress fields estimated for the pipes in Table 5.1 ranged between 6 and 11 ksi (41–76 MPa).

The mechanical properties of UOE pipe vary to some degree along the length because of variations in the plate (see Section 3.2). The plate axial yield stress is usually lower than the value in the transverse direction. The forming process further accentuates the

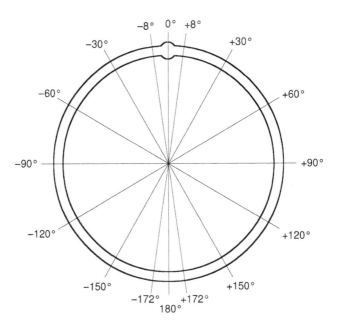

Figure 5.4 Angular positions at which diameter and thickness measurements are taken before the specimen is tested.

differences between the axial and circumferential properties, but more importantly, it lowers the circumferential compressive strength. In addition, the four manufacturing steps introduce some variation in properties around the circumference. Investigation of these differences led to the conclusion that the half of the pipe opposite the seam is the most representative of the average properties. For the tests in Table 5.1, two transverse specimens were extracted from a location diametrically opposite the seam. One was close to the inner surface, and the second was as close to the outer surface as was practical (see Figure A.2(b)). The specimens were tested in compression, producing stress–strain responses that extended up to a strain of approximately 1% (the specimens tend to buckle at higher strain values). For thinner pipes, a single specimen that encompasses most of the wall thickness will suffice.

The average diameter and wall thickness, the maximum ovality, the residual stress and the type of pressure loading were used in the 2-D model described in Section 4.3 (BEPTICO) to predict the collapse pressure of the 13 pipes tested. Integration points on the inner half of the pipe wall were assigned the stress–strain response from the inner specimen, and those in the outer half the one from the outer specimen. The predicted collapse pressures ($\hat{P}_{CO}$) are included in Table 5.1. They are seen to be in good agreement with the measured collapsed pressures, with the biggest deviation being on the order of 5%. This level of agreement has also been obtained in several additional full-scale tests, performed subsequently by industry not reported here. The good agreement between measured and calculated collapse pressures confirms the fidelity of the measurement protocol outlined above, as well as the adequacy of the idealizations introduced in the 2-D analysis performed.

5.3 IMPROVEMENT OF COMPRESSIVE PROPERTIES BY HEAT TREATMENT OF THE PIPE

The 30–40% degradation in collapse pressure induced by the UOE process illustrated above translates into significantly higher wall thickness requirements. This, in turn, results in not only higher material costs but also in higher offshore installation costs due to the increased line weight. In addition, there are limits in the plate wall thickness that can be formed into pipe by modern pipe mills. For example, these limits were challenged in the case of the Oman–India project, where the pipeline was designed to cross waters as deep as 3,350 m (11,000 ft [5.5]). Therefore, any scheme that improves the collapse pressure can be beneficial, in that it reduces the cost and can make some otherwise unfeasible projects possible.

In Section 5.1, it was demonstrated that heat-treating the pipe at normalizing temperatures returns the properties back to those of the original plate. This level of heating is very costly and can have detrimental effects on the welds and on corrosion resistance. Subsequent work showed that a more modest heat treatment process, involving heating to 225–275°C for a few minutes, can recover the compressive yield stress and cause a corresponding increase in collapse pressure [5.6]. This recovery is the result of strain aging of the material. The elevated temperature allows interstitial solute atoms such as carbon and nitrogen to diffuse around dislocations pinning them [5.7, 5.8]. Many pipelines are coated with epoxies for corrosion protection. Several of these coatings are fusion-bonded to the steel, requiring that the pipe be heated to 225–250°C for a short time (30–60 s;

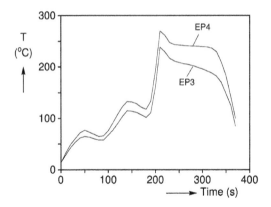

Figure 5.5   Temperature history of heat-treated 24-inch pipe [5.11].

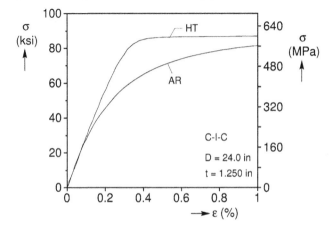

Figure 5.6   Comparison of transverse compressive stress–strain responses of 24-inch UOE pipe in the as-received state and after heat treatment [5.11].

e.g., Corro Coat EP-F 1003LD, 3M Scotchkote 226N/633, DUPOND NAP-GARD Mark X 7-2500, see manufacturer specifications). Thus, strain aging can be a byproduct of the coating process.

The potential benefits of this type of heat treatment on the collapse performance have been explored and proven in several testing programs [5.9–5.11]. The recovery in mechanical properties by this low temperature heat treatment was first confirmed using test coupons extracted from the pipe. Several pipes were subsequently heat treated as follows: the pipe was first heated in a furnace for about 200 s, reaching a temperature of about 150°C (see temperature history in Figure 5.5). It was subsequently passed through a 1.3 MW induction coil at nearly full power, increasing the temperature to about 225–250°C. After about 340 s the pipe was quenched. Figure 5.6 shows a comparison of circumferential compression stress–strain responses measured on coupons extracted from the pipe before and after heat treatment. Strain aging removes a significant part of the rounding induced by cold forming, and increases the yield stress dramatically.

Collapse pressures of as-received pipes and pipes heat-treated in the manner described above are compared in Table 5.2. The first four pipes were tested under the Blue Stream

Table 5.2   Pipe parameters and collapse pressures of UOE pipes tested as received and after heat treatment.

NO.	EP1	EP2	EP3	EP4	CT1	CT2	CT3	CT4
D in	23.734	24.048	24.018	23.998	17.951	18.015	17.939	18.013
t in	1.2608	1.2500	1.250	1.236	1.007	1.006	1.001	1.009
D/t	18.82	19.24	19.21	19.42	17.79	17.90	17.92	17.85
L/D	12.84	12.84	12.86	12.87	12.12	12.04	12.16	12.10
Δ_o % (max)	0.150	0.196	0.172	0.189	0.295	0.375	0.301	0.369
Condition	AR	AR	HT	HT	AR	AR	HT	HT
$|\sigma_R|$ ksi	9.34	3.85	3.84	3.74	12.8	6.1	9.5	17.6
σ_{oO} ksi	−51.33	−55.06	−69.91	−71.50	−52.0	−52.9	−68.1	−69.0
σ_{oI} ksi	−56.55	−62.89	−86.30	−82.82	−51.3	−55.8	−71.6	−72.2
HP/LP	HP	HP	HP	HP	HP	HP	HP	HP
P_{CO} psi	**5,075**	**5,365**	**6,525**	**6,540**	**5,196**	**5,213**	**7,294**	**6,981**
$\hat{P}_{CO}$ psi	4,790	4,959	6,576	6,583	5,045	5,146	7,211	6,785
$\hat{P}_{CON}$ psi	7,642	7,104	7,244	7,018	7,247	6,919	7,162	6,975

pipeline development program [5.11]. This involved nominally X65 grade, 24-inch pipes with approximately 1.250 in (31.8 mm) wall. The four Europipe pipes had relatively low ovalities that ranged between 0.15% and 0.2%. Two pipes were tested in the as-received (AR) state and two following heat treatment (HT). Included in the table are the measured yield stresses in compression tests in the circumferential direction. Although the yield stresses for the inner and outer specimens maintained a difference, heat treatment resulted in a significant increase in both values. The average collapse pressure of the AR pipes was 5,220 psi (360 bar) and the average of the HT pipes was 6,533 psi (450.6 bar). Thus, the heat treatment resulted in approximately a 25% increase in collapse pressure.

Included in the table are collapse pressures calculated using the individual geometric variables of the pipes and an elastic-perfectly plastic material response with a yield stress of 70 ksi (483 MPa) ($\hat{P}_{CON}$; measured plate properties ranged between 70 and 80 ksi). They are all above 7,000 psi, with an average value of 7,252 psi (500.1 bar). These values indicate that the heat treatment did not fully recover the material strength.

The second set of results comes from a development program undertaken by Corus Tubes [5.10]. It involved four nominally X65, 18-inch pipes with 1.000 in (25.4 mm) wall thickness. Again two were tested in the AR state and two following heat treatment of the type described above. The ovalities of these pipes were somewhat larger, ranging between 0.3% and 0.375%. The compressive yield stresses of the HT pipes are seen to have significantly higher values than the AR pipes. The average collapse pressure of the AR pipes was 5,205 psi (359.0 bar) while that of the HT was 7,138 psi (492.2 bar). So in this set the heat treatment has resulted in a 37% increase in collapse pressure. The larger improvement is attributed to the more uniform recovery in strength compared to the first group. Collapse pressures were calculated using the individual geometric variables of the pipes and an elastic-perfectly plastic material response with a yield stress of 65 ksi (448 MPa) ($\hat{P}_{CON}$). They are all around 7,000 psi (483 bar), with an average value of 7,076 psi (488 bar). These values are in agreement with the collapse pressures of the two HT pipes, confirming that for this set the heat treatment was more effective in recovering the material strength throughout the pipe wall.

Included in Table 5.2 are collapse pressures ($\hat{P}_{CO}$) calculated for each pipe using the geometric parameters listed and the measured stress–strain responses. All predictions are in good agreement with the test results, once more confirming the adequacy of the scheme adopted.

5.4 ONE-DIMENSIONAL MODEL OF UOE PIPE FORMING

Through modeling and simulation of UOE pipe forming, it became clear that the main cause of the degradation in compressive strength and the corresponding reduction in collapse performance illustrated in Sections 5.1 and 5.2 is the final expansion. This led Kyriakides *et al.* [5.4] to propose replacing expansion with compression (UOC). Furthermore, simulations showed that the UOE process can be optimized for collapse performance by altering some of the traditionally used forming parameters. This section presents a simple one-dimensional model of the process that can be used to illustrate the cause of the degradation in compressive yield strength, and to show that the alternative UOC process does away with this deficiency.

In this model, the plate is shaped by cylindrical bending and axisymmetric contraction and expansion, following steps 1–5 in Figure 5.7. In this manner, the finished pipe is circular with uniform properties around the circumference. The plate is assumed to be isotropic, stress-free and thin enough for the customary thin-walled assumptions to apply. Thus, the strain increments are given by

$$\left\{ \begin{array}{c} \dot{\varepsilon}_x \\ \dot{\varepsilon}_s \end{array} \right\} = \left\{ \begin{array}{c} \dot{\varepsilon}^o_{xx} \\ \dot{\varepsilon}^o_{ss} \end{array} \right\} + z \left\{ \begin{array}{c} \dot{\kappa}_{xx} \\ \dot{\kappa}_{ss} \end{array} \right\}, \tag{a}$$

where (x, s) are the axial and circumferential coordinates. The strains are assumed to be small, and the instantaneous stress–strain relationships are given by

$$\left\{ \begin{array}{c} \dot{\sigma}_x \\ \dot{\sigma}_s \end{array} \right\} = \left[\begin{array}{cc} C_{11} & C_{12} \\ C_{12} & C_{22} \end{array} \right] \left\{ \begin{array}{c} \dot{\varepsilon}_x \\ \dot{\varepsilon}_s \end{array} \right\}, \tag{b}$$

where $[C_{\alpha\beta}]$ come from an appropriate flow theory of plasticity model. It is essential that the plasticity model adopted has the capacity to accurately simulate the Bauschinger effect exhibited on unloading and reverse loading. This is achieved by adopting the two-surface nonlinear kinematic hardening model of Dafalias and Popov outlined in Section 13.4.2. The instantaneous constitutive matrix $D = C^{-1}$ is expressed as follows:

$$D = \frac{1}{E} \left[\begin{array}{cc} 1 + Q(2\hat{\sigma}_x - \hat{\sigma}_s)^2 & -\nu + Q(2\hat{\sigma}_x - \hat{\sigma}_s)(2\hat{\sigma}_s - \hat{\sigma}_x) \\ -\nu + Q(2\hat{\sigma}_x - \hat{\sigma}_s)(2\hat{\sigma}_s - \hat{\sigma}_x) & 1 + Q(2\hat{\sigma}_s - \hat{\sigma}_x)^2 \end{array} \right] \tag{c}$$

where

$$Q = \frac{E}{4\sigma_o^2 H(\delta, \delta_{in})}, \qquad \hat{\sigma}_i = (\sigma_i - \alpha_i),$$

and the plastic modulus H is evaluated from (13.35).

The example that will be presented involves a 24-inch pipe with 1.273 in (32.33 mm) wall. This pipe was formed from X70 grade steel plate. Figure 5.8 shows the measured

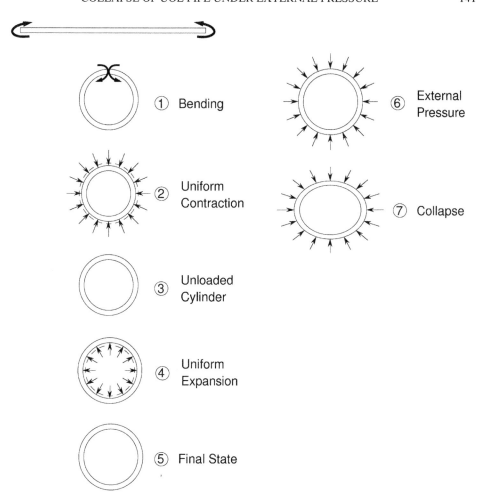

Figure 5.7 Deformation and loading steps used to simulate the UOE process followed by collapse under external pressure using the 1-D model of Ref. [5.4].

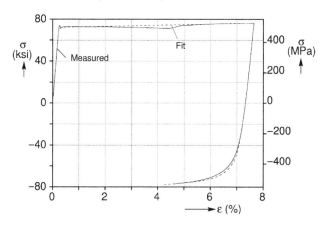

Figure 5.8 Uniaxial plate stress–strain response and the Dafalias–Popov fit used in the simulation.

Table 5.3 Constitutive model parameters for base case.

Part	E Msi (GPa)	ν	σ_o ksi (MPa)	σ_b ksi (MPa)	E_o^P ksi (MPa)	h Msi (GPa)
Monotonic	31 (214)	0.3	68.0 (469)	72.14 (498)	55.8 (385)	5.0 (34.5)
Hysteresis	24 (165)	0.3	52.9 (365)	86.40 (596)	189 (1303)	10 (69)

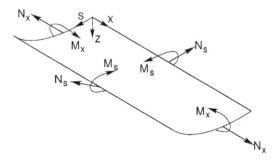

Figure 5.9 Definition of the stress and moment intensities for the 1-D model.

stress–strain response of this plate in the transverse direction. The specimen was pulled in tension to about 7.6% strain, followed by unloading and reverse loading. The response exhibits Lüders banding behavior that extends to about 4.7%. Subsequently, the material hardens and deforms uniformly. Reverse loading exhibits the Bauschinger effect.

The Dafalias–Popov model was calibrated to this response using the procedure described in Appendix E. The monotonic and hysteresis responses were modeled independently. The model parameters for each are given in Table 5.3, and the fit is drawn in Figure 5.8 with a dashed line. For numerical expediency, the Lüders banding is avoided by assigning a small positive constant slope to the monotonic part of the response.

Using Eqs. (a) and (b) the force and moment intensities are related to the membrane strains and curvatures as follows (see Figure 5.9):

$$\begin{Bmatrix} \dot{N}_{xx} \\ \dot{N}_{ss} \\ \dot{M}_{xx} \\ \dot{M}_{ss} \end{Bmatrix} = \int_{-t/2}^{t/2} \begin{bmatrix} C & zC \\ zC & z^2C \end{bmatrix} dz \begin{Bmatrix} \dot{\varepsilon}_{xx}^o \\ \dot{\varepsilon}_{ss}^o \\ \dot{\kappa}_{xx} \\ \dot{\kappa}_{ss} \end{Bmatrix}. \tag{d}$$

For cylindrical, plane strain forming

$$\dot{\varepsilon}_{xx}^o = \dot{\kappa}_{xx} = 0. \tag{e}$$

The simulation is conducted as follows (see Figure 5.7):

① The plate is bent into a cylinder by incrementally prescribing the curvature κ_s until the desired radius is reached.

② The circular cylinder is contracted circumferentially to a strain corresponding to that seen in the O-press.

③ The membrane compressive force developed, N_{ss}, is then released to zero. This approximately completes the simulation of the U and O processes.

Table 5.4 Main pipe shape parameters and collapse pressures (1-D model).

Parameters	UO	UOE	UOC	UOE-HT
$\varepsilon\,\%$	−0.2048	0.9838	−0.1712	–
$\Delta_o\,\%$	0.848	0.0967	0.192	0.0967
P_{CO} psi (bar)	4,063 (280.2)	5,112 (352.6)	7,449 (513.7)	7,477 (515.7)

④ Expansion is simulated by internally pressurizing the cylinder in an incremental fashion up to a strain corresponding to that seen during the expansion stage of the actual process.
⑤ The pressure is then removed and the pipe is unloaded.

When the pipe is finished by compression, steps ②–⑤ are replaced by the following: after step ①, the strain is circumferentially contracted to the desired strain, and the pipe is then unloaded incrementally.

During the simulation, the stresses are monitored at 12 points through the thickness, and integrations are performed by Gaussian quadrature. The main products of the simulation are the stresses, strains and their history parameters at the 12 integration points. Obviously, this model does not produce a pipe shape.

The collapse pressure of the pipe is evaluated using the custom collapse analysis described in Section 4.3 (BEPTICO). This version of BEPTICO uses the same two-surface plasticity model as the 1-D forming model outlined above. The input is the pipe diameter, wall thickness and initial ovality measured in an actual UOE or UOC pipe. In addition, the through-thickness stress–strain states produced by the 1-D model are assigned to all circumferential integration positions of the model. The pipe is then pressurized in a volume-controlled manner under plane strain loading conditions until it collapses. This aspect of the calculation follows along the lines described in Section 4.3.

The model performance is demonstrated by simulating the forming of a 24-inch pipe with 1.273 in (31.83 mm) wall and the mechanical properties listed in Table 5.3. The results are listed in Table 5.4. The circumferential membrane strains after UO, UOE and UOC are listed in the first row. The ovality of each pipe came from the 2-D simulation described in the next section, and is listed in row 2. The collapse pressure is listed in row 3. The UO pipe has a relatively high ovality and, as a consequence, a collapse pressure of only 4,063 psi (280.2 bar). Expansion reduces the ovality but simultaneously degrades the compressive properties. The net result is a collapse pressure of 5,112 psi (352.6 bar). Finishing the pipe by compression to a strain of −0.1712% results in an ovality of 0.192%, a value that is nearly double that of the corresponding UOE pipe. Despite this, the collapse pressure climbs to 7,449 psi (513.7 bar). Assigning the pipe the properties of the plate (see Figure 5.8) and the ovality of the UOE pipe yields a collapse pressure of 7,477 psi (515.7 bar). This can be assumed to be the collapse pressure yielded by this model for a fully heat-treated UOE pipe.

The results will be compared in the next section to corresponding predictions using the more complete 2-D model of the process and shown to be of sufficient accuracy. A weakness of this model is that it does not produce a shaped pipe. The diameter, thickness and ovality shape must be provided. In practice, these can come from measurements performed after each step at the pipe mill (this was done for the case analyzed in [5.4]). In the present calculations the geometric parameters were provided from the 2-D forming simulation described in the next section.

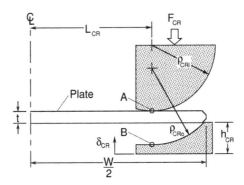

Figure 5.10 Schematic and parameters of the crimping press.

5.5 TWO-DIMENSIONAL MODELS OF UOE/UOC

The 1-D model outlined in Section 5.4 does not capture variations of properties around the circumference induced by the process, nor does it produce the final cross sectional shape. A more elaborate 2-D finite element model of the process that is capable of accurately simulating each forming step was presented in Herynk *et al.* [5.12]. The finished pipe can then be collapsed under external pressure. The model was used to develop an understanding of how each step influences the pipe shape and mechanical properties as well as the collapse pressure. This section summarizes some of the key findings of this work.

5.5.1 UOE/UOC Forming Steps

The UOE pipe manufacturing process involves many steps that are summarized in the flowchart shown in Figure 3.13 and described in Section 3.6. Here we review in more detail the four mechanical forming steps that the model simulates.

a. Crimping Press

The longitudinal edges of the plate are trimmed by milling, bringing the width to the exact required value. Simultaneously, the ends are beveled to later form v-grooves in the circular skelp to accommodate the welding. The first forming step involves crimping of the edges of the plate into circular arcs over a width of about one radius on each side. This is achieved by pressing the ends between two shaped dies as shown in Figure 5.10. Because of the large forces required, this is done in a step fashion involving lengths of one to four pipe diameters depending on the wall thickness. In order to accommodate different pipe thicknesses and diameters, each mill possesses sets of dies of different dimensions. For a given pipe, the dies with the most appropriate inner and outer radii (ρ_{CRi} and ρ_{CRo}) are selected. The relative horizontal positions of the dies can be adjusted to accommodate for mismatch between the die radii and the thickness of the plate. The width of plate to be crimped is set by the horizontal placement of the dies (L_{CR}). The values L_{CR} can be influenced by the thickness of the plate and the load capacity of the press.

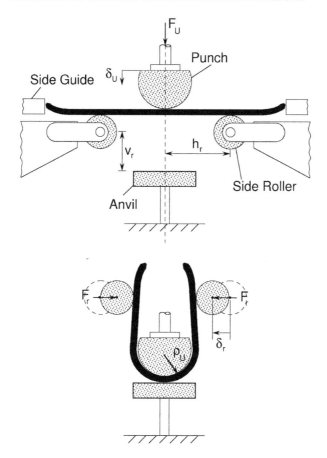

Figure 5.11 Schematic and parameters of the U-press.

b. U-Press

The plate moves next to the U-press, where it initially rests centered between a pair of side rollers that run along its entire length (Figure 5.11). The U-punch moves down and bends the entire plate through three-point bending. The radius of the punch ρ_U is selected so that the lower half of the plate acquires an outer radius near that of the final pipe (see Figures 5.11 and 3.12(a)). The U-punch stops when the plate contacts a series of anvils set at a predetermined height. The U-punch is then held in place, and the side rollers are moved inwards as shown in Figure 5.11. The horizontal position (h_r) and inward travel (δ_r) of the rollers are selected such that the final position of the straight arms of the U-shaped plate, or "skelp," is nearly vertical.

c. O-Press

The skelp is then conveyed to the O-press, which consists of two semi-circular (radius ρ_O) stiff dies as shown in Figures 5.12 and 3.12(b). The top die is actuated downwards, forcing the skelp into a nearly circular shape (Figure 3.12(c)). The O-forming

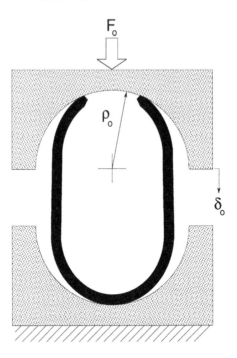

Figure 5.12   Schematic and parameters of the O-press.

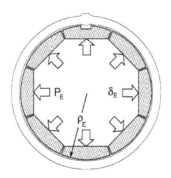

Figure 5.13   Schematic and parameters of the expansion process.

finishes by forcing the dies further together, producing a net compressive strain of 0.1–0.2%.

After leaving the O-press, the pipe seam is welded using submerged arc welding machines that weld it first on the inside and then on the outside. Extensive ultrasonic examination is performed on the weld before the pipe is expanded.

d. Expansion

Expansion is accomplished by an internal mandrel shown in Figures 5.13 and 3.12(d). The mandrel consists of eight, ten, or twelve segments. The segment radius (ρ_E) is chosen to be near the inner radius of the pipe. The mandrel is hydraulically actuated, and in one

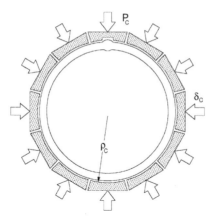

Figure 5.14   Schematic and parameters of the compression process.

step it typically expands a length of one-half to one diameter (depending on the wall thickness). Each step maintains some overlap between the expanded and unexpanded pipe sections. Expansion improves the roundness of the pipe and brings it to its desired final size. Traditionally, in order to achieve low ovality, the pipe is expanded 0.8–1.3% from its diameter after the O-step.

e. Compression

Compressive devices that would substitute the expanders do not exist for large-diameter pipes. A small number of such devices exist for smaller pipes made by the ERW process (US Patent 2,999,405 [5.13]). Figure 5.14 shows a schematic of such a device with 12 mandrel segments with radius ρ_C. It is envisioned to operate in a similar fashion as the expanders.

5.5.2   Numerical Simulation

a. Discretization

In the 2-D model, the plate is formed under plane strain conditions. In addition, symmetry about the plate mid-width is assumed. The problem is solved within the nonlinear FE code ABAQUS using a user-defined nonlinear kinematic hardening plasticity subroutine based on the two-surface model of Dafalias and Popov. The structure is discretized by linear, reduced integration plane strain continuum elements (CPE4R). Figure 5.15 shows the mesh adopted after convergence studies. There are seven elements through the thickness. Along the length, there are 110 elements in the main part of the plate (W_1), 10 elements in the edge of the straight part (W_2), and 15 elements in the section beveled to accommodate welding (W_3). The plate geometry is altered by prescribing new values for W, W_i, t, and t_i.

The plate is deformable, and the forming dies were modeled as analytical rigid surfaces. Contact between two surfaces is accomplished by a strict "master-slave" algorithm, in which rigid surfaces constitute the master surface in a contact pair. The contact pairs are allowed "finite sliding," and contact was assumed to be frictionless. Contact is made to be "soft" by user-defined variables.

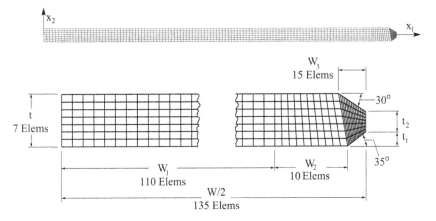

Figure 5.15 Finite element mesh of the UOE model.

b. Constitutive Model

As was the case for the 1-D model, the elastic–plastic behavior of the material is modeled through the two-surface nonlinear kinematic hardening model of Dafalias and Popov outlined in Section 13.4. The same example case will be analyzed using the plate material properties given in Table 5.3.

c. Collapse Under External Pressure

Once the forming is completed, the pipe is loaded by external pressure up to collapse. The pressurization is achieved by surrounding the pipe (one-half of the cross section) with a fluid-filled cavity made of two-node hydrostatic fluid elements (F2D2). Fluid is pumped into the cavity by using the cavity reference node as an inlet. This volume-controlled loading enables the tracing of the pressure maximum that corresponds to the collapse pressure ($\hat{P}_{CO}$).

5.5.3 An Example of UOE Forming

The model is now used to simulate the UOE forming of the same 24-inch pipe analyzed in Section 5.4 using the 1-D model. The pipe has a wall thickness of 1.273 in (32.33 mm) and the mechanical properties listed in Table 5.3. The process forming parameters used are listed in Table 5.5. These values are representative of values that a modern steel mill might use for this pipe.

Crimping is achieved by moving the outer die upwards (displacement δ_{CR}) while the inner die is stationary. The die is displaced until points A and B in Figure 5.10 are a distance of $(t + 0.02)$ in apart, as illustrated in the sequence of configurations shown in Figure 5.16. During most of the crimping process (configuration ①) the reaction force is nearly constant, but it ramps up sharply once the plate makes significant contact with the dies (configuration ②). The added gap of 0.020 in (0.5 mm) ensures that reaction forces will not become excessive. At the end, the dies are moved apart and the plate springs back elastically.

Table 5.5 Parameters of the base case UOE simulation.

	Var.	Description	Value
Plate	t	Thickness of the plate in (mm)	1.273 (32.33)
	W	Width of the plate in (mm)	71.00 (1803)
	X	Grade of the plate steel ksi (MPa)	70 (482)
Crimp	ρ_{CRi}	Inner crimping radius in (mm)	10.45 (265.4)
	ρ_{CRo}	Outer crimping radius in (mm)	11.75 (298.5)
	δ_{CR}	Under-closure of crimp die in (mm)	0.02 (0.5)*
	L_{CR}	Horizontal position of the dies in (mm)	26.64 (676.7)
	h_{CR}	Height of outer crimp die in (mm)	3.50 (88.9)*
U	ρ_U	Radius of U punch in (mm)	9.70 (246.4)
	δ_U	U punch vertical travel in (mm)	28.5 (724)
	δ_r	U roller horizontal travel in (mm)	4.00 (102)
	h_r	Horizontal position of the roller in (mm)	18.0 (457)*
	v_r	Vertical position of anvil in (mm)	28.5 (724)
O	ρ_O	Radii of the O dies in (mm)	11.959 (303.8)*
	δ_O	Overlap of the O dies' centers in (mm)	0.00
E	ρ_E	Radii of the expansion mandrels in (mm)	10.66 (270.8)
	δ_E	Radial expansion of mandrels in (mm)	0.14 (3.6)
	N_E	Number of expansion mandrels	8
C	ρ_C	Radii of compression mandrels in (mm)	11.88 (301.8)*
	δ_C	Radial compression of mandrels in (mm)	0.14 (3.6)
	N_C	Number of compression mandrels	12*

* Indicates constant values

In the U-press the punch first moves down (δ_U) until the skelp contacts the anvil (configurations ⓪–② in Figure 5.17). The load reaches a maximum at about half of the total displacement, and then gradually drops as the plate continues to bend. Next, the rollers move inward a distance δ_r, bending the plate "arms" usually past vertical (configuration ③). The rollers are then retracted and the plate springs back to its final U-shape, with the "arms" nearly vertical (configuration ④).

In the O-press, the upper die is moved down, forcing the skelp to gradually conform to the circular shape (configuration ① Figure 5.18). Initially, the lower part of the skelp bends to match the curvature of the lower die. The top edges of the skelp then slide along the top die until they meet. Subsequently, the skelp arms bend to match the curvature of the top die, and as a result the force rises. The pipe eventually nearly matches the shape of the dies, and further displacement compresses it circumferentially (configuration ②). Compression is terminated when the centers of the two-half dies become coincident. Because in this press the whole length of pipe is formed at once, the force required is very significant. Often this ends up being the step that defines the capacity of a given pipe mill. For this case, the pipe is compressed to $\varepsilon_{O1} = -0.288\%$ and unloads to a final value of $\varepsilon_O = -0.205\%$ after the dies are released (average membrane strains).

In practice, after the O-step, the pipe is welded using submerged arc welding. This is an important step that is given immense attention in the mill, including manual, ultrasonic and radiographic inspections over the whole length of the seam. In the simulation, welding

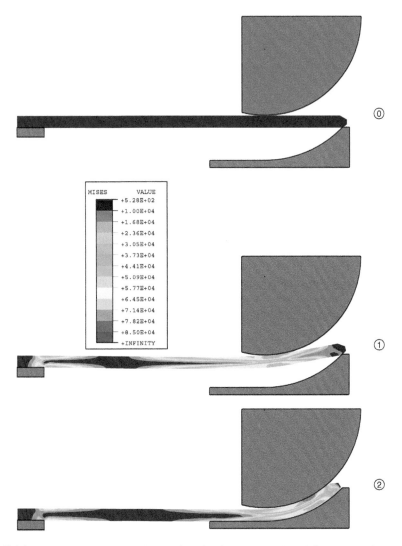

Figure 5.16   Three configurations during the crimping process (von Mises stress shown in color contours).

is modeled by replacing the elements in the beveled end with elements without the bevel that are initially stress-free. The end to which these elements are added is held fixed while this is done. When the replacement is completed, the constraint is removed and the new elements share whatever stress existed in the pipe.

The pipe is expanded by moving the reference nodes of the eight segments out radially a distance δ_E (configuration ① in Figure 5.19). When all mandrels have contacted the pipe the force required rises sharply as the pipe expands and plasticizes. This particular pipe was expanded to $\varepsilon_{E1} = 1.389\%$ and relaxes to $\varepsilon_E = 0.981\%$ on unloading.

Compression is applied in the place of expansion using a 12-segment mandrel with a radius $\rho_C = 11.88$ in (301.8 mm). The compression mandrels are modeled in a manner similar to the expansion ones. The radial displacement of the mandrels δ_C is prescribed.

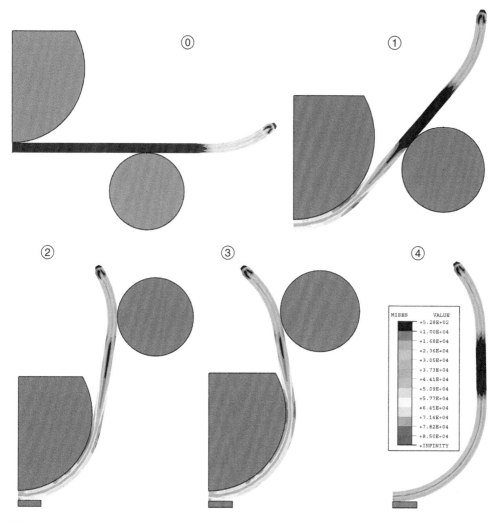

Figure 5.17 Sequence of configurations during the U-ing process (von Mises stress shown in color contours).

The pipe is compressed to a strain of $\varepsilon_{C1} = -0.390\%$ (configuration ① in Figure 5.20), and when unloaded springs back to $\varepsilon_C = -0.171\%$ (configuration ②).

a. Collapse Under External Pressure

The collapse pressure of a given pipe is governed by its D/t, its circularity and by the compressive mechanical properties of the material. Figure 5.21 shows the shapes of the calculated cross sections of the UO (a), UOE (b) and UOC pipes (c) (with the radial deviation from best-fit circles exaggerated). The shape is affected by each step of the process. Sector A is influenced mostly by the crimp. Sector B corresponds to the straight arms of the skelp. This section is the last to conform to the O-die, and it remains somewhat flatter than the rest of the cross section. Sector C depends on ρ_U, δ_r and ρ_O and is seen to be reasonably circular. Initial ovality as defined in Eq. (4.25) is the imperfection that

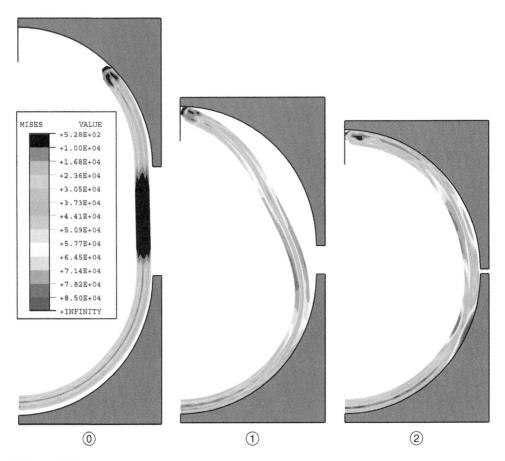

MISES	VALUE
	+5.28E+02
	+1.00E+04
	+1.68E+04
	+2.36E+04
	+3.05E+04
	+3.73E+04
	+4.41E+04
	+5.09E+04
	+5.77E+04
	+6.45E+04
	+7.14E+04
	+7.82E+04
	+8.50E+04
	+INFINITY

Figure 5.18 Initial and two deformed configurations during the O-ing process (color contours show von Mises stress).

mostly influences collapse under external pressure (see Section 4.4). The UO pipe has a relatively high value of ovality of $\Delta_o = 0.848\%$ that is also evident in Figure 5.21(a). Expanding the pipe by $\varepsilon_E = 0.984\%$ results in the shape shown in Figure 5.21(b). The pipe is now much more circular, with $\Delta_o = 0.097\%$, but sector B is still slightly more distorted than the rest of the cross section. The expansion has introduced slight thickness variations in the pipe, as some regions stretched more than others. These are seen greatly distorted in the figure because of the large magnification of w adopted. Concavities introduced by the eight mandrels can be seen on the inner surface. The shape of the actual UOE pipe being simulated is shown in Figure 3.15. The calculated shape is not exactly the same, primarily because of some differences in the process setting. However, the similarity between the two shapes is clearly visible.

Another evaluation benchmark for the simulation is its capacity to reproduce the changes to the circumferential stress–strain responses introduced by the UOE process. This check was performed as follows: two integration points near the through thickness positions where the actual specimens were extracted from (at 180° from weld) were selected, and the state variables locked into these points were recorded. Residual stresses

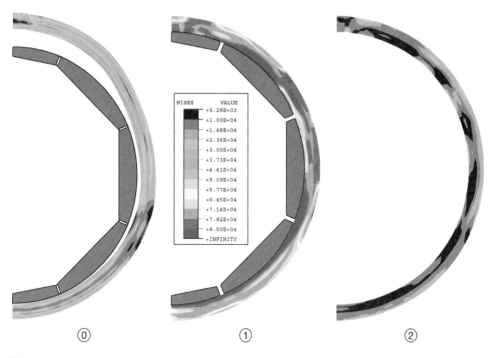

MISES	VALUE
	+5.28E+02
	+1.00E+04
	+1.68E+04
	+2.36E+04
	+3.05E+04
	+3.73E+04
	+4.41E+04
	+5.09E+04
	+5.77E+04
	+6.45E+04
	+7.14E+04
	+7.82E+04
	+8.50E+04
	+INFINITY

⓪   ①   ②

Figure 5.19   Three configurations during the expansion process (von Mises stress shown in color contour).

were incrementally reduced to zero, and each point was then loaded to a uniaxial compression strain of 1.0% in the circumferential direction. The predicted stress–strain responses are plotted in Figure 5.22. Included are corresponding responses measured experimentally. The comparisons are quite favorable.

Figure 5.21(c) shows the final shape of the UOC pipe in which $\varepsilon_C = -0.171\%$. The shape is somewhat less round than the UOE pipe ($\Delta_o = 0.192\%$) because of the smaller compressive strain used.

In the final step of the simulations, each of the three pipes was subjected to external pressure. The calculated pressure-change in volume $(P - \delta\upsilon)$ responses appear in Figure 5.23. Each response is initially relatively stiff. At higher pressures, their stiffness is gradually reduced because of inelastic action. Eventually a limit pressure is reached that represents the collapse pressure of the pipe. The collapse pressure of the UO pipe is 4,171 psi (287.7 bar) (see Table 5.6). Note that because of the assumed cross section symmetry, the pipe must collapse symmetrically, either in a vertical (0) or in a horizontal (O) mode. In this case, the pipe collapsed in the vertical mode.

The $P - \delta\upsilon$ response of the UOE pipe is overall somewhat stiffer, while the collapse pressure increased to 4,895 psi (337.6 bar). In this case, the ovality of the UO pipe was so much larger than that of the UOE pipe that the expansion resulted in a 17.4% increase in collapse pressure. In the next section it will be demonstrated that this is not always the case.

The $P - \delta\upsilon$ response of the UOC pipe is even stiffer, and the collapse pressure is now 6,843 ksi (471.9 bar), with a horizontal buckling mode. This particular UOC pipe has nearly double the ovality of the UOE pipe. Despite this, it collapses at a pressure that

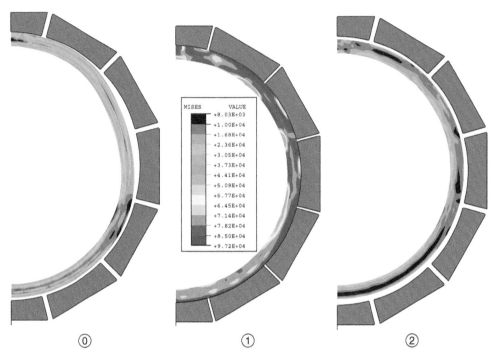

Figure 5.20   Three configurations during the compression process (von Mises stress shown in color contours).

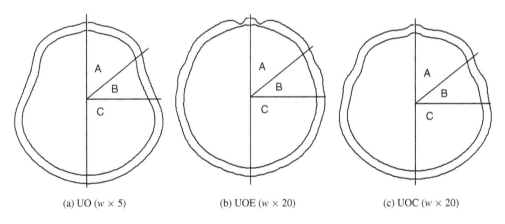

Figure 5.21   Pipe cross sectional shapes from simulations.

is 40% higher. This confirms the potential benefits of replacing E with C pointed out earlier [5.4].

Finally, a collapse calculation is conducted using the shape of the UOE pipe, assumed to be stress-free, and the initial monotonic part of the stress–strain response of the plate in Figure 5.8. The calculated response included in Figure 5.23 stays nearly linear until the limit pressure is reached at 7,461 psi (514.6 bar). This value is 52% higher than that of the UOE pipe, illustrating the significant benefit that heat-treating the pipe could lend.

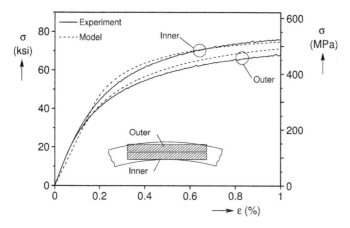

Figure 5.22 Comparison of measured and calculated compressive stress–strain responses at two locations through the pipe cross section.

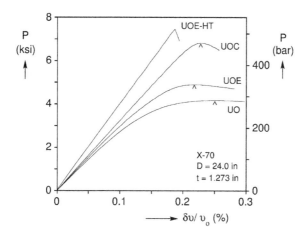

Figure 5.23 Calculated pressure-change in volume responses of UO, UOE, UOC and "heat-treated" pipes.

Comparing the collapse pressures in Table 5.6 with the corresponding ones calculated with the 1-D model in Table 5.4, it is clear that the 1-D model does very well. The biggest difference occurs for the UOC pipe, where the 2-D collapse pressure is 7% higher than the 1-D prediction. The main contribution of the 2-D model is the prediction of the exact shape of the cross section and variation of properties around the circumference. The good agreement between the 1-D and 2-D collapse pressures indicates that the ovality is the main imperfection influencing collapse, and that the variation of properties around the circumference plays a relatively minor role.

5.5.4 Parametric Study-Optimization of UOE/UOC

Each step of the UOE process has variables that influence the shape and collapse performance of the finished pipe. In an effort to quantify the influence of such parameters on pipe

Table 5.6 Predicted main pipe shape parameters and collapse pressures (2-D model).

Parameters	UO	UOE	UOC	UOE-HT
Circumference in (mm)	70.855 (1799.7)	71.552 (1817.4)	70.733 (1796.6)	71.552 (1817.4)
$\varepsilon \%$	-0.2048	0.9838	-0.1712	–
$\Delta_o \%$	0.848	0.0967	0.192	0.0967
$\hat{P}_{CO}$ psi (bar)	4,171 (287.7)	4,895 (337.6)	6,843 (471.9)	7,461 (514.6)
Mode	0	0	⊙	–

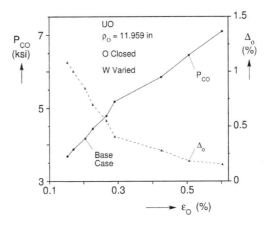

Figure 5.24 Effect of ε_O on the performance of UO pipe. The plate width is increased as ε_O increases.

performance, several of these variables were varied individually while keeping all other parameters at the values of the base case (Table 5.5). Unless otherwise stated, the pipe wall thickness is 1.273 in (32.33 mm), the plate properties are those given in Table 5.3, and the O- and E-strains are $\varepsilon_O \approx -0.20\%$ and $\varepsilon_E \approx 1.00\%$. Each pipe is collapsed by external pressure after the UO, and again after the UOE (or UOC) steps. Each parameter is varied within a range of values that is considered to be practical. Press load capacities are not considered in these simulations. The major findings of this study are summarized below. A more extensive evaluation of the manufacturing steps is reported in [5.12].

a. O-Strain

We first consider the possibility of finishing the pipe after the UO step by compressing it to increasingly higher values of ε_O. To achieve this, the starting width of the plate is varied somewhat so that the required strain is achieved on full closure of the two semicircular dies. Figure 5.24 shows that the ovality decreases with the ε_O, correspondingly P_{CO} increases nearly linearly with ε_O. At $\varepsilon_O = -0.602\%$, the collapse pressure was over 7,000 psi (483 bar), which once again demonstrates that compression increases P_{CO}. Because of press capacity limitations and other practical issues, it is rather doubtful that compression larger than -0.5% can be systematically applied in the O-press. Despite this, the results show that increasing the compression in the O-press can be beneficial to the collapse performance of the pipe.

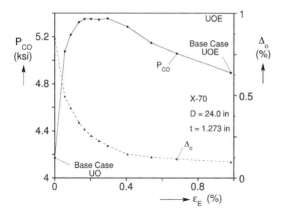

Figure 5.25   Effect of expansion on pipe ovality and collapse pressure.

b. Expansion

Current practice is to expand all UOE pipes by increasing the diameter by about 1% (0.8–1.2%). The effect of ε_E on ovality and collapse pressure is illustrated in Figure 5.25 for the same pipe. $\varepsilon_E = 0$ corresponds to the UO pipe with $\Delta_o = 0.848\%$ and $P_{CO} = 4{,}171$ psi (287.7 bar). As ε_E increases, the ovality initially drops sharply, causing P_{CO} to rise sharply. Further increase in ε_E produces increasingly less improvement in ovality and more degradation of the compressive material properties. This results in P_{CO} reaching a maximum of 5,356 psi (369.4 bar) at $\varepsilon_E = 0.30\%$, with a corresponding ovality of 0.193%. Further increase in ε_E progressively reduces P_{CO}. At $\varepsilon_E = 0.98\%$, the ovality is 0.096%, but P_{CO} has dropped to 4,895 ksi (337.6 bar).

The same exercise is now repeated for pipes compressed in the O-press to three additional ε_O values: 0.269%, 0.425%, and 0.506% (achieved by varying the plate width). The pipes are expanded to different strain levels in the range of 0–1% and then pressurized to collapse. The final ovality and collapse pressure of these pipes are plotted against ε_E in Figures 5.26(a) and 5.26(b), respectively. The cases with $\varepsilon_E = 0$ represent the four pure UO pipes. As ε_O increases, the ovality of the UO pipes decreases, with a corresponding increase in collapse pressure. For all cases, expansion further reduces the ovality, but most of the benefit occurs for $\varepsilon_E < 0.3\%$. This reduction in ovality has a corresponding positive impact on P_{CO}. However, as was the case for the results in Figure 5.25, the positive impact is limited to smaller values of ε_E. Indeed, as ε_O increases, the peak collapse pressure of each case moves to the left, indicating that increasingly less expansion is needed for an optimally performing UOE pipe. At even higher O-strains, the peak collapse pressure corresponds to the UO pipe. Such pipes are already very round and strong, and expansion can only degrade the compressive material strength and lower P_{CO}.

c. Compression

The advantages of replacing expansion with compression are further illustrated in Figure 5.27. The base case pipe is formed up to the O-step as before. The pipe is then compressed different amounts (mandrel radius $\rho_C = 11.88$ in –301.8 mm), resulting in

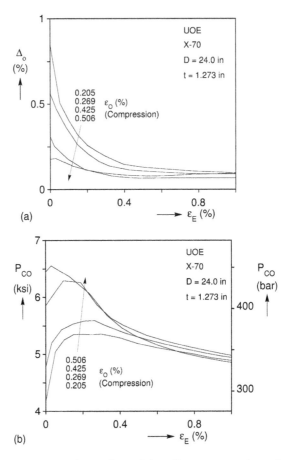

Figure 5.26   Effect of ε_E on (a) pipe ovality and (b) collapse pressure for various values of ε_O.

net compressive strains (ε_C) ranging from 0 to -0.625%. The calculated pipe ovalities and collapse pressures are plotted vs. ε_C in Figures 5.27(a) and (b), respectively.
Included in each plot for comparison are the corresponding results from the expanded
pipe. Figure 5.27(a) shows that equal amounts of expansion or compression result in
pipes with similar ovality levels. At the same time, the collapse pressure of UOC pipe
is significantly higher than that of UOE pipe (Figure 5.27(b)). In the case of UOC, P_{CO}
first increases nearly linearly with ε_C, reaching a value of 6,843 psi (471.9 bar) at the
relatively low value of $|\varepsilon_C| = 0.171\%$. At this point, the collapse mode changes its ori-
entation from vertical to horizontal, the ovality temporarily increases somewhat, and
additional compressive strain is needed until the collapse pressure begins to rise again.
Around 0.35% compressive strain, P_{CO} starts increasing again with ε_C, reaching a value of
8,025 psi (553.4 bar) at $|\varepsilon_C| = 0.625\%$. Indeed, the collapse pressure can increase further
if the pipe is compressed to higher values of ε_C. The differences from the corresponding
UOE results are very striking. For example, if we compare pipes expanded by 0.625%
and contracted by the same amount, UOC results in a collapse pressure that is 58%
higher.

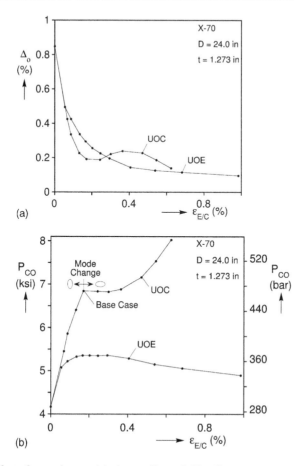

Figure 5.27 Effect of ε_E and ε_C on (a) pipe ovality and (b) collapse pressure.

d. Expansion and Compression of Pipe of Various *D/t* Values

The results presented this far are for a pipe with $D/t = 18.85$. We now examine how the expansion and compression strains introduced in the last manufacturing step affect the ovalities and collapse pressures of pipes of different D/t values. The material properties used are those in Table 5.3. The pipe diameter is kept at 24 in (610 mm), but the thickness is varied to get pipes with the following D/t values: 17.45, 18.85, 21.33, 24.00 and 27.44. In order to optimally form these pipes, the width of the plate (W), the radius of the crimp inner die (ρ_{CRi}), the U-punch radius (ρ_U) and roller displacement (δ_r), and the radius of the expansion mandrels (ρ_E) had to be adjusted. The remaining variables are kept at the levels given in Table 5.5 (thus $\varepsilon_O = -0.2\%$ for all cases).

Ovality and collapse pressure results are plotted against ε_E for the five D/t values considered in Figures 5.28(a) and (b), respectively. Expansion has a positive effect on ovality for all D/t values. However, the reduction in ovality is more pronounced at lower D/t values. This is because thinner pipes deformed to the same ε_O have smaller ovality after the O-step. In all cases, most of the reduction in ovality occurs for $\varepsilon_E < 0.35\%$.

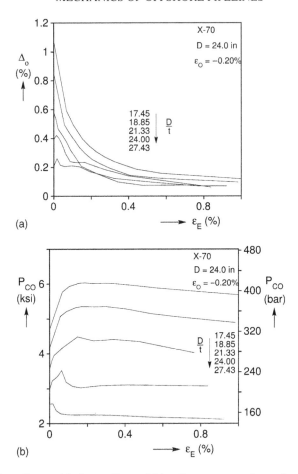

Figure 5.28 Effect of ε_E on (a) pipe ovality and (b) collapse pressure for various D/t pipes.

The collapse pressure of the three thicker pipes is seen to increase with ε_E, reaching a maximum around 0.3% expansion. At higher expansion strains, the collapse pressure progressively decreases. For $D/t = 24$, the collapse pressure reaches its maximum at $\varepsilon_E = 0.064\%$. Higher expansions do not improve the collapse pressure above the value of the UO pipe. For $D/t = 27.43$, expanding the pipe to any strain reduces the collapse pressure to a level below that of the UO pipe. In this case, the UO pipe has an ovality of only 0.2%. Expansion only moderately reduces the ovality. It simultaneously degrades compressive properties, which reduces P_{CO}. In addition, as the D/t increases, the horizontal collapse mode replaces the vertical. For the three thicker pipes ($D/t \leq 21.33$), all the cases collapsed in the vertical mode. For $D/t = 24.00$, all of the cases right of the peak collapsed in the horizontal mode. For $D/t = 27.43$, all of the cases collapsed in the horizontal mode.

A similar set of collapse calculations are performed for the same five D/t pipes as a function of the compressive strain (UOC). Ovality and collapse pressure are plotted against ε_C in Figures 5.29(a) and (b), respectively. For the three thicker pipes, finishing the pipe with compression increases the collapse pressure significantly. As D/t increases,

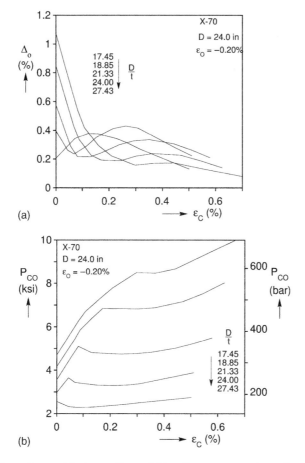

Figure 5.29 Effect of ε_C on (a) pipe ovality and (b) collapse pressure for various D/t pipes.

the benefit of ε_C is reduced and essentially disappears for the thinnest pipe considered. The sharp local peaks in Figure 5.29(b) correspond to collapse mode changes: cases left of the peak collapse in the vertical mode, while to the right, they collapse in the horizontal mode. Figure 5.29(a) reveals that these mode changes correspond to a temporary increase in ovality, and the higher the D/t value, the larger the increase. For thinner pipes ($D/t \geq 21.33$), this increase in ovality is high enough to cause P_{CO} to drop and give results comparable to the UOE process. As D/t increases, the mode change occurs at increasingly lower compressive strains, until (for $D/t = 27.43$) all of the cases collapse in the horizontal mode. For this pipe, compression is detrimental until relatively high strains ($\varepsilon_C > 0.40\%$) are reached, and the ovality is reduced. For higher D/t pipe this trend is, however, similar to what was observed for expanded pipe.

5.6 CONCLUSIONS AND RECOMMENDATIONS

UOE pipe, as typically manufactured at the writing of this book, is known to have lower collapse pressure than seamless pipe of the same D/t and steel grade. This deficiency in collapse pressure has been demonstrated using results from several full-scale collapse programs undertaken during the last decade. The lower performance has been shown to be the result of the cold forming process, and in particular the final cold expansion of the pipe. Expansion improves the tensile circumferential strength but lowers the compressive strength, due to rounding of the stress–strain response associated with the Bauschinger effect. The degradation in collapse pressure has been shown to exceed 30%.

It has been shown that normalizing the pipe stress relieves it and results in the recovery of its mechanical strength. The high energy required as well as potential detriments to welds make normalizing impractical. An alternative milder heat treatment process, where the pipe is heated to about 250°C for about a minute has been shown to recover compressive strength through strain aging. Since most pipes are corrosion-coated with epoxies that require heating the pipe to 225–250°C, it is possible that the strength recovery can be a byproduct of the coating. For the strength recovery to be effective it is imperative that proper quality and control processes be instituted to guarantee that the whole length of each pipe is properly heat-treated. Furthermore, strain aging is affected by alloy content. Thus, a heat treatment procedure must be custom designed to the needs of each project.

The parametric study of UOE conducted using the 2-D forming model of [5.12] demonstrated that the collapse pressure of UOE pipe can be improved to some degree by optimizing the UOE process parameters. For example, this can be achieved by increasing the compressive strain induced in the O-press (ε_O) as much as possible while simultaneously reducing the expansion strain (ε_E). For a chosen value of ε_O, an optimum expansion exists that will balance ovality improvement and material degradation effects to maximize P_{CO}. Simulations were conducted for pipes with $17.4 < D/t < 27.5$. Best collapse performance was found to require $\varepsilon_E < 0.3\%$.

Finishing the pipe by compression instead of expansion reduces the pipe ovality and increases its mechanical compressive properties. One way of inducing the required compression is in the O-press. Because the O-press deforms the whole length of the pipe simultaneously, this is limited by press capacity and by other factors. A more attractive alternative is to add a new component to the process that will compress the pipe radially in a step-by-step fashion. We envision a segmented female mandrel operating on the outer surface of the pipe (essentially the opposite of the current expansion mandrel). The beneficial effects of such a device on the collapse pressure of lower D/t pipes were clearly demonstrated in the simulations. At similar values of strain, UOC yields pipes with similar ovality as UOE but with much higher collapse pressures. High enough compression can increase P_{CO} to values greater than that of the heat-treated pipe, as the material is hardened past the yield stress of the original plate.

We expect the commissioning of a compression device to become an enabling technology for deep-water pipeline applications. What is envisioned is that present expanders will continue to be used for land pipelines, where expansion increases the tensile strength in the hoop direction, and for high D/t offshore pipelines. The compression device would be used for deep-water pipes, where an increase in collapse pressure would result in significant savings in materials as well as installation costs. Indeed, for future ultra deep-water applications, the use of the compression device may be the only viable alternative.

REFERENCES

5.1. Hunter, A.F., Zimmer, R.A., Wang, W.-J., Bozeman, J.D., Adams, C.J. and Rager, B.L. (1990). Designing the TLWP. *Proc. Offshore Technology Conference* **3**, OTC 6360, 147–158.

5.2. Lohr, C.J., Bowen, K.G., Calkins, D.E. and Kipp, R.M. (1994). Design, fabrication and installation of Auger LMS and tendons. *Proc. Offshore Technology Conference* **2**, OTC 7622, 553–563.

5.3. Enze, C.R., Brasted, L.K., Arnold, P., Smith, J.S., Breaux, J.N. and Luyties, W.H. (1994). Auger TLP design, fabrication, and installation overview. *Proc. Offshore Technology Conference* **3**, OTC 7615, 379–387.

5.4. Kyriakides, S., Corona, E. and Fischer, F.J. (1991). On the effect of the UOE manufacturing process on the collapse pressure of long tubes. *Proc. Offshore Technology Conference* **4**, OTC 6758, pp. 531–543. Also, *ASME J. Engineering for Industry* **116**, 93–100, 1994.

5.5. Stark, P.R. and McKeehan, D.S. (1995). Hydrostatic collapse research in support of the Oman-India gas pipeline. *Proc. Offshore Technology Conference* **2**, OTC 7705, 105–120.

5.6. Al-Sharif, A.M. and Preston, R. (1996). Improvements in UOE pipe collapse resistance by thermal aging. *Proc. Offshore Technology Conference* **2**, OTC 8211, 579–588.

5.7. Wilson, D.V. and Russell, B. (1960). The contribution of atmosphere locking to the strain-ageing of low carbon steels. *Acta Metallurgica* **8**, 36–45.

5.8. Hall, E.O. (1970). *Yield Point Phenomena in Metals and Alloys*. Plenum Press. New York.

5.9. DeGeer, D., Marewski, U., Hillenbrand, H.-G., Weber, B. and Crawford, M. (2004). Collapse testing of thermally treated line pipe for ultra-deepwater applications. *Proc. 23rd Int'l Conf. Offshore Mechanics and Arctic Engineering*, June 20–25, Vancouver, BC, Canada, Paper OMAE2004-51569.

5.10. Fryer, M., Tait, P., Kyriakides, S., Timms, C. and DeGeer, D. (2004). The prediction and enhancement of UOE–DSAW collapse resistance for deepwater pipelines. *Proc. 5th Biennial International Pipeline Conference* **3**, October 4–8, 2004, Calgary, AL, Canada, 1961–1966.

5.11. DeGeer, D., Timms, C. and Lobanov, V. (2005). Blue Stream collapse test program. *Proc. 24th Int'l Conf. Offshore Mechanics and Arctic Engineering*, June 12–17, Halkidiki, Greece, Paper OMAE2005-67260.

5.12. Herynk, M.D., Kyriakides, S., Onoufriou, A. and Yun, H.D. (2007). Effects of the UOE/UOC manufacturing process on pipe collapse pressure. *Int'l J. Mechanical Sciences*, **49**, 533–553.

5.13. Ewart, J.C. (1961). *Apparatus for Radially Compressing Articles*. US Patent 2,999,405, A.O. Smith Corporation.

6

Collapse of Dented Pipes Under External Pressure

Collapse of as-received pipe under external pressure and the main factors that govern it have been discussed in Chapters 4 and 5. Here we consider the collapse resistance of pipes that have suffered local damage in the form of a dent. Offshore pipelines are relatively large scale structures installed in hostile environments. Damage in the form of dents is one of the most common causes of pipeline failures [6.1, 6.2]. They occur due to impact by foreign objects and other causes during transportation of the pipe, during installation and trenching, and during the operation of the line. Dents reduce the local collapse capacity of the pipe, and can be the initiators of propagating buckles [6.3, 6.4]. In addition, for gas lines under high internal pressure, dents, sometimes accompanied by gouges, are often the prelude to burst failures [6.5]. Once a dent is detected, the operator must decide on the integrity of the structure. Repairs are generally complicated and costly. Thus, the ability to assess the integrity of the dented pipe analytically is very important.

Results from experimental studies of the collapse performance of dented pipes are reviewed first. The results are used to develop a practical measure of the "magnitude" of the dent with regard to collapse resistance. The experiments are complemented by numerical results of denting followed by collapse. The chapter ends by the discussion of *universal collapse resistance curves* proposed by Park and Kyriakides [6.6] as a design tool for dented pipes.

6.1 DENT CHARACTERISTICS

Kyriakides *et al.* [6.4] measured the collapse pressures of seamless, carbon steel 1018 tubes pre-dented to various degrees. Their D/t values ranged between 43 and 33. The tubes were indented by rigid indentors of different dimensions with "point," "plate," and "knife" shapes. The dented tubes were subsequently collapsed under external pressure. The study concluded that despite the complex three-dimensional nature of such dents, the collapse pressure of the dented pipes best correlated to the "ovalization" of the most deformed cross section of the damaged region. Other details of the geometry of the dents were found to play a secondary role. The dent ovalization (Δ_{od}) is defined as follows:

$$\Delta_{od} = \frac{D_{\max} - D_{\min}}{D_{\max} + D_{\min}} \tag{6.1}$$

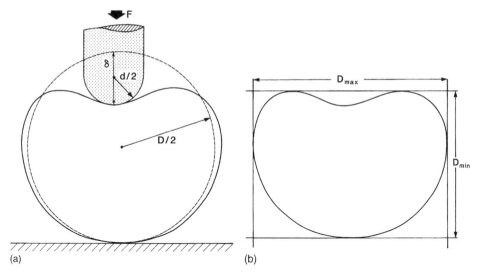

Figure 6.1 (a) Indention geometric parameters. (b) Geometric parameters of most deformed cross section of a dent.

where D_{max} is the maximum distance across the deformed cross section and D_{min} is the minimum distance across the convex part of the deformed cross section, as defined in Figure 6.1(b).

Guided by this finding, a second set of experiments was conducted by Park and Kyriakides [6.6] in which denting was performed only by spherical indentors of different diameters. The tubes used in this set of experiments had generally lower D/t values. The experimental procedures and results of this study are outlined in the next section.

6.2   DENTING AND COLLAPSE EXPERIMENTS

6.2.1   Indention

This set of experiments was conducted on commercially available stainless steel 304 seamless tubes with nominal diameters of 1.25 in (31.8 mm) and nominal D/t values of 18.9, 24.2 and 33.6. The test specimens were typically about 30 tube diameters long. The geometric and material properties of each tube were measured, and are reported in Tables 6.1–6.3 of Ref. [6.6]. The average values of the relevant properties of each D/t group are listed in Table 6.1.

Each test specimen was dented at mid-length using a rigid indentor with a spherical head of diameter d, as shown schematically in Figure 6.1(a). The indention was carried out in a standard, electromechanical testing machine with the test specimen resting on a rigid, flat plate. A thin rubber pad was placed between the tube and the rigid plate to help prevent rotation of the tube during the early stages of the indention. Test specimens from each of the three tube families were indented to final indention depths in the range of $\delta_o < 0.4D$ (where δ_o is the maximum dent depth after unloading as shown in Figure 6.1(a)). Indentors with diameters $d = 0.4D$ and $1.6D$ were employed. For each tube D/t, a sufficient number of tubes (8–12) were indented using the indentor with $d = 0.4D$ to

Table 6.1   Mean geometric and material parameter of tubes used in the study.

D in (mm)	D/t	E Msi (GPa)	σ_o ksi (MPa)	σ_y ksi (MPa)	n
1.252 (31.80)	18.90	30.7 (212)	37.97 (262)	33.80 (233.0)	10.4
1.249 (31.72)	23.25	29.9 (206)	54.08 (373)	50.00 (344.7)	12.6
1.251 (31.78)	34.01	28.0 (193)	46.32 (319)	42.80 (295.1)	13.0

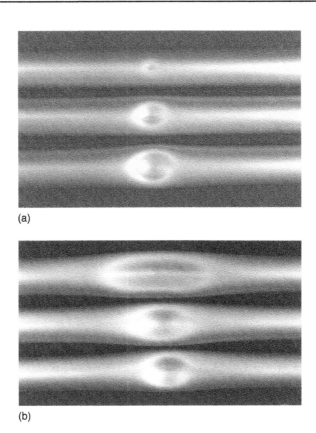

(a)

(b)

Figure 6.2   (a) Dents of different amplitude produced by an indentor with $d = 1.6D$ ($D/t = 18.9$). (b) Photographs of tubes in Figure 6.2(a) collapsed under external pressure.

ensure that the trend of the behavior was captured with accuracy. An additional three tubes from each D/t group were indented to varying degrees using the indentor with $d = 1.6D$. Figure 6.2(a) shows three tubes indented to varying degrees with the larger diameter indentor.

An example of an indention force–displacement record, from a tube with D/t of 24.2, is shown in Figure 6.3(a). In this case, the tube was indented to a final value of $\delta_o = 0.33D$. (The small nonlinearity observed in the unloading part of this response to a large extent is caused by the rubber pad.)

The geometry of each dent was characterized by taking several circumferential scans along the length of the tube as described in Ref. [6.6]. The major geometric parameters of each dent are its maximum depth δ_o and the ovality of the most deformed cross section,

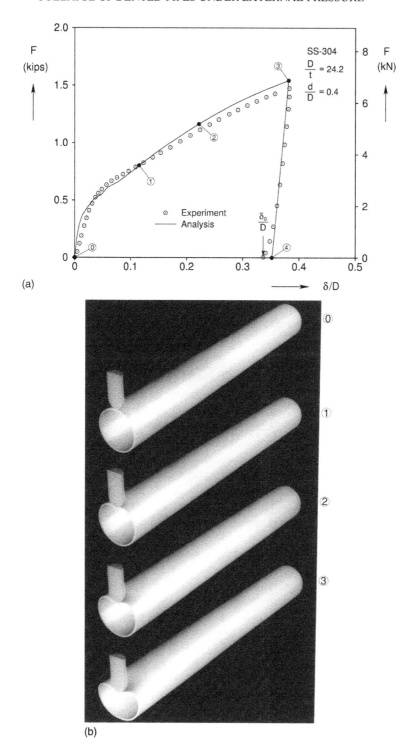

(a)

(b)

Figure 6.3 (a) Indention force-maximum dent depth response (indentor diameter $d = 0.4D$) (b) Sequence of calculated deformed configurations during indention.

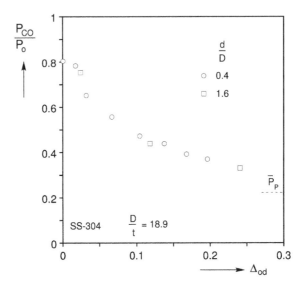

Figure 6.4   Measured collapse pressures as a function of dent parameter Δ_{od} ($D/t = 18.9$).

Δ_{od}, defined in Eq. (6.1). The measured values of these dent parameters are included in Tables 6.1–6.3 of the reference.

6.2.2   Collapse experiments

The dented tubes were sealed at both ends and placed in the 10 ksi (690 bar) capacity pressure vessel shown schematically in Figure 4.10. The vessel was filled with water and pressurized with a high pressure precision metering pump at a quasi-static rate. The pressure was monitored by a pressure transducer and dial gages. Collapse was sudden and catastrophic, resulting in a sudden drop in pressure. The nature of the collapse is illustrated in Figure 6.2(b), which shows photographs of the collapsed regions of the three tubes shown at the dented stage in Figure 6.2(a). The sequence of the tubes in the two pictures is the same. Because of the high stiffness of the pressurizing system and the nearly volume-controlled pressurization scheme used, collapse is local, affecting sections of the tubes which are only a few pipe diameters long. Despite its high stiffness, the pressurizing system does store some energy during pressurization, primarily due to the compressibility of the pressurizing medium (water) and the compliance of the specimen itself. During collapse, this energy is dumped into the collapsing section of the tube. Tubes with relatively small dents have higher collapse pressures, and as a result more energy is transferred from the system to the collapsing tube. Thus, the extent of the collapse varies inversely with the dent size; in other words, the collapsed section is larger for smaller dents, as illustrated in the figure.

The collapse pressures recorded, normalized by either the elastic buckling pressure (P_C) or the yield pressure (P_o) of each tube, are plotted against the dent parameter Δ_{od} in Figures 6.4–6.6. For each tube family studied, one tube was collapsed without denting. In this case, Δ_{od} is assigned the value of the maximum initial ovality of the tube. In each set, the results from the small diameter indentor are indicated with circular symbols

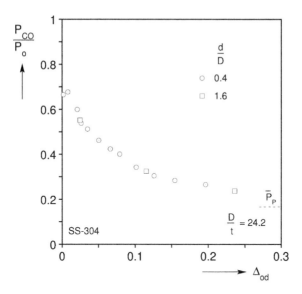

Figure 6.5   Measured collapse pressures as a function of dent parameter Δ_{od} ($D/t = 24.2$).

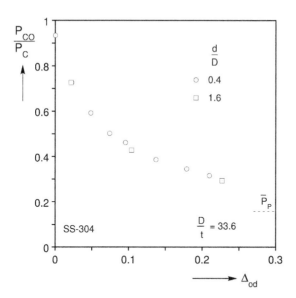

Figure 6.6   Measured collapse pressures as a function of dent parameter Δ_{od} ($D/t = 33.6$).

and those from the large diameter indentors with square symbols. The results have the following general features:

1. In all cases, denting is seen to reduce the collapse pressure significantly. Dents with values of Δ_{od} of approximately 0.1 reduce the collapse pressure by approximately 40%, 50%, and 54%, respectively, for D/t values of 18.9, 24.2, and 33.6. The trend is that as Δ_{od} increases, the collapse pressure tends to approach the propagation pressure

($\bar{P}_p$) of the tube (marked with horizontal dashed lines in the figures). A similar trend was reported in [6.4].

2. In all cases, the correlation of the results from the small diameter and the large diameter indentors is excellent. In view of the large difference in the two indentor diameters, this indicates that Δ_{od}, as defined in Eq. (6.1) and Figure 6.1(b), is a successful measure of the severity of the dent as it affects the collapse pressure. If we couple this with similar observations made in [6.4] for a much broader range of dent types and geometries, this geometric measure of the severity of dents can be considered to have a rather broad applicability.

6.3 MODELING OF DENTING AND COLLAPSE

The processes of denting and collapse described in the experimental section were formulated through appropriately nonlinear shell theories and simulated numerically using the FE computer code ABAQUS (version 5.2). The shell deformation is assumed to be symmetric about the mid-span and about the plane passing through the bottom of the dent and containing the axis of the tube. Thus, the section analyzed consists of one fourth of the shell. The shell was discretized with four-noded, curved shell elements (S4RF). They are based on finite strain kinematics which, in addition to finite rotations of and about the normals to the mid-surface, allow for finite membrane stretching. The elements have hourglass control and reduced integration. Five integration points were used through the thickness.

The hemispherical head of the indentor was assumed to be rigid and the IRS4 interface element was used to model contact of the indentor and the shell. The contact was assumed to be frictionless. During the indention process, the tube was assumed to be in smooth contact with a rigid flat plane. The same interface elements were used to model contact between the tube and this plane. Contact between the inner surfaces of the collapsing tube was again assumed to be smooth, and was modeled with INTER4 interface elements.

The tube section analyzed had 60 elements in the axial direction and 40 elements in the circumferential direction. A finer mesh was used to discretize the section of the tube, approximately one tube diameter long, which contained the dent.

The material was assumed to be a J_2-type elastoplastic, finitely deforming solid with isotropic hardening. The stress–strain response used is the true stress–logarithmic strain version of the engineering stress–strain response represented by the Ramberg–Osgood parameters given in Table 6.1. For each of the calculations that follow, the geometric parameters were those of the experiment being simulated.

The tube was indented by prescribing the displacement of the rigid indentor incrementally. When the dent reached the required depth, the indentor was incrementally retracted, and the tube was loaded by external pressure. Collapse is initiated when a limit load instability develops in the response. Riks' path-following method was used to capture the features of the post-limit load response.

A sample simulation of the indention process is shown in Figure 6.3(a), where the indentor force–displacement response calculated for one of the tubes with $D/t = 24.2$ is compared with the corresponding measured response. For this case, the indentor had a diameter of $d = 0.4D$ and the tube was dented to a depth of $\delta_o = 0.33D$, which corresponds to one of the larger dents induced to this tube D/t. The calculated response is seen to be in good agreement with the one measured. Small differences between the two responses are due to the presence of the rubber pad between the tube and the rigid base and the unknown amount of friction between the indentor and the tube. Both factors were present in the experiments but were neglected in the analysis precisely because their effect is small.

Figure 6.3(b) shows the undeformed configuration and a sequence of three deformed configurations calculated using the model. Their positions on the response are identified with numbered bullets. Configuration ⓪ shows the tube in its initial undeformed state. Configurations ①, ② and ③ show the tube at increasing stages of deformation as the indentor displacement increases. The nature of deformation is quite clear from the shape acquired by the most deformed cross section seen in the pictures. The rather local nature of the deformation, alluded to in the experimental section, is also clear from these pictures.

Figure 6.7 shows the progression of events during the pressurization process. The calculated pressure-change in volume response is shown in Figure 6.7(a), and several deformed configurations corresponding to it are shown in Figure 6.7(b). The permanent deformation caused by indention, obtained after unloading the indentor (point ④ in Figures 6.3(a) and 6.7(a)), is depicted as configuration ④ in Figure 6.7(b). Initially, the response is stiff and stable and the deformation of the tube is limited. This is terminated by a pressure maximum. Beyond the pressure maximum, the tube is seen in Figure 6.7(b) to collapse with most of the deformation taking place in the neighborhood of the dent. The collapse process involves further ovalization of the dented section (see ⑤). Despite the one-sided nature of the dent, as the collapse progresses the deformation becomes increasingly more doubly symmetric (see ⑥). At first contact between the opposite walls of the tube, the most deformed cross section acquires a nearly perfectly doubly symmetric cross section (see ⑦). Contact of the walls locally stabilizes the collapsing section and stops further radial deformation. First contact corresponds to the local pressure minimum in the response. The local strengthening due to contact results in a small increase of the pressure required for further deformation. Collapse is now forced to occur along the axis of the tube, spreading to the sections that up to this point have been relatively unaffected by the collapse. Indeed, under the quasi-static, volume-controlled pressurization loading scheme used here, the deformation will tend to spread at the propagation pressure of the tube, which is indicated with a dashed line in Figure 6.7(a). So, as expected, the dent reduces the local collapse resistance of the tube and initiates a propagating buckle.

The ovalization of the most deformed cross section is a key measure of the magnitude of the dent. It is thus important to evaluate the accuracy with which the numerical model can reproduce the most deformed cross sections. Figure 6.8 shows a comparison of calculated and measured shapes of most deformed cross sections for dents with three different indention depths. Although the measured and predicted shapes have some differences, in general they compare well with each other for all three cases.

6.3.1   Prediction of Collapse Pressure of Dented Tubes

The finite element model was used to develop additional data on the effect of denting on the collapse pressure of the three families of tubes used in the experiments. The geometric and material parameters used are those in Table 6.1. The collapse pressures calculated are plotted against the dent parameter Δ_{od} together with the experimental results in Figures 6.9–6.11.

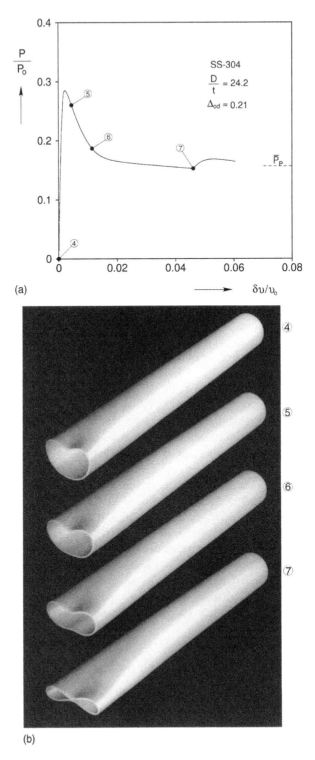

(a)

(b)

Figure 6.7 (a) Calculated pressure-change in volume response of dented tube (Figure 6.3). (b) Collapse configurations correspond to response in Figure 6.7(a).

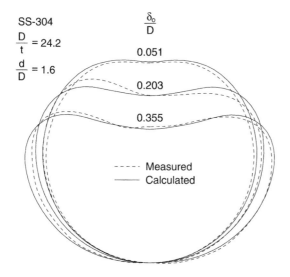

Figure 6.8   Comparison of measured and calculated shapes of dent most deformed cross sections for three indention depths.

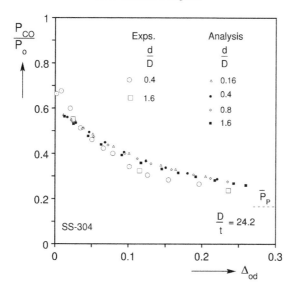

Figure 6.9   Comparison of measured and calculated collapse pressures as a function of dent parameter Δ_{od} ($D/t = 24.2$).

The most extensive parametric study was carried out for the tube with $D/t = 24.2$ for which indentors with four different diameters were used ($d/D = 0.16, 0.4, 0.8$ and 1.6, Figure 6.9). For each indentor, dents with Δ_{od} values ranging from approximately 0.012 to approximately 0.26 were generated. The first observation that can be made from this plot is that the correlation between the results from the four indentors is excellent. This gives further support for the choice of Δ_{od} as an appropriate measure of the magnitude of the dent. The second observation is that for values of Δ_{od} larger than approximately

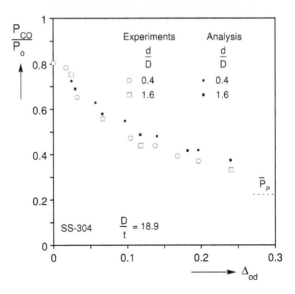

Figure 6.10 Comparison of measured and calculated collapse pressures as a function of dent parameter Δ_{od} ($D/t = 18.9$).

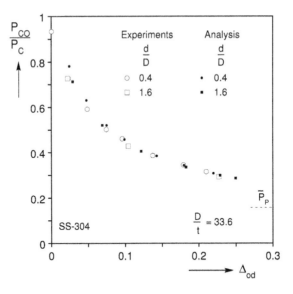

Figure 6.11 Comparison of measured and calculated collapse pressures as a function of dent parameter Δ_{od} ($D/t = 33.6$).

0.015, the predicted collapse pressures are in good agreement with the measured values. Small differences can be attributed to small variations in wall thickness and in material properties from tube to tube, and from the idealizations inherent to the J_2 flow theory with isotropic hardening used in the analysis. In the case of very small dents, the difference between theory and analysis is larger. This will be further discussed later in the chapter.

Results from a similar parametric study for dented tubes with $D/t = 18.9$ involving only the two indentors used in the experiments, appear in Figure 6.10. The correlation between the results from the two indentors is again very good, and the agreement between the experiments and the predictions is also good.

Results for dented tubes with $D/t = 33.6$, again involving the two indentors used in the experiments, appear in Figure 6.11. The same conclusions can be drawn for this case as those drawn for the other two tube families. Interestingly, these results exhibit the best correlation between results from different indentors and between experiment and analysis. This is primarily due to the fact that the geometric and material properties of the tubes in this family exhibited the smallest variation.

The results presented above clearly demonstrate that, for a given tube and indentor, the indention process and the collapse of the dented section can be simulated numerically with good accuracy. Furthermore, the experimental results generated under well controlled laboratory conditions, as well as the analytical results produced, quantify the parametric dependence of the collapse pressure of dented tubes with three different D/t values. We will now return to the problem as it occurs in the field, and in the light of these results, try to establish how best to assess the danger of collapse of a given dented offshore pipeline.

6.4 UNIVERSAL COLLAPSE RESISTANCE CURVES FOR DENTED PIPES

It is almost certain that the shape of a dent found in a pipeline will differ from the dents generated by our spherical and smooth indentors. In addition, the field dent will probably have been generated by impact by an unidentified foreign object. Thus, the main evidence available to the field engineer will probably be a picture of the dented section, which for deeper waters will come from a ROV camera. The engineer has to decide if the pipeline, when depressurized, is in danger of local collapse and of developing a propagating buckle.

As has been demonstrated, the severity of a local dent is governed by the ovality parameter Δ_{od} of its most deformed cross section. The shape and magnitude of the concave region, do not affect the collapse pressure significantly. This implies that for any pipe, it might be possible to generate a collapse pressure–dent ovality curve that is applicable to a broad variety of dent geometries. Under this scenario, the problem is reduced to finding the most convenient way of generating this curve. Once generated, this universal dent collapse resistance curve can be used to establish the severity of any field dent and plan the most appropriate course of repair.

6.4.1 Localization of Collapse Under External Pressure

Clues as to how such a universal dent collapse resistance curve might be generated came from a review of the mechanism of collapse of an intact pipe first discussed in [6.7]. Here we will perform this 3-D collapse calculation using a FE model. The tube analyzed is SS-304 with $D/t = 25.3$ and has an axially uniform initial ovality (Δ_o) of 0.05%. The calculated pressure-change in volume response is shown in Figure 6.12(a). (Nominal yield stress $\sigma_o = 37.8$ ksi (261 MPa), $\sigma_y = 34$ ksi (234 MPa), $n = 13$ up to $\varepsilon = 1.5\%$, then linear hardening). Four deformed configurations corresponding to the numbered points on the response are shown in Figure 6.12(c). The response has an initial stiff and stable part during

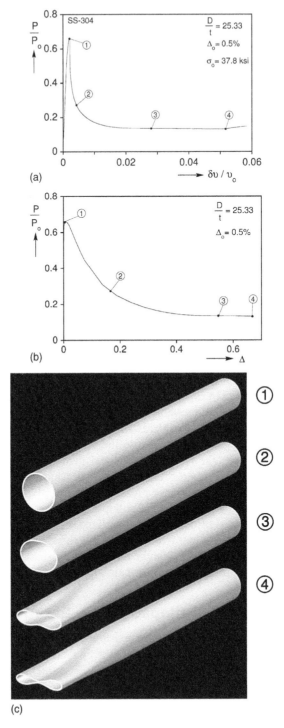

Figure 6.12 Simulation of local collapse of a tube with uniform initial ovality. (a) Pressure-change in volume response, (b) pressure maximum ovality response and (c) sequence of deformed configurations.

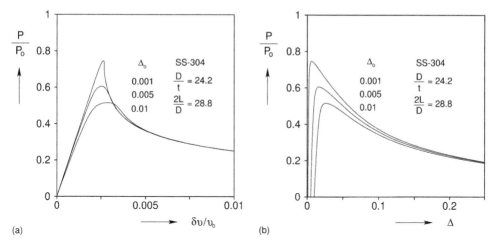

Figure 6.13 Collapse of tubes with three different values of initial ovality. (a) Pressure-change in volume response and (b) pressure maximum ovality response.

which the tube progressively ovalizes uniformly. Soon after the pressure maximum (① in Figure 6.12), the deformation localizes (see configurations ②–④) to a section that is a few tube diameters long. With the pressure required for equilibrium dropping, deformation and stresses in this local section continue to grow as the remainder of the tube unloads. The pressure drops sharply to a local minimum, which corresponds to first contact between the walls of the collapsing tube. At this stage, collapse of the crown point is arrested. Subsequent pressurization results in the spreading of the collapse to the part of the tube, which up to this point has remained intact [6.8]. The pressure is also plotted against the ovality of the most deformed cross section in Figure 6.12(b) (Δ as defined by Eq. (6.1) and Figure 6.1(b)). Such plots will be useful in the proposed scheme that follows.

An important conclusion from these results is that localized collapse has a well-defined mode of deformation, with a distinct variation in the axial direction covering a length of a few tube diameters. It has been observed that irrespective of how collapse is initiated, the deformation of the collapsing section reverts to this mode. Thus, the postbuckling response of a circular tube and the deformed shape associated with it can be viewed as fundamental to this process. The last point is illustrated in Figure 6.13(a), which shows a comparison of $P - \delta v$ responses for the $D/t = 24.2$ tube in Table 6.1, but with initial (uniform) ovalities of 0.1%, 0.5% and 1%. The initial ovality is seen to affect the prebuckling response, the limit load and the initial part of the post-limit load response. However, soon after the limit load, the three responses are seen to converge to the same path. Furthermore, it was shown in [6.7] that unloading from the descending part of the response arrests collapse. Reloading is stable until the original post-limit load response is intersected when collapse resumes following the original path (see Figures 6.15 and 6.17 in reference).

6.4.2 The Universal Collapse Resistance Curve

In view of the fundamental nature of the postbuckling response of tubes, we examine its relationship, if any, to the collapse performance of dented tubes. The first step in this

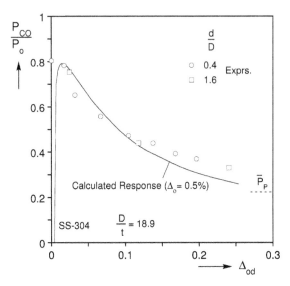

Figure 6.14 Comparison of the measured collapse pressures as a function of dent parameter Δ_{od} and the calculated pressure–ovality response of a tube with an initial uniform ovality of 0.5% ($D/t = 18.9$).

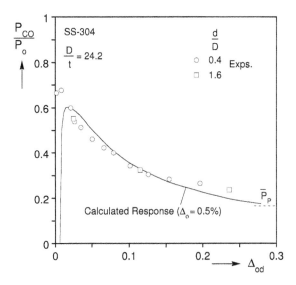

Figure 6.15 Comparison of the measured collapse pressures as a function of dent parameter Δ_{od} and the calculated pressure–ovality response of a tube with an initial uniform ovality of 0.5% ($D/t = 24.2$).

endeavor is to establish the *natural* pressure–maximum ovality (Δ) response of the intact tube. Figure 6.13(b) shows such responses for the same cases analyzed in Figure 6.13(a). Again, the prebuckling and initial postbuckling parts of the three responses differ but, as collapse progresses, they tend to converge to the same path. Such responses were generated with the properties in Table 6.1 for the three families of tubes used in the experiments. The three families of responses were then compared with the corresponding collapse

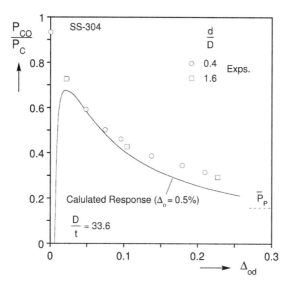

Figure 6.16 Comparison of the measured collapse pressures as a function of dent parameter Δ_{od} and the calculated pressure–ovality response of a tube with an initial uniform ovality of 0.5% ($D/t = 33.6$).

pressure–dent ovality plots obtained from the experiments (see Figures 6.14–6.16). In all three cases, it was found that the descending part of the response corresponding to initial uniform ovality of $\Delta_o = 0.5\%$ matched, in the main, the $P_{CO} - \Delta_{od}$ results. Some minor discrepancies between the calculated responses and the collapse pressures of the dented tubes are observed for relatively small dent ovalities, whereas for larger values of Δ_{od} the calculated responses bound the results from below. In view of this success, the descending part of each of the three calculated $P - \Delta$ responses will be designated as its *Universal Collapse Resistance Curve* (UCRC), with the word *Universal* meant to apply to dents.

Since for each tube or pipe the proposed UCRC is generated by analyzing a cylindrical shell with an initial uniform ovality of 0.5%, it can only be used for dents with $\Delta_{od} > 0.5\%$. The experimental results generated in this study, as well as those in [6.4], indicate that the collapse pressures of tubes with small, localized imperfections ($\Delta_{od} < 0.5\%$) do not correlate well with those of tubes with larger dents. This is due to the fact that such small dents only affect the geometry of the pipe in the vicinity of the dent, and do not significantly increase the local ovality of the pipe. As a result, Δ_{od} is not a suitable geometric measure of the magnitude of such dents. Very small local dents can in general be neglected in the field, at least from the point of view of their effect on the collapse performance of the pipe. If a pipe is designed with an assumed uniform initial ovality of 0.5% (usual recommended value), then the design collapse pressure can be expected to be lower than the collapse pressure of the same pipe with a local dent with $\Delta_{od} < 0.5\%$. (Ways of analyzing shallow dents that affect a section of the pipe of a few diameters in length are discussed in [6.7].)

The problem has been tackled from the point of view that the pipe is dented in the absence of external pressure. The dented pipe is subsequently pressurized externally, resulting in its collapse. Offshore pipelines can be dented while in service, in which case the denting is augmented by external pressure. This problem has been investigated in [6.9].

6.5   CONCLUSIONS AND RECOMMENDATIONS

It has been demonstrated that denting a pipe can reduce its collapse resistance quite significantly. A dent can best be characterized by the maximum ovality of its most deformed cross section, Δ_{od}, defined in Eq. (6.1) and Figure 6.1(b). The detailed shape of the concave part of the dent is of secondary importance. From the experiments performed, it was demonstrated that Δ_{od} values of about 0.1 result in a reduction in collapse pressure of about 50%. Local dents with $\Delta_{od} < 0.005$ were found to be relatively benign, at least from the point of view of the pipe collapse under external pressure. Their effect on other performance characteristics of pipelines, such as corrosion, burst, pigging, etc., should be evaluated independently.

The denting and collapse resistance can be modeled numerically along the lines described in Section 6.3. This process requires knowledge of the body impacting the pipe. A *Universal Collapse Resistance Curve* for dented pipes has been proposed as an alternative to case-specific custom analyses. This curve can be generated for each pipeline during the design stage. The curve can be used to evaluate the potential of collapse of any dent encountered during the lifetime of the pipeline. The UCRC is generated from a calculated pressure–maximum ovality response (three-dimensional) of a pipe with the geometric and material properties of the pipeline and with an assumed uniform initial ovality of 0.5%. For each of the three tube D/t values analyzed, it was shown that the descending part of this response agrees well with the plot of collapse pressures of dented tubes against the dent parameter Δ_{od}.

REFERENCES

6.1. Demars, K.R., Nacci, V.A. and Wang, W.D. (1977). Pipeline failure: A need for improved analyses and site surveys. *Proc. Offshore Technology Conference* 4, OTC2966, 63–70.

6.2. Strating, J. (1981). A survey of pipelines in the North Sea. Incidents during installation, testing and operation. *Proc. Offshore Technology Conference* 3, OTC4069, 25–32.

6.3. Kyriakides, S. and Babcock, C.D. (1982). On the initiation of a propagating buckle in offshore pipelines. *Proc. 3rd International Conference on the Behavior of Offshore Structures* 2, Boston, MA, 187–199.

6.4. Kyriakides, S., Babcock, C.D. and Elyada, D. (1984). Initiation of propagating buckles from local pipeline damages. *ASME J. Energy Resour. Technol.* 106, 79–87.

6.5. Lancaster, E.R. and Palmer, S.C. (1993). Assessment of mechanically damaged pipes containing dents and gouges. *Service Experience and Life Management: Nuclear, Fossil, and Petrochemical Plants*, ASME PVP 261, 61–68.

6.6. Park, T.-D. and Kyriakides, S. (1994). On the collapse of dented cylinders under external pressure. *Intl J. Mechanic. Sci.* 38, 557–578.

6.7. Dyau, J.-Y. and Kyriakides, S. (1993). On the localization of collapse in cylindrical shells under external pressure. *Intl J. Solids Struct.* 30, 463–482.

6.8. Dyau, J.-Y. and Kyriakides, S. (1993). On the propagation pressure of long cylindrical shells under external pressure. *Intl J. Mechanic. Sci.* 35, 675–713.

6.9. Karamanos, S.A. and Elefteriadis, C. (2004). Collapse of pressurized elastoplastic tubular members. *Intl. J. Mechanic. Sci.* 46, 35–56.

7

Buckling and Collapse Under Combined External Pressure and Tension

In several important offshore and onshore oil exploration and production applications, long, tubular structures are held suspended in either vertical or catenary-shaped positions in a water head that often extends for several thousand feet. In such configurations, the submerged weight of the pipe must be supported at the top, resulting in significant build-up of axial tension in the structure. Although the tensile force decreases as the depth increases, there can be sections that simultaneously experience significant external pressure and axial tension. For commonly used line grade steels and pipe D/ts, tension can reduce the buckling and collapse pressures. Accordingly, these critical pressures must be reevaluated including the effect of tension.

For example, a pipeline being installed in deep water using the J-lay method (Figure 2.19) can have sections in which the effect of tension on the collapse pressure can be significant. The same is true for either straight or catenary production risers shown in Figures 2.5–2.9. Tubular tendons used in tension-leg platforms (Figure 2.6) are similarly affected, in some cases even more critically.

Even more importantly, collapse under combined axial tension and external pressure is the governing design loading for casing installed down-hole, both onshore and offshore. Figure 7.1 shows schematically an example of an offshore well with its many tubular components. The well shown starts at a water depth of about 3,000 ft (914 m) and goes down to 17,000 ft (5,200 m) below sea level. Telescoping tubulars with progressively decreasing diameters comprise the well. A production riser runs from the mudline to the platform at sea level. It encloses the production tieback, which in turn contains the production tube and other tubes carrying production fluids. Close to the mudline, 36-inch casing surrounds the wellhead and the well. Inside it, a 20-inch casing grouted with concrete runs down to 4,500 ft (1,372 m). A 16-inch liner is hung from the 20-inch casing and goes down to 5,500 ft (1,676 m) and is similarly grouted. Inside these liners is a 13.625-inch tube, which hangs from the wellhead to 7,200 ft (2,195 m). An 11.75-inch liner hangs from it and goes down to 13,000 ft (3,962 m). Inside these liners is the production tieback (8.625 in), which runs from about 13,000 ft to the wellhead, where it is partially hung, and continues on up to the platform. Attached to it at the bottom is a

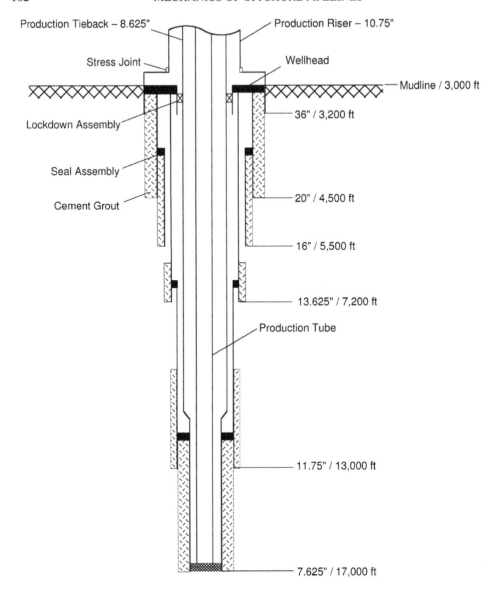

Figure 7.1 Schematic of an offshore production well.

smaller diameter liner (7.625 in) that ends at the well and is also several thousand feet long.

Tension builds up in the two hanging sections because of their weight. During construction and later during operation, either drilling mud or a mixture of mud and sea water will be on either side of the tubes. Despite this, they are often designed to be capable of sustaining the full external pressure corresponding to their depth. The internal tieback tube is held partly at the platform and partly at the wellhead. Significant external pressure can be seen by the tieback tube if the drilling mud is accidentally lost. The larger diameter liners closer to the mudline that are installed as drilling progresses to maintain the integrity of the hole, can also experience pressure in the presence of tension. Thus, in general, for both down-hole as well as above the mudline, drilling and production tubulars can have stringent loading requirements of combined external pressure and tension. This has motivated full-scale studies of the problem, such as those reported in [7.1–7.3], as well as the development of custom testing facilities at pipe mills capable of full scale tests under this type of combined loading. An example of such a facility is shown schematically in Figure 4.17(b).

In this chapter, elastic and then elastic–plastic bifurcation buckling of a long tube under combined external pressure and axial tension are first discussed. The more practical problem of estimating the effect of tension on the collapse pressure of tubes with small initial geometric imperfections is considered in Section 7.3. The problem is modeled through an extension of the nonlinear formulation presented for the corresponding external pressure problem (Chapter 4). The extension is correspondingly reflected in the custom numerical code BEPTICO [7.4]. In Section 7.4, the performance of this model is evaluated by comparison of predictions with experimental results from Madhavan *et al.* [7.5].

Pipes suspended in a column of water experience an *effective tension* (T_e) given by

$$T_e = T + P\frac{\pi D^2}{4},$$ (7.1)

where T is the local tension acting on the cross section and P is the external pressure [7.6]. T_e is the value that must be used for combined loading design considerations. Because of the way the experiments that follow were conducted, in most of the results the tension quoted will be simply T.

7.1 ELASTIC BUCKLING

The bifurcation buckling pressure of a long circular elastic cylinder of radius R and wall thickness t, acted upon simultaneously by external pressure P and tension T, remains the same as that of the pure pressure case given in Eq. (4.6). The tensile stress starts to affect the buckling pressure once the combined loading starts to yield the pipe. In the presence of tension the yield pressure, modified in accordance to the von Mises yield criterion, is P_{oT} given by:

$$P_{oT} = P_o\left(-\frac{\tau}{2} + \sqrt{1 - \frac{3}{4}\tau^2}\right), \quad \tau = T/T_o, \quad T_o = \pi\sigma_o D_o t, \quad P_o = \frac{2\sigma_o t}{D_o}.$$ (7.2)

The elastic buckling pressure (4.6) holds provided

$$P_C \leq P_{oT}.$$ (a)

For a uniformly ovalized pipe, the formulation in Section 4.1.1 also remains applicable at least up to yielding, and the displacements (v, w) are related to the external pressure through (4.9). The collapse pressure can again be associated with the onset of yielding in the manner suggested by Timoshenko, by taking into account in (4.11) the lowering

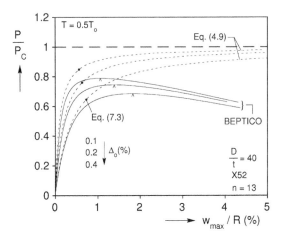

Figure 7.2 Pressure–maximum displacement responses for an imperfect tube at a tension of $T = 0.5T_o$. Results from linear elastic analysis and from BEPTICO for elastic–plastic material.

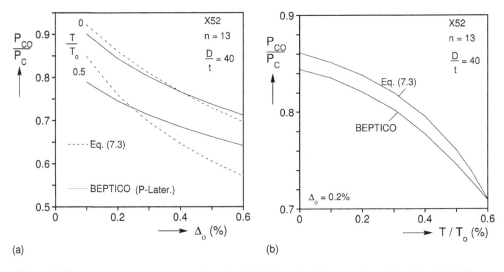

Figure 7.3   (a) Collapse pressure as a function of initial ovality for two values of tension. (b) Collapse pressure as a function of tension.

of yield stress in the circumferential direction by the tension. The derivation is similar to the one in Section 4.1.1 for pure external pressure, and results in replacing P_o by P_{oT} in Eq. (4.11) as follows:

$$P_{CO}(T) = \frac{1}{2}\left\{(P_{oT} + \psi P_C) - [(P_{oT} + \psi P_C)^2 - 4P_{oT}P_C]^{1/2}\right\}, \qquad (7.3)$$

where

$$\psi = \left(1 + 3\Delta_o \frac{D_o}{t}\right).$$

A demonstration of the use of this formula is shown in Figure 7.2 for an X52 pipe with $D/t = 40$. The applied tension is $0.5T_o$. Shown are $P - w_{max}$ results corresponding to elastic buckling for different values of initial ovality (use $\Delta_o = a/R$ in (4.9)). The onset of yielding according to Eq. (7.3) is identified with solid dots marked on the elastic responses. The corresponding pressure levels enriched with additional results are plotted against Δ_o in Figure 7.3(a). Included in the same figure are corresponding results for $T = 0$. Comparing the two, we conclude that tension lowers the collapse pressure while its dependence on ovality with and without tension is similar. Figure 7.3(b) shows how tension lowers the collapse pressure for a fixed value of initial ovality ($\Delta_o = 0.2\%$).

7.2   PLASTIC BUCKLING

When the combined loads yield the perfect pipe, the bifurcation pressure must be evaluated through the corresponding elastoplastic bifurcation formulation. The inelastic response of the pipe is evaluated incrementally using the J_2 flow theory as customized for the present problem in Eq. (4.14), with

$$\sigma_e = (\sigma_\theta^2 - \sigma_\theta\sigma_x + \sigma_x^2)^{1/2}, \tag{b}$$

where

$$\sigma_\theta = -\frac{PR}{t} \quad \text{and} \quad \sigma_x = \frac{T}{2\pi Rt}.$$

If the material exhibits anisotropic yielding, Eqs. (4.14) are replaced by (13.21b) suitably reduced. The effect of biaxial loading is demonstrated through an example in which tension $T < T_o$ is applied first, held fixed and the external pressure is increased. In this case $\dot{\sigma}_x = 0$ and Eqs. (4.14) reduce to

$$\dot{\varepsilon}_x = \frac{1}{E}[-v + Q(2\sigma_x - \sigma_\theta)(2\sigma_\theta - \sigma_x)]\dot{\sigma}_\theta,$$

$$\dot{\varepsilon}_\theta = \frac{1}{E}[1 + Q(2\sigma_\theta - \sigma_x)^2]\dot{\sigma}_\theta. \tag{c}$$

Figure 7.4 shows a set of pressure–radial displacement results corresponding to different values of tension for an X52 ($n = 13$, see Appendix C) material with $D/t = 20$. Tension reduces the yield stress in the circumferential direction and, correspondingly, the pressure that can be sustained. We note that such results can be affected by the loading path followed. The use of the flow theory of plasticity is thus essential.

The bifurcation pressure can be established once more from Eq. (4.16a) with C_{22} and Ω appropriately calculated using the incremental J_2 deformation theory Eqs. (4.15) with the biaxial state of stress of the present problem (or by the anisotropic version given in (13.28)). For a stress-controlled problem like the one associated with the results in

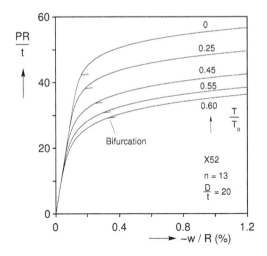

Figure 7.4 Pressure–circumferential strain responses of an X52 pipe at prescribed tension values. Indicated are the bifurcation buckling points.

Figure 7.4, the critical stress can be evaluated by gradually increasing the pressure and checking for bifurcation, as described in Section 4.2.1. Bifurcation pressures calculated in this manner for the X52 pipe considered above are marked on the $P - w$ responses in Figure 7.4 with the symbol $\vdash$. All occur at strains lower than 0.5%. If the loading path is not stress-controlled, then the instantaneous stresses must first be evaluated via the flow theory Eqs. (4.14). The bifurcation check is again conducted incrementally by inserting the calculated stress state into the deformation theory given in Eqs. (4.15) at every loading step considered.

7.3 NONLINEAR FORMULATION

As was the case for the purely external pressure loading, a more complete analysis of plastic buckling and collapse of long pipes under combined external pressure and axial tension requires numerical treatment. The problem is once more assumed to be sufficiently represented by a long tube or pipe with uniform geometric and material properties along the length. It can thus be treated with the same basic formulation developed in Section 4.3 by adding the contribution of tension to the work term in the PVW (4.22) as follows (BEPTICO [7.4]):

$$\delta \dot{W}_e = -\hat{P}R \int_0^{2\pi} \left[\delta \dot{w} + \frac{1}{2R}(2\hat{w}\delta \dot{w} + 2\hat{v}\delta \dot{v} + \hat{w}\delta \dot{v}' + \hat{v}'\delta \dot{w} - \hat{v}\delta \dot{w}' - \hat{w}'\delta \dot{v}) \right] d\theta$$
$$+ \hat{T}\delta \dot{\varepsilon}^o, \tag{7.4}$$

where $\dot{\varepsilon}^o$ is the mean strain in the axial direction. The performance of the model will be evaluated in the next section, where predictions are compared to experimental results.

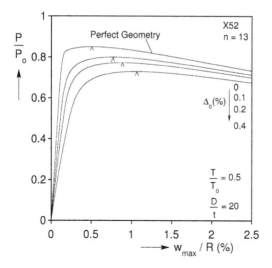

Figure 7.5 Pressure–maximum displacement responses at $T = 0.5T_o$ for perfect and imperfect tubes with $D/t = 20$. Perfect case bifurcates plastically.

7.3.1 Examples

An illustration of the effect of tension on the collapse of a pipe which buckles elastically and collapses plastically is shown in Figure 7.2 for an X52 ($n = 13$) pipe with $D/t = 40$ and $T = 0.5T_o$. Pressure-maximum deflection responses are shown for the same cases analyzed in Section 7.1 using the elastic closed form solution. In the presence of an imperfection, each response initially follows the corresponding elastic one. With inelastic action, rigidity is gradually reduced and a pressure maximum develops. The pressure maximum, marked on the response with "∧", corresponds to the collapse pressure. The sensitivity of P_{CO} to initial ovality is similar to that exhibited in the pure external pressure case, but the collapse pressures are lower. This sensitivity is further illustrated in Figure 7.3(a), where similar results for the pure pressure case are included (lateral pressure loading is used in this case). For pure external pressure, the Timoshenko formula is seen to be in very good agreement with the numerical results, as pointed out in Chapter 4. For the combined loading case, results from Eq. (7.3) are in reasonable agreement with the numerical results, but the collapse pressure is seen to drop more precipitously with ovality than the more accurate numerical results. This difference is influenced by the shape of the stress–strain response and by the level of tension used. For stress–strain responses with very low hardening, Eq. (7.3) tends to be uniformly conservative.

We now repeat the calculations of P_{CO} as a function of T for a fixed value of initial ovality ($\Delta_o = 0.2\%$) numerically. The results are compared to the corresponding ones from Eq. (7.3) in Figure 7.3(b). The analytical results are somewhat higher, but follow the general trend of the numerical values. Again, this difference depends on the shape of the stress–strain response. Thus, in general Eq. (7.3) is a useful formula, but should be used with some caution.

Results characteristic of pipes that buckle in the plastic range are shown in Figure 7.5 for an X52 ($n = 13$) material and $D/t = 20$. $P - w$ responses for various values of initial ovality and tension $T = 0.5T_o$ are shown. The results exhibit a similar trend as the corresponding ones for pure external pressure in Figure 4.8. In each case, a pressure maximum is attained which will be associated with the onset of collapse. Comparison of the two sets of responses illustrates that the presence of tension lowers the collapse pressures. A more extensive demonstration of this reduction is given in the next section.

7.4 COLLAPSE UNDER EXTERNAL PRESSURE AND TENSION

The problem was studied through full-scale experiments by Kyogoku *et al.* [7.1] and Tamano *et al.* [7.2, 7.3]. The combined tension/external pressure testing facility of Sumitomo described by Kyogoku *et al.* is shown schematically in Figure 4.17(b). Madhavan *et al.* [7.5] studied the problem parametrically through an extensive experimental program involving small scale tubes. The investigation was part of the larger study on factors affecting pipe collapse reported in [7.7]. The experimental procedures and results of [7.5] are summarized below.

The experiments involved SS-304 seamless tubes with D/t values in the range of 13 to 39. They were conducted in a custom combined tension/external pressure testing facility shown schematically in Figure 7.6. It consisted of a pressure chamber mounted in a test frame equipped with a hydraulic actuator used to apply tension. The chamber had an inside diameter of 2.5 in (64 mm), an internal length of 42 in (1,070 mm) and a pressure capacity of 10,000 psi (690 bar). The actuator had a load capacity of 20 kips (89 kN) and could be operated either in load or displacement control using a closed-loop servo controller.

The test specimens had nominal diameters in the range of 1.25–1.375 in (32–35 mm). Their ends were sealed by bonding to solid end-plugs, resulting in a test section at least 20D long. One of the end-plugs was threaded and connected into the blind end-cap at the bottom. The other one passed through the top end-cap and connected to the actuator through a universal joint. The chamber was filled and pressurized with water using a positive displacement pump. In the process, the specimen was free to expand axially. A data acquisition system was used to monitor and record the pressure, axial tension, and axial elongation during the tests.

Small geometric imperfections in the test specimens were measured prior to each test. The maximum ovality recorded is reported in [7.7] along with mean values of D and t. The mechanical properties of the tube materials were measured using axial test specimens extracted from each length of the tube used in the experiments. The measured stress–strain responses were fitted with to the Ramberg–Osgood expression (Appendix C) and the results are given in the same report. Yield anisotropy was established for some tubes in separate experiments conducted in the manner described in Appendix B.

Experiments were conducted in which tension was prescribed and the pressure was increased until collapse was recorded (designated as $T \rightarrow P$ loading path). In this case, collapse was associated with the attainment of a pressure maximum, resulting in local collapse similar to that shown in Figure 4.9. Alternatively, pressure was applied first and fixed at a chosen value by adding an accumulator to the pressurizing system. The

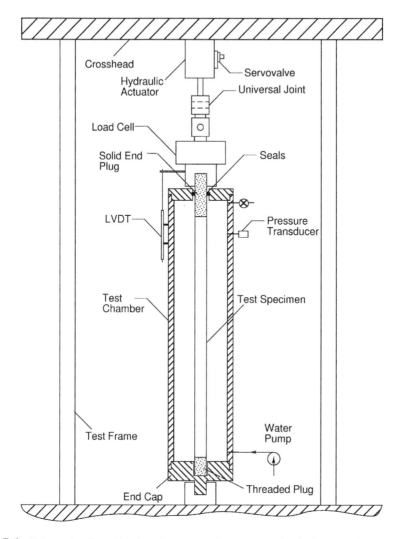

Figure 7.6 Schematic of combined tension–external pressure testing facility used in [7.5].

specimen was subsequently tensioned to collapse (designated as $P \rightarrow T$ loading path). Because of the presence of the accumulator, in this case the initially local collapse spread to the whole specimen.

Additional sets of experiments were conducted in which the effect of initial ovality on collapse under this combined loading was examined. In these tests, the tubes were first uniformly ovalized by crushing between rigid plates and then collapsed in the manner described above. Ovality values up to $\Delta_o = 4\%$ were induced in this manner.

In the case of some of the lower D/t tubes (19.2 and 13.2) tested under $T \rightarrow P$ loading, as the tension level approached the yield tension, pressure tended to uniformly elongate the tube. Such experiments were terminated when the actuator bottomed out. Similar

Table 7.1 Mean values of geometric and material parameters of tubes used in the collapse tests under combined tension and external pressure.

D in (mm)	$\dfrac{D}{t}$	Δ_o %	E Msi (GPa)	σ_o ksi (MPa)	σ_y ksi (MPa)	n	S	P_C psi (bar)	P_o psi (bar)	T_o lbf (kN)
1.253 (31.83)	19.68	0.07	27.1 (187)	46.04 (318)	41.2 (284)	10.1	0.85	–	4,929 (340)	10,950 (48.7)
1.251 (31.78)	25.22	0.11	28.9 (199)	50.00 (345)	45.7 (315)	12.0	1.0	–	4,129 (285)	9,361 (41.6)
1.242 (31.55)	28.41	0.31	28.0 (193)	53.69 (370)	49.8 (343)	12.8	0.77	2,988 (206)	–	8,829 (39.3)

pressure-induced flow of the material was also reported in [7.8] for buckle propagation experiments under high tension and external pressure.

7.4.1 Experimental Results and Numerical Predictions

The results of this experimental program are reported in detail in [7.7] and are also summarized graphically in [7.5]. Here we will show three sets of the experimental results plotted in a manner consistent with similar results in other chapters in this book dealing with collapse. The mean values of the geometric and material properties of each set considered are listed in Table 7.1.

Figure 7.7 shows the collapse pressure–tension interaction envelope formed by 12 experiments on tubes with D/t value of 25.2 (for experimental points P_C, P_o and T_o are based on individual specimen variables and usually differ somewhat from the values listed in Table 7.1). $T \rightarrow P$ loading was used in these experiments. Tension, which interacts with pressure mainly through plasticity, is seen to reduce the collapse pressure significantly. At tensions approaching the yield tension, the collapse pressure is seen to ease into a plateau. Included in the figure is the interaction collapse envelope predicted using BEPTICO [7.4]. The predictions are based on the mean values of the measured variables listed in Table 7.1, Ramberg–Osgood fits of axial stress–strain responses and the measured yield anisotropy. Because the actual stress–strain data were not available for larger strain values, for $\varepsilon_x \geq 1.5\%$ the stress–strain response was assumed to maintain a constant tangent modulus corresponding to the value of the Ramberg-Osgood fit at $\varepsilon_x = 1.5\%$. The predicted envelope is seen to be in good agreement with the experimental one. It is also seen to reproduce the pressure plateau at higher tensions. The modest over-prediction of the collapse pressure observed at higher values of tension is mainly due to the use of mean geometric and material properties and to the extrapolation of the stress–strain response adopted. The axial strains corresponding to the collapse pressures were measured and are reported in [7.7]. Generally they are also in agreement with predicted values.

A similar interaction collapse envelope for the same loading path for tubes with $D/t = 19.7$ is shown in Figure 7.8. The general trend of this envelope is similar to that in Figure 7.7. The figure includes the corresponding numerical predictions in which the

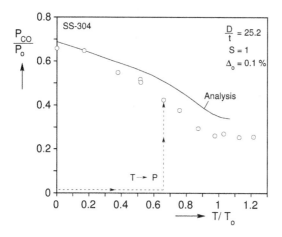

Figure 7.7   Experimental and predicted $T \to P$ collapse envelopes for $D/t = 25.2$.

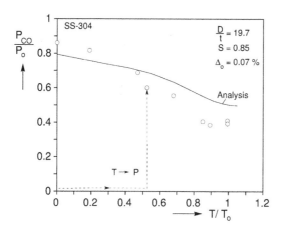

Figure 7.8   Experimental and predicted $T \to P$ collapse envelopes for $D/t = 19.7$.

mean properties given in Table 7.1 were used. The agreement between the two is good, with some deviation again at higher tension levels.

The third interaction collapse envelope, shown in Figure 7.9, is for tubes with $D/t = 28.4$ tested under the $P \to T$ loading. This is a thinner tube, which under pure external pressure would be expected to buckle elastically. The measured collapse pressures are thus normalized by P_C. However, the yield stress in the circumferential direction was significantly lower than the axial yield stress, and this results in the relatively low normalized collapse pressures seen in the plot. The general trend in the data is similar to those of the other two. Numerical predictions based on the mean geometric and material properties are seen to follow the experimental results very closely. Included for comparison is the corresponding interaction collapse envelope for the $T \to P$ loading path. It is seen to be very close to the $P \to T$ envelope, with some deviation occurring at tension levels above

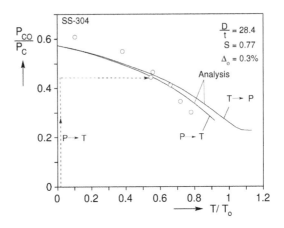

Figure 7.9   Experimental and predicted $P \rightarrow T$ collapse envelopes for $D/t = 28.4$.

$0.8T_o$. Based on this and similar results from lower D/t tubes, [7.5] concluded that $P_{CO}–T$ interaction collapse envelopes are not significantly affected by the loading path followed.

The critical tension–pressure combinations of tubes which were initially ovalized were predicted with similar accuracy. Thus, it can be concluded that the uniform tube formulation is both sufficient and appropriate for the problem.

7.5 ADDITIONAL PARAMETRIC STUDY

Collapse under tension and pressure can be affected by the same set of parameters as those discussed for the pure external pressure problem. Of these, initial ovality remains very important. This is illustrated in Figure 7.10, which shows how the collapse pressure of an X52 grade pipe with $D/t = 20$ at a tension of $0.5T_o$ is affected by ovality. When the ovality is increased from 0.02% to 1%, the collapse pressure is reduced by about 25%, which is very significant. Included in the same figure is the corresponding plot for pure hydrostatic pressure loading. In this case the reduction with ovality is even more severe.

Another variable of importance is the shape of the stress–strain response. Figure 7.11 shows pressure–tension interaction collapse envelopes for pipes with $D/t = 20$, but with three different stress–strain responses all categorized as grade X52 (see Figure C.2). In this case, the collapse pressures are normalized by a yield pressure ($P'_o = 2\sigma'_o t / D_o$) that is based on the API yield stress $\sigma'_o = \{\sigma | \varepsilon = 0.5\%\}$ and is the same for all three responses. As the transition from elastic to plastic behavior becomes sharper, the collapse pressure under zero tension increases, as seen earlier in Figure 4.21. At the same time, in the presence of tension, the collapse pressure drops more precipitously for the EP and $n = 30$ materials than it drops for the $n = 13$ material. The results indicate that an interaction relationship based strictly on the yield stress (e.g. [7.9]) can deviate from the true envelope, and for some cases can be unconservative.

As in the case of collapse under pure external pressure, it is useful to examine the effect of circumferential bending-type residual stresses (Eq. (A.3)) on the pressure–tension interaction collapse envelope. We will examine this effect for the SS-304 steel tubes with

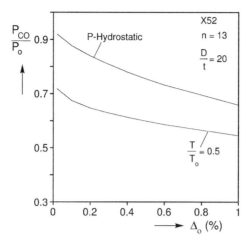

Figure 7.10   Effect of initial ovality on collapse for hydrostatic pressure loading and for $T = 0.5T_o$.

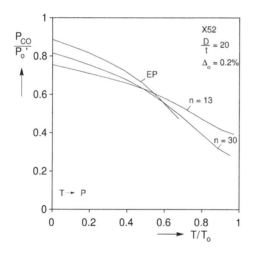

Figure 7.11   Effect of material hardening on $T \rightarrow P$ collapse envelope for X52 steel ($D/t = 20$).

$D/t = 28.4$ studied earlier in Figure 7.9 using the mean properties given in Table 7.1. The $T \rightarrow P$ loading path will be used. Figure 7.12 shows how this type residual stress with amplitude of $|\sigma_R| = 0.4\sigma_o$ affects the interaction collapse envelope. At lower tension levels, σ_R reduces the collapse pressure nearly the same amount it does for pure pressure. At higher tension levels, the reduction in the collapse pressure decreases. For tensions higher than $0.7\sigma_o$, the effect is insignificantly small. This trend was found to be repeated for other D/t values, although it can be affected to some degree by the shape of the stress-strain response.

In summary, the tension–pressure interaction collapse envelope is affected by initial ovality, by the shape of the stress–strain response, to a lesser extent by circumferential

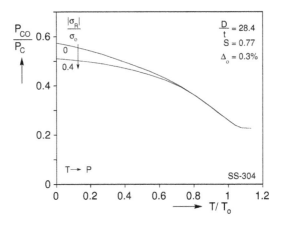

Figure 7.12   Effect of residual stress σ_R on $T \rightarrow P$ collapse envelope for $D/t = 28.4$.

bending-type residual stresses, and by other factors. These can be easily taken into account in a uniform tube type formulation such as the one described in Section 7.3. Attempts to represent such interaction collapse envelopes by simple formulae which do not account for such parameters are bound to be either overly conservative or erroneous, at least for pipes or casing that buckle plastically ($P_o < P_C$).

7.6   CONCLUSIONS AND RECOMMENDATIONS

Buckling and collapse of long tubes under combined axial tension and external pressure are influenced by the same problem parameters as the pure external pressure problem. They mainly depend on the material elastic modulus and yield stress, on the tube D/t, and on initial ovality. The main effect of tension is that it interacts with the circumferential stress (Eq. (7.2)) and lowers the pressure at which the pipe yields. For tubes thin enough to buckle elastically, the elastic buckling formula (4.6) remains applicable. In this case, the influence of ovality can be captured through the modified Timoshenko formula in (7.3). For lower D/t tubes that buckle plastically, the plastic bifurcation pressure in Eq. (4.16a) is still applicable, but in this case the biaxial state of stress must be used to evaluate the incremental deformation theory moduli in (4.15). Plastic collapse can be evaluated numerically via a uniform ovality model (BEPTICO). The following additional points should be helpful in designing against collapse for this combined loading problem.

- Tension generally reduces the collapse pressure, primarily through plastic interaction with the circumferential stress associated with external pressure.
- As is the case for pure external pressure, ovality can reduce the combined loading collapse pressure significantly.
- Although the deformation induced to the pipe is influenced by the loading path followed (in the Tension–Pressure plane), the influence of loading path on the collapse envelope is generally small.

- The shape of the stress–strain response of the material as well as anisotropic yielding can affect collapse under this combined loading, and must be included in modeling.
- The effect of bending-type residual stresses is modest, provided that σ_R is kept below $0.3\sigma_o$.
- Wall thickness eccentricity does not influence the collapse pressure significantly if kept below values of 10% (Ξ_o).

REFERENCES

7.1. Kyogoku, T., Tokimasa, K., Nakanishi, H. and Okazawa, T. (1981). Experimental study of axial tension load collapse strength of oil well casing. *Proc. Offshore Technology Conference* **3**, OTC4108, 387–395.

7.2. Tamano, T., Mimura, H. and Yanagimoto, S. (1982). Examination of commercial casing collapse strength under axial loading. ASME, *Proc. 1st Offshore Mechanics/Arctic Engineering/Deepsea Systems Symposium* **1**, 113–118.

7.3. Tamano, T., Inoue, Y., Mimura, H. and Yanagimoto, S. (1981). Examination of commercial casing collapse strength under axial loading. *ASME J. Energy Resourc. Technol.* **104**, 343–348.

7.4. Kyriakides, S., Dyau, J.-Y., and Corona, E. (1994). *Pipe Collapse Under Bending, Tension and External Pressure* (BEPTICO). Computer Program Manual, Engineering Mechanics Research Laboratory Report No. 94/4.

7.5. Madhavan, R., Babcock, C.D. and Singer, J. (1993). On the collapse of long, thick-walled tubes under external pressure and axial tension. *ASME J. Press. Ves. Technol.* **115**, 15–26.

7.6. Sparks, C.P. (1980). Mechanical behavior of marine risers mode of influence of principal parameters. *ASME J. Energy Resourc. Technol.* **102**, 214–222.

7.7. Kyriakides, S., Corona, E., Madhavan, R., and Babcock, C.D. (1987). *Factors affecting pipe collapse – Phase II*. Final Report to the American Gas Association, Engineering Mechanics Research Laboratory Report No. 87/8.

7.8. Kyriakides, S. and Chang, Y.-C. (1992). On the effect of axial tension on the propagation pressure of long cylindrical shells. *Int. J. Mechanic. Sci.* **34**, 3–15.

7.9. Murphey, C.E and Langner, C.G. (1985). Ultimate pipe strength under bending, collapse and fatigue. *Proc. 4th International Conference on Offshore Mechanics and Arctic Engineering* **1**, 467–477.

8
Inelastic Response, Buckling and Collapse Under Pure Bending

Bending is a primary loading experienced by pipelines during installation and operation. For typical pipe D/t s and line grade steels, bending and the associated limit states are affected by the plastic characteristics of the material. For example, plastic bending is experienced when a pipeline is spooled onto a reel, and again later when it is unspooled at an installation site. Significant bending in the presence of tension is experienced during installation by the S-lay method, as the pipe conforms to the curvature of the stinger and beyond in the overbend region (see Figure 2.15). Bending in the presence of external pressure is experienced in the sagbend of all major installation methods (e.g., reeling, J-lay, S-lay, see Chapter 2) as well as in free-spans on the sea floor. Land pipelines can experience bending due to subsidence of the foundation, as illustrated in Figure 8.1. Subsidence can result from earthquake activity, from permafrost degradation and from other causes. Advances in directional drilling can also result in bending of tubulars. Figure 8.2 shows a pipeline being pulled through an underground river crossing. The borehole was pre-drilled by directional drilling. The pipeline conforms to the curvature of the hole, bending plastically in some cases. This chapter deals with the mechanics of inelastic bending and the associated limit states. The presentation will be extended in subsequent chapters to bending in the presence of tension and to combined bending and external pressure.

8.1 FEATURES OF INELASTIC BENDING

A long pipe bent into the plastic range, to first order, behaves as a beam. As such, its moment–curvature response can be evaluated by integrating the material stress–strain

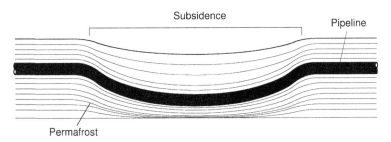

Figure 8.1   Bending of a pipeline due to subsidence of permafrost foundation.

response over the undeformed cross section (see Section 8.3). The actual response differs from the beam theory predictions by the fact that bending induces ovalization of the cross section (Brazier effect [8.1]). The growing ovalization, in turn, causes a progressive reduction in bending rigidity that eventually leads to a moment maximum in the response. For tubes with low D/t values made of typical structural metals that exhibit relatively modest hardening, the ovalization localizes in a zone a few diameters long, where it continues to grow rapidly as the overall moment drops. Thus, the moment maximum is the limit state of this structure.

An example of such a moment–curvature (M–κ) response is shown in Figure 8.3(a). It came from a pure bending test on an SS-304 tube with D/t of 26.1. The corresponding ovalization–curvature response represented by the change of diameter in the plane of bending (ΔD) is shown in Figure 8.3(b). The moment maximum, marked on the response with a caret ($^\wedge$), occurred at a normalized curvature of 1.505. At the maximum moment, the change in diameter reaches a value of about 6%. Subsequent to the moment maximum, the tube developed a diffuse local collapse like the one shown in Figure 8.4(a).

In the case of higher D/t pipes, wrinkling and shell-type buckling leading to localized collapse can precede the natural limit load of the Brazier effect. Figure 8.5(a) shows the moment–curvature of an aluminum 6061-T6 tube with $D/t = 35.7$. Figure 8.5(b) shows the change in diameter measured at two locations along the length of the tube. The tube initially ovalized uniformly, but at normalized curvatures larger than about 0.80 some localized ovalization occurred (compare $\Delta D/D$ at points A and B). At the point on the response indicated by ($\downarrow$), short wavelength wrinkles were observed in the overly-ovalized region (e.g., see pocket of wrinkles in Figure 8.5(c)). Soon after the wrinkles appeared, a limit load developed on the response (identified by "$^\wedge$"), and the tube buckled locally at a dropping moment shortly thereafter. Figure 8.4(b) shows an example of a local kink failure triggered by the wrinkles.

For even thinner pipes, wrinkling and shell-type buckling can occur even earlier. It can be followed by more complex diamond shell buckling modes, such as the one shown in Figure 8.4(c). An extensive study of how ovalization and wrinkling interact and lead to local collapse as the tube D/t changes can be found in Kyriakides and Ju [8.2]. A sample of their results is presented next.

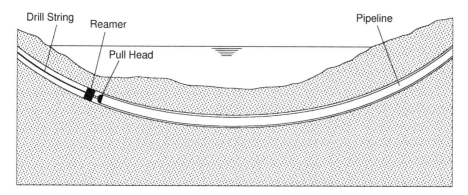

Figure 8.2   Pipeline bent as it is pulled through an underground borehole drilled by directional drilling.

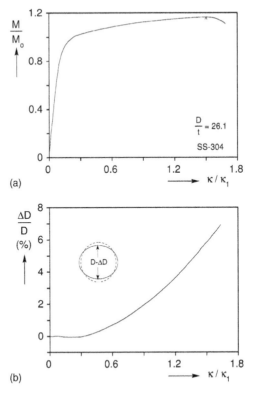

Figure 8.3 Pure bending response of an SS-304 tube with $D/t = 26.1$: (a) moment–curvature and (b) ovalization–curvature.

8.2 BENDING EXPERIMENTS

Inelastic bending experiments on tubes and pipes have been reported in [8.2–8.8]. The experiments involved large diameter UOE pipes, seamless pipes, small scale tubes and pipes of steel and aluminum alloys manufactured in several different ways. Both the material response and the manufacturing process can affect the bending response and its limit states. For this reason, correlating the results of these studies is not attempted. Instead, we will use primarily results from controlled experiments on small scale aluminum and steel tubes to illustrate the physics of the problem.

Figure 8.6 shows a four-point bending testing machine developed for testing tubes with diameters of 1–2 inches (25–50 mm). This machine was used in several studies that will be referred to here [8.2, 8.7, 8.8]. It consists of two free-turning sprockets mounted on two stiff support beams at a maximum distance apart of 50 in (1270 mm). Heavy chains which run around the sprockets are connected to two actuators and load cells to form a closed loop (Figure 8.6(b)). This arrangement allows both bending and reverse bending to be prescribed. For monotonic bending tests, the curvature capacity is extended by removing the lower cylinder (see Figure 8.6(a)) and extending the top one to its full stroke of 10 in (254 mm). Solid extension rods that closely fit into the ends of the tube to be tested are used to engage rollers mounted on the sprockets (Figure 8.6(c)). The rollers allow free axial motion of the end-rods. The arrangement allows maximum specimen length

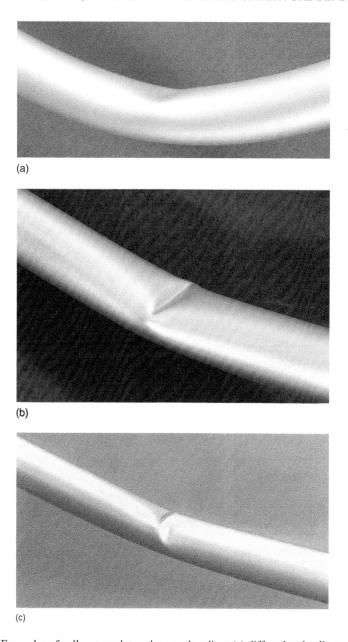

(a)

(b)

(c)

Figure 8.4 Examples of collapse modes under pure bending: (a) diffuse local collapse characteristic of low D/t s, (b) local kink for intermediate D/t s and (c) diamond-mode for high D/t s.

of 41 in (1041 mm). Bending is achieved by contracting one cylinder, causing rotation of the sprockets. The rotation of each sprocket is monitored by an LVDT displacement transducer as follows. Thin inextensional cables that run over the hub of each sprocket are connected to the cores of the LVDTs. Rotation of the sprockets causes linear motion of the cores, producing proportional changes in the output voltages. While the deformation of

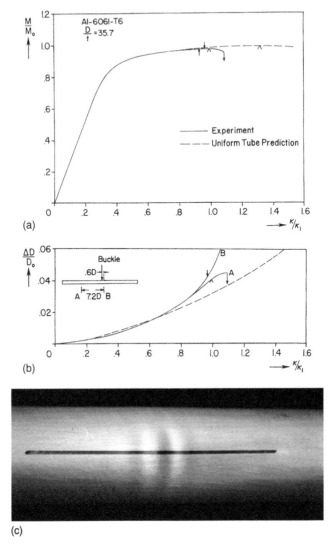

Figure 8.5 Experimental and predicted bending response of an Al-6061-T6 tube with $D/t = 35.7$: (a) moment–curvature, (b) ovalization–curvature and (c) pocket of wrinkles developed in an Al-6061-T6 ($D/t = 36.1$) tube that later precipitated a local kink.

the tube is uniform, its curvature (κ) is proportional to the sum of the angles of rotation of the two sprockets. The moment (M) is directly proportional to the tensile load measured by the load cell. In the set-up described, bending is essentially curvature-controlled. The machine was designed to be stiff compared to the specimens. These characteristics allow tracking of the response of a specimen past limit moments and other instabilities.

Several additional instruments that mount directly onto the tube are used to monitor the ovality of the cross section and its axial evolution if required. Figure 8.7 shows the simplest of these instruments. It consists of a stiff frame that carries a sliding cross beam. The instrument is weakly spring-loaded and contacts the tube through knife edges. The

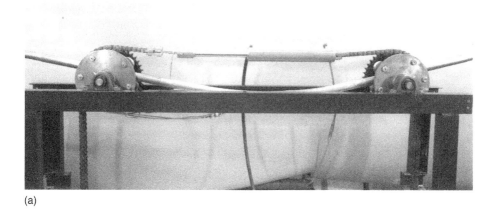

(a)

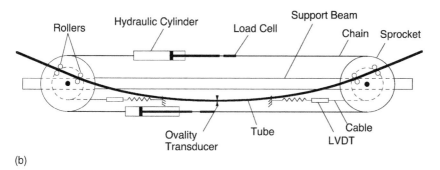

(b)

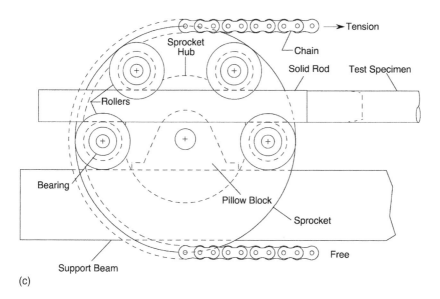

(c)

Figure 8.6   Pure bending testing facility: (a) photograph, (b) schematic and (c) load transfer assembly.

Figure 8.7 Instrument for measuring ΔD at a fixed axial location.

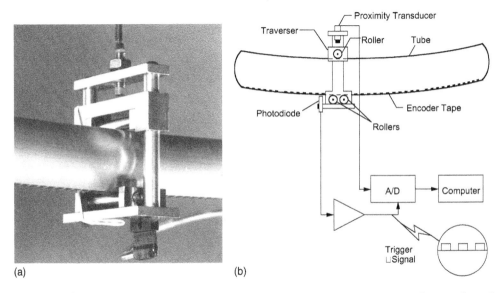

Figure 8.8 Instrument for measuring ΔD along the specimen length: (a) photograph and (b) schematic.

relative displacement between the two knife edges is monitored with a displacement transducer and corresponds to the change in diameter (ΔD) in the plane of bending.

The outputs of the instruments are monitored through a computer-operated data acquisition system which is programmed to record them whenever any one of the variables $\{M, \kappa, \Delta D\}$ changes by preset amounts. The moment–curvature and change in diameter–curvature results in Figure 8.1 constitute a typical set of results from such an experiment. The M, κ and ΔD are respectively normalized by the following parameters

$$M_o = \sigma_o D_o^2 t, \quad \kappa_1 = \frac{t}{D_o^2}, \quad \text{and} \quad D. \qquad (8.1)$$

$\Delta D/D$ is approximately equal to the induced ovality and will be referred to as such.

Figure 8.8 shows a second transducer that contacts the tube with rollers. The instrument is periodically rolled along the length of the tube to generate axial scans of the ovality.

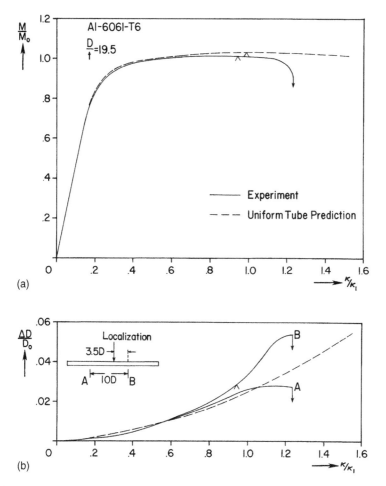

Figure 8.9 Experimental and predicted bending response of an Al-6061-T6 tube with $D/t = 19.5$: (a) moment–curvature and (b) ovalization–curvature.

Its axial position is established by an encoder system. A flexible tape with black–white markings is bonded to the lower surface of the tube. An emitter/receiver photodiode mounted on the transducer produces a signal whenever a white line is encountered. The signal triggers the A/D converter, which records the ovalization at that point.

Figure 8.9(a) shows the moment–curvature response of an Al-6061-T6 tube with $D/t = 19.5$, which exhibits a limit load instability. Figure 8.9(b) shows the ovalization at two locations. Instrument A is outside the zone of localized deformation, and as a result after the moment maximum the recorded ovality ceases to grow. Instrument B, which is in the neighborhood of localized deformation, records accelerated ovality growth in the curvature range of the limit moment and beyond. A sequence of axial scans of ovalization taken at different values of curvature with the instrument in Figure 8.8 is shown in Figure 8.10 (where x represents the axial coordinate). The ovalization remains essentially uniform up to a curvature of $\kappa/\kappa_1 = 0.75$. Subsequently, a section approximately $8D$ long around

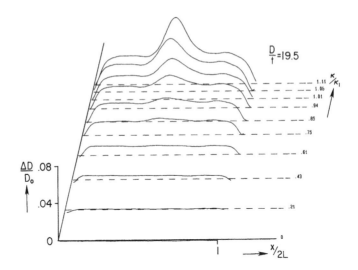

Figure 8.10 Axial scans of ΔD for an Al-6061-T6 tube with $D/t = 19.5$ at different curvatures.

the mid-span of the tube starts to ovalize at a faster rate. After the limit moment is reached at $\kappa/\kappa_1 = 0.94$, the rate of growth of local ovality accelerates and the tube eventually collapses. The last ovality scan shown was taken just before the tube collapsed.

In the case of thinner tubes, wrinkles that initially have small amplitude can develop on the compressed side of the tube. A more sensitive instrument shown in Figure 8.11 with a higher axial resolution was used to monitor the evolution of such wrinkles over a span of 4 in (102 mm). It consists of an axial guide with a fine lead screw used to move a vertical traversing system along the bent tube. The guide mounts on the tube through v-shaped end-plates which straddle the tube. The device is held in place by elastic springs. The traverser contacts the top and bottom of the tube through spring loaded, small diameter rollers. It is moved along the length of the tube by turning the lead screw. The two transverse beams that carry the rollers are free to slide vertically to accommodate changes in the geometry of the tube. Changes in the distance between the rollers are monitored by a non-contacting proximity transducer. The axial position of the traverser is monitored by a simple rotary encoder consisting of an emitter/receiver photodiode and a circular disk with six equally spaced slits mounted on the lead screw. Whenever the photodiode light beam is obstructed, the encoder sends a pulse that triggers the data acquisition system to record the proximity signal. The axial resolution of the system is 192 measurements per inch.

Figure 8.12 shows axial scans of ovality corresponding to the experiment in Figure 8.5. At a curvature of $\kappa/\kappa_1 = 0.81$ the first signs of localization of ovalization were observed. This grew with curvature to cover a region about $8D$ long. At a curvature of $\kappa/\kappa_1 = 0.96$, a pocket of short wavelength wrinkles became visible in this region. An axial scan of the wrinkles taken at a curvature of 0.99 with the instrument in Figure 8.11 is included in Figure 8.12. The short wavelength wrinkles are clearly visible. These wrinkles precipitated the failure in the form of one sharp local kink.

Table 8.1 summarizes results from 11 bending experiments, covering the D/t range of 19.5–60.5. Listed are the geometric and material parameters of the tubes, the critical curvatures and average half wavelengths (λ) of wrinkles when present. The curvature at

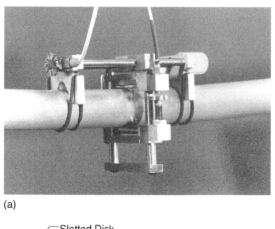

(a)

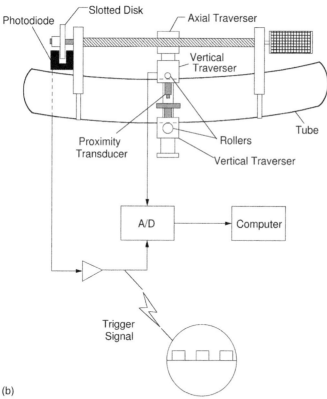

(b)

Figure 8.11 Ovalization transducer with high axial resolution for monitoring evolution of short
wavelength wrinkles: (a) photograph and (b) schematic.

which axial wrinkling was first observed is identified by κ_b, and the one corresponding
to the maximum moment by κ_L. The bending strains corresponding to $\{\kappa_b, \kappa_L\}$ based
on $\varepsilon = \kappa D/2$ are listed under $\{\varepsilon_b, \varepsilon_L\}$. For the two higher D/t tubes, failure was sudden
and catastrophic, which did not allow measurement of the wrinkle wavelengths. In these
cases, κ_b was taken to be the curvature at collapse. For the next three, a limit load was not

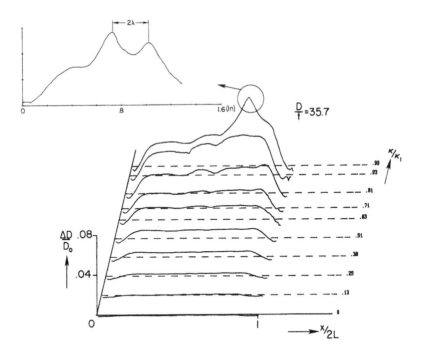

Figure 8.12   Axial scans of ΔD for an Al-6061-T6 tube with $D/t = 35.7$ at different curvatures. The inset shows short wavelength wrinkles that developed.

recorded, as they failed at an increasing curvature soon after the onset of wrinkling. The following three developed both wrinkles and a limit moment, and both are reported. The final three did not develop wrinkles, but instead developed a moment maximum followed by local, diffuse collapse.

The critical strains $\{\varepsilon_b, \varepsilon_L\}$ are plotted against D/t in log–log scales in Figure 8.13. They fall on a nearly linear trajectory, indicating a powerlaw relationship between the critical strain and D/t. For the middle D/t values, the onset of wrinkling and the subsequent limit load were not separated significantly. For these the bifurcation wrinkling strain, which is lower, can be taken as the critical state. If this is adopted, then the critical strain fits to the following power law with a correlation coefficient of $R^2 = 0.997$

$$\varepsilon_{CR} = 0.69018 \left(\frac{t}{D}\right)^{1.0893}. \tag{a}$$

It must be noted that the power of the fit as well as the multiplying coefficient depend on the stress–strain response of the material. The transitional D/t values from bifurcation buckling to a natural limit load instability are similarly dependent on the material stress–strain response.

Ju and Kyriakides [8.9] presented a sequence of models for analyzing this interaction and tracking the tube response into the postbuckling regime. In the following sections, we outline first a pure bending/axially uniform ovality formulation capable of capturing the natural limit load instability. This is then supplemented with a wrinkling bifurcation

Table 8.1 Tube geometric and material parameters of pure bending tests on Al-6061-T6 tubes.

D in (mm)	$\dfrac{D}{t}$	Ξ_o (%)	Δ_o (%)	E Msi (GPa)	σ_o ksi (MPa)	σ_y ksi (MPa)	n	S	$\dfrac{\kappa_b}{\kappa_1}$	$\dfrac{\kappa_L}{\kappa_1}$	ε_b (%)	ε_L (%)	$\dfrac{\lambda}{R}$
1.253 (31.82)	60.5	1.7	0.06	10.0 (68.9)	43.4 (299)	–	28	–	0.93	–	0.79	–	–
1.500 (38.10)	52.6	1.4	0.13	10.1 (69.6)	43.4 (299)	–	33	–	1.0	–	0.95	–	–
1.000 (25.40)	50.0	3.9	0.12	10.3 (71.0)	44.5 (307)	44.6	29	0.90	0.91	–	0.97	–	0.2877
1.251 (31.78)	44.0	4.5	0.12	9.75 (67.2)	44.1 (304)	44.0	25	0.92	0.95	–	1.13	–	0.3681
1.378 (35.0)	38.5	0.0	0.07	10.3 (71.0)	41.6 (287)	41.4	25	–	0.95	–	1.26	–	0.2935
1.250 (31.75)	35.7	0.86	0.05	9.77 (67.4)	41.1 (283)	40.9	28	0.91	0.96	0.99	1.42	1.49	0.2469
1.127 (28.63)	32.2	2.9	0.05	10.3 (71.0)	41.7 (288)	41.5	26	0.95	0.94	0.94	1.58	1.58	0.3172
0.998 (25.34)	28.2	1.0	0.06	9.60 (66.2)	44.1 (304)	44.0	35	–	0.89	0.90	1.69	1.72	0.2804
1.252 (31.80)	25.3	1.2	0.06	10.0 (69.0)	41.5 (286)	–	30	0.93	–	0.95	–	2.03	–
1.251 (31.78)	21.1	3.2	0.09	10.3 (71.1)	41.4 (285)	–	28	–	–	0.98	–	2.55	–
1.251 (31.78)	19.5	3.5	0.05	9.96 (68.7)	44.8 (309)	–	37	–	–	0.95	–	2.79	–

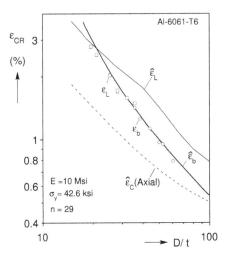

Figure 8.13 Experimental and predicted critical strains as a function of D/t for Al-6061-T6 tubes under pure bending.

buckling check for establishing the onset of wrinkling in the case of thinner pipes. These two levels of analysis cover sufficiently most of the needs of pipeline design.

8.3 FORMULATION

As was the case for external pressure loading, we again consider a long pipe of radius R and wall thickness t, bent under pure bending (moment M) about the x_2-axis to a curvature κ as shown in Figure 8.14. In the process, the cross section ovalizes in the sense that it flattens to some degree in the plane of bending. The axial strain is then

$$\varepsilon_x = \varepsilon^o + \zeta\kappa, \tag{8.2}$$

where ε^o is the strain of the axis of the pipe. For relatively thick pipes bent to a relatively small curvature ($\sim$ yield curvature $\kappa_o \approx 2\sigma_o/DE$), the ovalization can be neglected, $\varepsilon^o = 0$, $\zeta = R(1 + z/R)\cos\theta$, and the moment is calculated in the Saint-Venant sense as follows:

$$M = \int_A \sigma_x \zeta \, dA = 2R^2 \int_0^\pi \int_{-t/2}^{t/2} \sigma_x \cos\theta \left(1 + \frac{z}{R}\right)^2 dz \, d\theta. \tag{8.3}$$

For general elastic–plastic material stress–strain responses, the integral is evaluated numerically using an appropriate quadrature scheme.

Inelastic pure bending that includes the nonlinearity of cross sectional ovalization was first considered by Ades [8.10], by assuming the tube cross section to deform into an elliptical shape. Gellin [8.11] developed a more accurate solution to the problem by using improved kinematics. Both formulations used the J_2 deformation theory of plasticity to model the inelastic behavior of the material. Gellin's solution was further refined by Shaw and Kyriakides [8.12], who adopted more general kinematics and the more appropriate J_2 flow theory of plasticity. This formulation will be presented here with some extensions

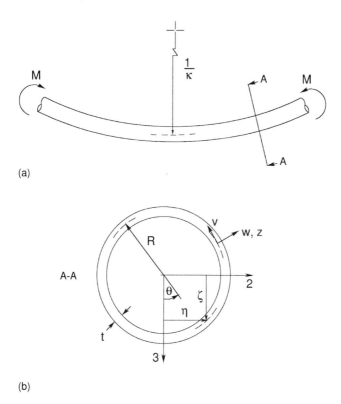

Figure 8.14   Problem parameters: (a) global view and (b) cross sectional parameters.

that allow the inclusion of more general initial geometric imperfections and wall thickness variations ([8.13–8.15] are additional relevant references).

The cross section is allowed to have the same axially uniform, small initial geometric imperfection, $\overline{w}(\theta)$, defined in Eq. (4.18) and the wall thickness is allowed to vary around the circumference in the sense of Eq. (4.19). The pipe is constrained to bend to a curvature κ in the x_1–x_3 plane. As a result, it will develop moments M_2 and M_3 related to the axial stress σ_x as follows

$$M_2 = R \int_0^{2\pi} \int_{-t/2}^{t/2} \sigma_x \zeta \left(1 + \frac{z}{R}\right) dz\, d\theta \quad \text{and} \quad M_3 = R \int_0^{2\pi} \int_{-t/2}^{t/2} \sigma_x \eta \left(1 + \frac{z}{R}\right) dz\, d\theta,$$

(8.4)

where ζ and η are defined in Figure 8.14(b) and t will in some cases imply $t(\theta)$.

8.3.1 Kinematics

The kinematical relationships are an extension of the small strain/finite rotations strain–displacement equations of Section 4.3. The circumferential strain is given by

$$\varepsilon_\theta = (\varepsilon_{\theta\theta}^o + z\kappa_{\theta\theta}) - \overline{\varepsilon}_\theta, \quad -\frac{t}{2} \leq z \leq \frac{t}{2},$$

(8.5a)

where $\bar{\varepsilon}_\theta$ is the apparent strain of the initial imperfection. The membrane component of (8.5a) is

$$\varepsilon_{\theta\theta}^o = \left(\frac{v' + w}{R}\right) + \frac{1}{2}\left(\frac{v' + w}{R}\right)^2 + \frac{1}{2}\left(\frac{v - w'}{R}\right)^2, \tag{8.5b}$$

while the local curvature is

$$\kappa_{\theta\theta} = \left(\frac{v' - w''}{R^2}\right) \Big/ \sqrt{1 - \left(\frac{v - w'}{R}\right)^2}. \tag{8.5c}$$

The distances ζ and η are given by

$$\zeta = (R + w)\cos\theta - v\sin\theta + z\cos\theta \tag{8.6a}$$

and

$$\eta = (R + w)\sin\theta + v\cos\theta + z\sin\theta. \tag{8.6b}$$

8.3.2 Constitutive Behavior

The pipe material is modeled as an elastic–plastic solid using the flow theory of plasticity with isotropic hardening (see Section 13.2). (Note that this can easily be replaced by an appropriate nonlinear kinematic hardening model when the needs of the problem so dictate.) The yield surface will be allowed to exhibit Hill-type anisotropy as represented by Eq. (13.11) with $\sigma_{x\theta} = 0$,

$$f = \left[\sigma_x^2 - \left(1 + \frac{1}{S_\theta^2} - \frac{1}{S_r^2}\right)\sigma_x\sigma_\theta + \frac{1}{S_\theta^2}\sigma_\theta^2\right]^{1/2} = \sigma_{e\,\mathrm{max}}. \tag{a}$$

8.3.3 Principle of Virtual Work

As in the experiments, the pipe is bent by incrementally prescribing the curvature, κ. Under this loading, the Principle of Virtual Work (*PVW*) can be expressed as

$$R\int_0^{2\pi} \int_{-t/2}^{t/2} (\hat{\sigma}_x \delta\dot{\varepsilon}_x + \hat{\sigma}_\theta \delta\dot{\varepsilon}_\theta)\left(1 + \frac{z}{R}\right) dz\, d\theta = 0, \tag{8.7a}$$

where $(\dot{\bullet})$ denotes a variable increment and $(\hat{\bullet}) \equiv (\bullet + \dot{\bullet})$. The problem domain is discretized in the same fashion as for the external pressure problem, using the following series expansions for the two displacements

$$w \cong R\left[a_o + \sum_{n=1}^{N}(a_n\cos n\theta + b_n\sin n\theta)\right] \quad\text{and}\quad v \cong R\sum_{n=2}^{N}(c_n\cos n\theta + d_n\sin n\theta).$$
$$\tag{8.7b}$$

Terms containing the coefficients c_0, c_1 and d_1 are omitted in order to preclude rigid body motion. When (8.7b) are substituted into Eqs. (8.5)–(8.7a) a system of $4N$ nonlinear

algebraic equations results. These equations are solved for the unknown coefficient increments and $\dot{\varepsilon}^o$ using the Newton–Raphson method. The integrals in (8.7a) are evaluated
via a Gauss integration scheme. At the end of each converged solution, the moments are
evaluated from (8.4). The number of terms in the series (8.7b) and the number of integration points around the circumference depend on the features of the initial imperfection
that need to be modeled.

For the most common imperfection case of simple initial ovality ($\overline{w}/R = -\Delta_o \cos 2\theta$),
the deformation becomes symmetric about the plane of bending and as a result $M_3 = 0$
(and $M_2 \equiv M$). In this case, $b_n = c_n = 0$ $\forall n$ and the number of equations reduces to
$2N + 1$. Five to seven integration points through the thickness of the pipe and twelve
around the half circumference are usually sufficient ([8.16]; note that the number of integration points usually must be increased when the stress–strain response exhibits low
hardening). The formulation presented above is implemented in the numerical model
BEPTICO that will be used to conduct uniform ovality bending calculations.

8.3.4 Bifurcation Buckling Under Pure Bending

For thinner pipes, bifurcation buckling leading to wrinkling of the compressed side of
the pipe can precede the natural limit load discussed above. The elastic version of the
problem was considered by Fabian [8.17] and Axelrad [8.18] among others. The elastic–
plastic problem was considered by Gellin [8.11], Fabian [8.13] and by Ju and Kyriakides
[8.9, 8.19], who went on to study the evolution of the wrinkles. In this problem, the
prebuckling state is a plastically bent cylinder with uniform ovalization. This makes it
significantly more complex than problems with simpler uniaxial, or at least proportional
loading, prebuckling states.

Here we present the formulation for a bifurcation check that can be conducted in conjunction with the prebuckling solution from BEPTICO. In particular, we have a long pipe
that is initially circular with radius R and wall thickness t. Deformations will be symmetric about the plane of bending, and thus $b_n = c_n = 0$ $\forall n$ in the displacement expansions
(8.7b). At any calculated equilibrium state, a bifurcation check can be made by assuming
that there exist two possible incremental solutions given by u^1 and u^2. We denote their
difference by $\{\sim\}$ and associate it with the buckling mode given by

$$\tilde{w} = R \cos \frac{\pi x}{\lambda} \sum_{n=0}^{N_w} c_n \cos n\theta,$$

$$\tilde{v} = R \cos \frac{\pi x}{\lambda} \sum_{n=1}^{N_v} d_n \sin n\theta, \qquad (8.8)$$

$$\tilde{u} = R \sin \frac{\pi x}{\lambda} \sum_{n=0}^{N_u} e_n \cos n\theta,$$

where $\{\tilde{u}, \tilde{v}, \tilde{w}\}$ are displacement components being measured from a circular toroidal
reference shell shown in Figure 8.15(a) and 2λ is the axial wavelength of wrinkles. This

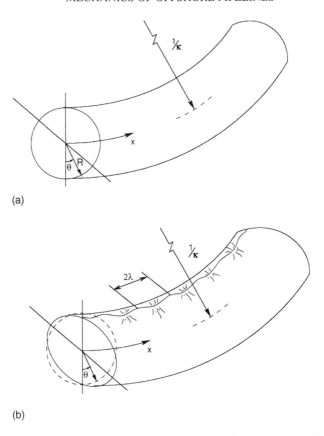

(a)

(b)

Figure 8.15   (a) Toroidal reference surface for bifurcation calculations and (b) wrinkled shell.

reference shell has principal curvatures

$$\kappa_1 = \frac{\kappa \cos \theta}{1 + \kappa R \cos \theta} \quad \text{and} \quad \kappa_2 = \frac{1}{R}. \tag{8.9}$$

The expected wrinkling-type buckling mode is shown in Figure 8.15(b). It is convenient to identify Eqs. (8.8) with the vector of unknown coefficients

$$\tilde{q} = [c_0, c_1, \ldots c_{N_w}, d_1, d_2, \ldots d_{N_v}, e_0, e_1, \ldots e_{N_u}]^T. \tag{a}$$

Since $\{\sim\}$ is the difference of two solutions, it must satisfy the incremental form of the PVW, which is then expressed as

$$R \int_0^\lambda \int_0^{2\pi} \left\{ \left[\tilde{N}_{\alpha\beta,j} \tilde{\varepsilon}^o_{\alpha\beta,i} + \tilde{M}_{\alpha\beta,j} \, \tilde{\kappa}_{\alpha\beta,i} + N_{xxo}(\tilde{\phi}_{1,i}\tilde{\phi}_{1,j} + \tilde{\phi}_{,i} \, \tilde{\phi}_{,j}) \right. \right.$$

$$\left. \left. + N_{\theta\theta o}(\tilde{\phi}_{2,i}\tilde{\phi}_{2,j} + \tilde{\phi}_{,i} \, \tilde{\phi}_{,j}) \right] \tilde{q}_i \tilde{q}_j \right\} dx \, d\theta = H_{ij}\tilde{q}_i\tilde{q}_j = 0, \quad i,j = 1, 2, \ldots (N_w + N_v + N_u),$$

$$\tag{8.10}$$

where α, $\beta = x$ or θ repeated, indices imply summation over their range and $(\bullet)_{,i} \equiv (\bullet)_{,\tilde{q}_i}$. $N_{xxo}(\theta)$ and $N_{\theta\theta o}(\theta)$ are the prebuckling stress intensities given by

$$N_{xxo} = \int_{-t/2}^{t/2} \sigma_{xo} \left(1 + \frac{z}{R}\right) dz \text{ and } N_{\theta\theta o} = \int_{-t/2}^{t/2} \sigma_{\theta o} dz. \tag{b}$$

The stresses in (b) come from the prebuckling solution. In deriving Eq. (8.10), Sanders' shell equations in Appendix D, appropriately linearized, are used. They are given by

$$\tilde{\varepsilon}_{xx}^o = \frac{1}{1 + \kappa R \cos\theta}[\tilde{u}_{,x} - \tilde{v}\kappa \sin\theta + \tilde{w}\kappa \cos\theta],$$

$$\tilde{\varepsilon}_{\theta\theta}^o = \frac{1}{R}[\tilde{v}_{,\theta} + \tilde{w}] + \phi_{2o}\,\tilde{\phi}_2,$$

$$\tilde{\varepsilon}_{x\theta}^o = \frac{1}{2(1 + \kappa R \cos\theta)}\left[\tilde{u}_{,\theta}\left(\frac{1}{R} + \kappa \cos\theta\right) + \tilde{v}_{,x} + \tilde{u}\kappa \sin\theta\right] + \frac{1}{2}\phi_{2o}\,\tilde{\phi}_1,$$

$$\tilde{\kappa}_{xx} = \frac{1}{1 + \kappa R \cos\theta}\left[\tilde{\phi}_{1,x} - \tilde{\phi}_2\kappa \sin\theta\right], \tag{c}$$

$$\tilde{\kappa}_{\theta\theta} = \frac{1}{R}\,\tilde{\phi}_{2,\theta},$$

$$\tilde{\kappa}_{x\theta} = \frac{1}{2(1 + \kappa R \cos\theta)}\left[\tilde{\phi}_{1,\theta}\left(\frac{1}{R} + \kappa \cos\theta\right) + \tilde{\phi}_{2,x} + \tilde{\phi}_1\kappa \sin\theta + \frac{\tilde{\phi}}{R}\right],$$

where

$$\tilde{\phi}_1 = \frac{1}{1 + \kappa R \cos\theta}[-\tilde{w}_{,x} + \tilde{u}\kappa \cos\theta], \quad \tilde{\phi}_2 = \frac{1}{R}[-\tilde{w}_{,\theta} + \tilde{v}],$$

$$\tilde{\phi} = \frac{1}{2(1 + \kappa R \cos\theta)}\left[\tilde{v}_{,x} - \tilde{u}_{,\theta}\left(\frac{1}{R} + \kappa \cos\theta\right) + \tilde{u}\kappa \sin\theta\right],$$

$$\phi_{2o} = \frac{1}{R}[-w_{o,\theta} + v_o]$$

and again, the notation $(\bullet)_o$ refers to prebuckling solution values. The following constitutive equations are adopted in (8.10):

$$\begin{Bmatrix} \tilde{N}_{xx} \\ \tilde{N}_{\theta\theta} \\ \tilde{N}_{x\theta} \\ \tilde{M}_{xx} \\ \tilde{M}_{\theta\theta} \\ \tilde{M}_{x\theta} \end{Bmatrix} = \int_{-t/2}^{t/2} \begin{bmatrix} 1 & z \\ z & z^2 \end{bmatrix} C_d\, dz \begin{Bmatrix} \tilde{\varepsilon}_{xx}^o \\ \tilde{\varepsilon}_{\theta\theta}^o \\ \tilde{\varepsilon}_{x\theta}^o \\ \tilde{\kappa}_{xx} \\ \tilde{\kappa}_{\theta\theta} \\ \tilde{\kappa}_{x\theta} \end{Bmatrix}, \tag{d}$$

where $\boldsymbol{C}_d$ are the instantaneous deformation theory moduli corresponding to the state of stress induced by uniform bending (inverse of $\boldsymbol{D}_d$ in (13.26) or the anisotropic counterpart (13.28) with $\sigma_{x\theta} = 0$). Equation (8.10) is a quadratic form $\tilde{\boldsymbol{q}}^T \boldsymbol{H} \tilde{\boldsymbol{q}}$. If the solution is unique, $\boldsymbol{H}$ is positive definite. At the bifurcation point, the determinant of $\boldsymbol{H}$ is zero. Thus, the bifurcation check at every converged prebuckling solution involves calculating the determinant of $\boldsymbol{H}$ for a set of trial values of λ in (8.8). The critical half wavelength (λ_b) is the one for which the determinant changes sign at the lowest curvature.

8.4 PREDICTIONS

Ju and Kyriakides [8.19] used the uniform ovality formulation to calculate the moment–curvature and ovalization–curvature responses of each of the 11 experiments in Table 8.1. Predictions for two cases are included in Figures 8.5 and 8.9 along with the experimental results. In Figure 8.5, the calculated responses are in agreement with the measured ones up to the onset of localized ovalization and the onset of wrinkling. Because wrinkling precipitated failure, the calculated limit moment occurs at a significantly higher curvature than the actual moment maximum. However, the calculated onset of bifurcation, marked on the M–κ response with ($\uparrow$), is seen to be close to the point at which wrinkling was first observed experimentally, indicated by ($\downarrow$).

Figure 8.9 includes similar predictions for the tube with $D/t = 19.5$ that failed by localized diffuse collapse after attaining a limit moment. The predicted M–κ and ΔD–κ responses are in excellent agreement with the ones measured up to the limit moment. The analysis is thus capable of establishing this limit state. Figure 8.16 includes similar predictions for the experiment on an SS-304 tube presented in Figure 8.3. The predictions, including the curvature at the limit moment, are very close to the measured results. Included in Figure 8.16(a) is the moment–curvature corresponding to Saint-Venant bending (8.3), which precludes ovalization and, as a result, is monotonically increasing. It replicates the initial stages of the measured response well, but departs from it at higher curvatures because of the increasing effect of ovalization.

Results from similar calculations were conducted for all experiments, and the general trend is similar. Calculations were also conducted using the average material properties of the 11 tubes tested. The average stress–strain response is represented by the following Ramberg–Osgood parameters

$$E = 10^4 \, \text{ksi} \, (69.0 \, \text{GPa}), \quad \sigma_y = 42.6 \, \text{ksi} \, (294 \, \text{MPa}), \quad n = 29, \quad \nu = 0.32.$$

Predictions of the bifurcation ($\hat{\varepsilon}_b$) and limit strains ($\hat{\varepsilon}_L$) of tubes with D/t values ranging from 15 to 100 are included in Figure 8.13. The calculated bifurcation strains are significantly lower than the corresponding limit strains for a significant part of the D/t range. For $D/t < 22$ the limit load precedes the bifurcation instability. The correlation of predicted bifurcation strains with experimental results is very good. Since the differences between the strains recorded at bifurcation and the limit load were relatively small, $\hat{\varepsilon}_b$ provides a good lower bound for the bending capacity. We note that, as in other plastic bifurcation problems, the bifurcation strains predicted by using the J_2 flow theory of plasticity are unreasonably high. Use of deformation theory in such bifurcation checks should once more be preferred.

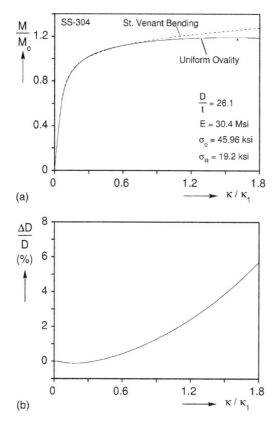

Figure 8.16 Predicted bending responses for the experiment shown in Figure 8.3: (a) moment–curvature and (b) ovalization–curvature.

The predicted transitional D/t at which the limit instability starts to precede wrinkling bifurcation is somewhat lower than the experimental value (no wrinkles were observed in the experiments for tubes with $D/t < 25.7$). This difference is partly due to the use of average material properties in the calculations (properties of individual tubes given in Table 8.1).

A lower bound for the critical bifurcation strain for shells in bending is the bifurcation strain corresponding to axisymmetric wrinkling under uniform axial compression. In other words, the strain corresponding to σ_C given in Eq. (11.4b). This can be calculated quite easily by again adopting the J_2 deformation theory incremental moduli as described in Chapter 11. The critical strains calculated in this manner for the aluminum properties given above are included in Figure 8.13 ($\hat{\varepsilon}_C$). Uniform compression is a more severe loading than bending, and as a result the predictions are quite conservative compared to the experimental results.

One area of disagreement with experiments noted in [8.9] was in the wavelength predicted by the bifurcation check model. The predicted wavelengths were uniformly longer than the measured values listed in Table 8.1 (see Figure 5 in [8.9]). A similar overprediction of measured wavelengths was also reported in [8.5]. Models of the postbuckling

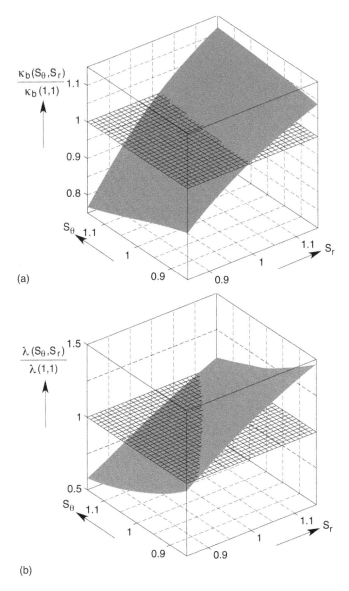

Figure 8.17   (a) Bifurcation curvature and (b) wrinkle half-wavelength as functions of the anisotropy
variables (Al-6061-T6, $D/t = 36.13$).

behavior of such tubes that account for the evolution of wrinkles, their localization and
subsequent events require the wrinkle wavelength as input. Because of this discrepancy,
the accuracy of such postbuckling calculations can be questioned.

This issue was recently revisited using experiments and analyses [8.20]. It was shown
that plastic anisotropy can significantly influence the predicted wavelength. The effect of
anisotropy variables S_r and S_θ on the predicted bifurcation curvature and half-wavelength
is illustrated in Figure 8.17 for an Al-6061-T6 tube with $D/t = 36.13$ (where $\kappa_b(1, 1)$ and
$\lambda(1, 1)$ are the calculated values for the isotropic material). κ_b is relatively insensitive

to S_θ, while it increases modestly for $S_r > 1$ and decreases with $S_r < 1$. By contrast, the predicted λ is influenced significantly by both anisotropy variables, increasing for $S_\theta < 1$ and $S_r > 1$ and decreasing for $S_\theta > 1$ and $S_r < 1$. Indeed, this variation of λ is quite similar to that observed for wrinkling under axial compression shown in Figure 11.17(b). Clearly, in addition to its influence on other pipe mechanical behavior effects, anisotropy can affect the predicted bifurcation buckling variables and should be incorporated in such calculations. This can be done by substituting the anisotropic incremental deformation theory constitutive equations (13.28) in the place of the isotropic ones in the bifurcation check model in Section 8.3.4.

Ju and Kyriakides [8.9] presented more complex custom models for following the evolution of wrinkling, its localization, and for checking for the possibility of a second bifurcation into a local shell mode (see also [8.21–8.23]). The onset and growth of diffuse local collapse, such as the one seen in Figure 8.10, was also modeled successfully. Such models are useful for predicting post-failure behavior of pipes and for establishing their sensitivity to geometric imperfections corresponding to these modes [8.9].

Postbuckling simulations of the type in [8.9] can also be performed using appropriate nonlinear finite element codes. A sample of such results involving an Al-6061-T6 tube with $D/t = 36.13$ is shown in Figure 8.18. A section of tube of length L is assigned an axisymmetric initial imperfection with wavelength 2λ given by

$$\bar{w} = -R\left[a_{oi} + a_i \cos\left(\frac{\pi x}{N\lambda}\right)\right]\cos\left(\frac{\pi x}{\lambda}\right), \quad -\frac{L}{2} \le x \le \frac{L}{2}, \quad 2N\lambda = L. \quad (8.11)$$

The imperfection has a small bias towards the center span ($a_{oi} = 0.0025$ and $a_i = 0.0005$). For the case presented, $\lambda = 0.329R$ was established from the bifurcation model including the appropriate anisotropies [8.20]. The length of the model was $L = 22\lambda$. Symmetry about the plane of bending allows modeling one half of the tube. The domain was discretized with 27-node brick elements with full integration (ABAQUS C3D27 – note that these can be substituted with suitable shell elements). The element distribution was 62 along the length (with a small bias in the central section 4λ long), 36 around the circumference and 2 through the thickness. The material was modeled as a J_2-type, elastoplastic, finitely deforming solid that hardens isotropically. The Ramberg–Osgood parameters of the engineering stress–strain response used are the averages of the parameters measured in this material group. In addition, the following anisotropies were adopted: $S_r = 0.887$ and $S_\theta = 1.015$ [8.20].

Initially, the calculated moment–curvature response shown in Figure 8.18(a) is very close to the one produced by the uniform ovality model discussed above (dashed line). Figure 8.18(b) shows a plot of the change in diameter ($\Delta D/D$) vs. the tube curvature in the plane of bending at mid-span ($x = 0$) and near the end of the tube ($x = L/2 - \lambda$). Figure 8.18(c) shows a set of configurations of the wrinkled tube. At first the tube ovalizes essentially uniformly, and ΔD at the two sites coincide with that of the uniform ovality model. During this early phase of bending, the amplitude of the axial wrinkles grows very gradually. As the bifurcation curvature of the uniform tube is approached (indicated by ↓), the growth of the wrinkles accelerates and a limit moment develops at $\kappa = 0.763\kappa_1$. This limit curvature is smaller than the one yielded by the uniform ovality model, and agrees well with the corresponding experimental one [8.20]. Configuration ① shows the wrinkles at the limit moment. In the scale of the drawing, they seem to have grown nearly uniformly. In reality, in the neighborhood of the limit moment the growth of the mid-span

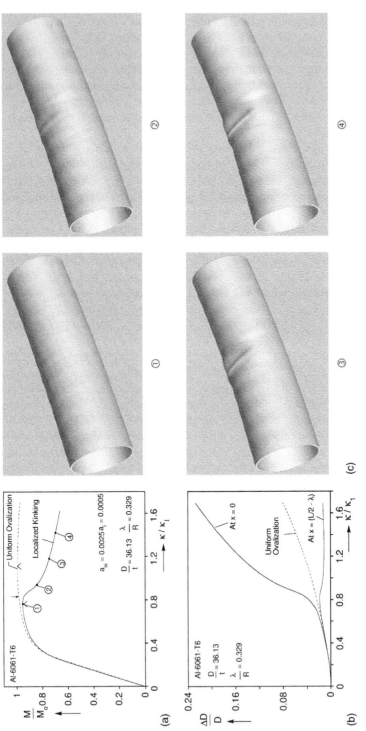

Figure 8.18 Predicted buckling and postbuckling response of an Al-6061-T6 tube with $D/t = 36.13$: (a) moment–curvature, (b) ovalization–curvature at two locations and (c) sequence of bent tube configurations corresponding to response in Figure 8.18(a) illustrating localization of wrinkling leading to a kink.

wrinkle accelerates and continues to grow at a high rate after the limit moment. This acceleration is reflected in the $\Delta D/D-\kappa/\kappa_1$ plot. The localization of the mid-span wrinkle is seen in configuration ② to lead to a sharp inward kink. At even higher curvatures, the drop in moment becomes more precipitous, and the kink becomes deeper as shown in configurations ③ and ④. By contrast, away from the mid-span ΔD decreases due to unloading, as illustrated in Figure 8.18(b) by the results corresponding to the end of the tube.

Configurations ②, ③ and ④ represent the early stages of development of a kink similar to the one shown in Figure 8.4(b). In the experimental setup described in Section 8.2, the tubes tested were typically 22D–25D long. In addition, the facility, although relatively stiff, has some compliance. As a result, soon after reaching the limit load, the tube kinks in a dynamic fashion and the moment drops significantly. Thus, the experimental kinks are at more severe stages of development than in the calculated configurations in Figure 8.18(c). At the same time, following numerically the kink development deeper into the postbuckling regime requires local refinement of the mesh, which was not pursued. Such post-buckling calculations primarily help understand the failure mode. Clearly, the critical state of interest is the limit curvature. This is influenced to some degree by the amplitude of the initial imperfection (8.11). However, in view of the closeness of the limit curvature to the bifurcation curvature discussed in Section 8.2, the design needs of most pipelines can be adequately served with the two simpler analyses outlined in Section 8.3.

8.5 PARAMETRIC STUDY

We now use the uniform ovality model and the bifurcation check algorithm of Section 8.3 to perform a limited parametric study for isotropic X52 steel pipes ($n = 13$, Appendix C). Figure 8.19 shows a set of moment–curvature and ovalization–curvature responses calculated with the uniform ovality model. Marked on the $M-\kappa$ responses are the limit moments and the onset of wrinkling bifurcations. The normalization used in these plots masks the significant difference introduced by D/t to the moment–curvature responses and the critical states. The buckling mode calculated for $D/t = 40$ is shown in Figure 8.20. It consists of a periodic wrinkling mode that affects the compressed side of the tube. The wavelength of the wrinkles is $0.8R$. It is worth noting that this value does not differ significantly from the corresponding wavelength for uniform axial compression of $0.83R$.

The critical strains corresponding to these responses are plotted in log–log scales against D/t in Figure 8.21. The limit strain varies from about 1.2% for $D/t = 60$ to about 5% for $D/t = 15$. The bifurcation strains are lower than the limit strains down to D/t of about 18. For even lower D/t s, ε_L is lower than ε_b, and it becomes the critical state.

Similar results for the same grade steel but with a hardening exponent of $n = 30$ are also included in Figure 8.21. For this material hardening, both critical states occur at significantly lower strains. For example, the limit strain now varies from about 0.97% for $D/t = 60$ to about 3.6% for $D/t = 15$. Another difference is that ε_L is now lower than ε_b for D/t values of about 24 and lower. The differences in the two sets of results illustrate the significant effect that hardening can have on these critical variables. It is thus essential that hardening be included in design considerations.

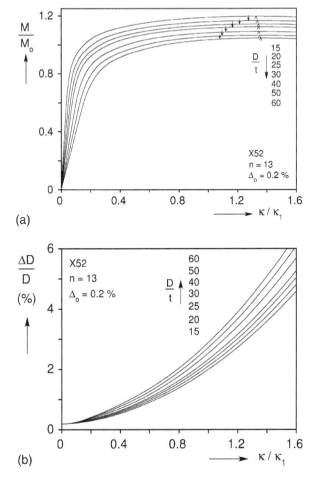

Figure 8.19 Predicted bending responses for X52 steel pipes of various D/t s: (a) moment–curvature and (b) ovalization–curvature.

In the interest of expediency, some design criteria (e.g., [8.6]) suggest the following empirical estimate of the limit state of pipes under pure bending:

$$\varepsilon_{CR} \approx \frac{t}{2D}.\qquad\qquad(8.12)$$

Predictions based on (8.12) are included in Figure 8.21. Such simplified expressions by their nature are incapable of differentiating between the two modes of instability. In addition, the effects of parameters such as initial ovality, material hardening, yield anisotropy and residual stresses cannot be captured. Furthermore, as demonstrated in the figure for the two X52 materials, the expression can be unconservative, as is the case for the higher D/t results for the $n = 30$ material. (Note also the difference between Eq. (8.12) and the powerlaw fit (a) in Section 8.2 of the experimental data in Table 8.1.) Despite these deficiencies, the simplicity of Eq. (8.12) makes it a reasonable candidate for use in

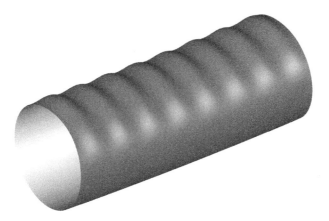

Figure 8.20   Pure bending wrinkling buckling mode for an X52 pipe with $D/t = 40$.

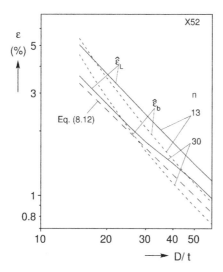

Figure 8.21   Predicted critical bending strains as a function of D/t for X52 steel pipes with two hardening exponents.

preliminary design as a tool for estimating the bending capacity. The design can then be refined by using numerical analyses such as the ones presented in this chapter.

It is worthwhile to compare once more the bifurcation strains under bending and axial compression. Figure 8.22 shows plots of bifurcation bending strains ($\hat{\varepsilon}_b$) and axial buckling strains ($\hat{\varepsilon}_C$) for a range of D/t values for the two X52 materials under consideration. As pointed out earlier, axial compression is a more severe loading than bending because the axial stress is uniformly distributed around the circumference. As a result, the cylinder bifurcates into axisymmetric wrinkling at significantly smaller strains than it does under pure bending for both material models. As illustrated in Chapter 11, the wrinkles grow stably, and significantly higher strain levels can be attained before failure in the form of localized collapse takes place. Under bending, the bifurcation is delayed while failure is in

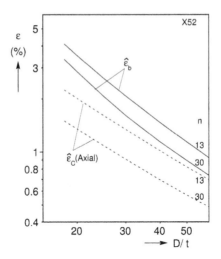

Figure 8.22   Wrinkling bifurcation strains vs. D/t for bending and axial compression for X52 steel
pipes with two hardening exponents.

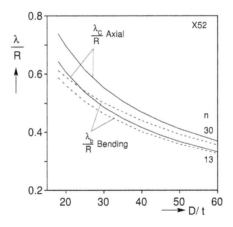

Figure 8.23   Wrinkle half-wavelengths vs. D/t for bending and axial compression for X52 steel pipes
with two hardening exponents.

the form of local kinking or diffuse localized collapse. Both are influenced by ovalization
and occur at strains that are not very much higher than ε_b.

Finally, we compare the wavelengths of wrinkles induced by bending and axial compression. Having the correct wavelength is important in postbuckling calculations like
the one in Figure 8.18, those in [8.9, 8.20–8.23] and in Chapter 11. Figure 8.23 shows
calculated half-wavelengths (λ) for the two loadings and the two X52 materials over a
range of D/t s. For axial loading, λ/R is proportional to $\sqrt{t/R}$ (11.4a). For bending, the
relationship is somewhat more complicated, but follows a similar trend. Increasing the
value of n increases the wavelength for both loadings. The bending wavelengths are somewhat lower than those for axial compression. The difference between the two increases
for lower D/t values.

The moment–curvature response, the induced ovalization and, as a result, the limit and bifurcation curvatures and strains can be affected to varying degrees by additional factors. These include anisotropic yielding of the material, circumferential bending-type residual stresses, wall thickness variations and geometric imperfections other than initial ovality. The effect of several of these parameters will be considered in the next chapter, which involves bending in the presence of external pressure.

8.6 SUMMARY AND RECOMMENDATIONS

The moment–curvature response of a long circular tube bent in the plastic range differs from that of a corresponding inelastic beam because of the bending-induced ovalization of the tube cross section. The growing ovalization causes a progressive reduction in bending rigidity, eventually resulting in the attainment of a moment maximum (limit moment). The response can be followed beyond the maximum by controlling the curvature (numerically or experimentally). However, soon after the limit moment, deformation localizes and the structure collapses either by developing diffuse local ovalization or by local kinking. Thus, the curvature at the moment maximum is one of the limit states of the structure. The nonlinear response and limit load instability can be evaluated by custom uniform ovality analyses (e.g. BEPTICO) or by appropriately nonlinear finite element models.

In the case of higher D/t tubes, wrinkling precedes the natural limit load. Although wrinkling initially grows in a stable manner, it contributes to further reduction in bending rigidity and leads to local buckling and collapse in the form of a local, often sharp, kink. Because of the proximity of the onset of wrinkling and failure, the former can be treated as the second limit state of the structure. The onset of wrinkling can be evaluated using a dedicated bifurcation check that works in conjunction with a uniform ovality nonlinear model. Given the wavelength of the wrinkles, more complex 3-D numerical models can be used to follow the evolution of wrinkling, its localization and the ensuing local kinking.

The curvature (or bending strain) at the limit moment as well as that at the onset of wrinkling are mostly influenced by the tube D/t and the material stress–strain response (hardening). Factors such as yield anisotropy, residual stress fields, wall thickness eccentricity, etc. can also influence these limit states.

REFERENCES

8.1. Brazier, L.G. (1927). On the flexure of thin cylindrical shells and other thin sections. *Proc. Roy. Soc. London* A **116**, 104–114.

8.2. Kyriakides, S. and Ju, G.-T. (1992). Bifurcation and localization instabilities in cylindrical shells under bending. Part I. Experiments. *Int. J. Solid. Struct.* **29**, 1117–1142.

8.3. Jirsa, J.O., Lee, F.K., Wilhoit, J.C. and Merwin, J.E. (1972). Ovaling of pipelines under pure bending. *Proc. Offshore Technology Conference* **I**, OTC1569, 573–578.

8.4. Sherman, D.R. (1976). Tests of circular steel tubes in bending. *ASCE J. Struct. Div.* **102**, ST11, 2181–2195.

8.5. Reddy, B.D. (1979). An experimental study of the plastic buckling of circular cylinders in pure bending. *Int. J. Solid. Struct.* **15**, 669–685.

8.6. Murphey, C.E. and Langner, C.G. (1985). Ultimate pipe strength under bending, collapse and fatigue. *Proc. 4th International Conference on Offshore Mechanics and Arctic Engineering*, **1**, 467–477.

8.7. Kyriakides, S. and Shaw, P.-K. (1987). Inelastic buckling of tubes under cyclic bending. *ASME J. Pressure Vessel Technol.* **109**, 169–178.

8.8. Aguirre, F., Kyriakides, S. and Yun, H.D. (2004). Bending of steel pipes with Lüders bands. *Int. J. Plast.* **20**, 1199–1225.

8.9. Ju, G.-T. and Kyriakides, S. (1992). Bifurcation and localization instabilities in cylindrical shells under bending. Part II. Predictions. *Int. J. Solid. Struct.* **29**, 1143–1171.

8.10. Ades, C.S. (1957). Bending strength of tubing in the plastic range. *J. Aeronaut. Sci.* **24**, 605–620.

8.11. Gellin, S. (1980). The plastic buckling of long cylindrical shells under pure bending. *Int. J. Solid. Struct.* **10**, 397–407.

8.12. Shaw, P.-K. and Kyriakides, S. (1985). Inelastic analysis of thin-walled tubes under cyclic bending. *Int. J. Solid. Struct.* **21**, 1073–1100.

8.13. Fabian, O. (1981). Elastic–plastic collapse of long tubes under combined bending and pressure load. *Ocean Eng.* **8**, 295–330.

8.14. Bushnell, D. (1981). Elastic-plastic bending and buckling of pipes and elbows. *Comput. Struct.* **13**, 241–248.

8.15. Kyriakides, S., Dyau, J.-Y. and Corona, E. (1994). *Pipe Collapse Under Bending, Tension and External Pressure (BEPTICO)*. Computer Program Manual, Engineering Mechanics Research Laboratory Report No. 94/4, January 1994.

8.16. Corona, E. and Kyriakides, S. (1988). On the collapse of inelastic tubes under combined bending and pressure. *Int. J. Solid. Struct.* **24**, 505–535.

8.17. Fabian, O. (1977). Collapse of cylindrical, elastic tubes under combined bending, pressure and axial loads. *Int. J. Solid. Struct.* **13**, 1257–1270.

8.18. Axelrad, E.L. (1980). Flexible shells. *Proc. 15th IUTAM Congress*, Toronto, Canada, 45–56.

8.19. Ju, G.-T. and Kyriakides, S. (1991). Bifurcation versus limit load instabilities of elastic–plastic tubes under bending and pressure. *ASME J. Offshore Mech. Arctic Eng.* **113**, 43–52.

8.20. Corona, E., Lee, L.-H. and Kyriakides, S. (2006). Yield anisotropy effects on buckling of circular tubes under bending. *Int. J. Solid. Struct.* **43**, 7099–7118.

8.21. Ju, G.-T. and Kyriakides, S. (1991). On the Effect of local imperfections on the stability of elastic plastic shells in bending. *Proc. International Colloquium: Buckling of Shell Structures on Land, in the Sea and in the Air*, Lyon, France, September 17–19, 1991, Ed. J.F. Jullien, Elsevier Applied Science, 370–380.

8.22. Karamanos, S.A. and Tassoulas, J.L. (1996). Tubular members. I. Stability analysis and preliminary results. *ASCE J. Eng. Mech.* **122**, 64–71.

8.23. Karamanos, S.A. and Tassoulas, J.L. (1996). Tubular members. II. Local buckling and experimental verification. *ASCE J. Eng. Mech.* **122**, 72–78.

9

Buckling and Collapse Under Combined Bending and External Pressure

Bending in the presence of external pressure is experienced by pipelines during their installation and also subsequently during their operation. In installation methods such as S-lay (Figure 2.15), J-lay (Figure 2.19), and reeling (Figure 2.4), the pipe is bent under relatively high external pressure in the sagbend [9.1, 9.2]. The pipeline is also bent under external pressure as it conforms to surface undulations on the seafloor. Bending at the sea floor can also be experienced due to snaking resulting from pipe expansion caused by the passage of hot hydrocarbons (in some cases purposely induced, in others accidentally). It is also a condition that can develop in case of upheaval buckling of a section of a buried pipe. Bending ovalizes the pipe cross section, which of course reduces its resistance to external pressure. This interaction through ovalization is aggravated by inelastic material effects, which usually come into play in lower D/t pipes used for offshore applications. This chapter deals with the mechanics of inelastic bending in the presence of external pressure and the associated limit states. The development of this interaction starts with the behavior under pure bending and pure external pressure. The reader would benefit from exposure to Chapters 4 and 8 before embarking on the following material.

9.1 FEATURES OF INELASTIC BENDING OF TUBES UNDER EXTERNAL PRESSURE

The moment–curvature response of a long pipe with $D/t = 26.1$ bent into the plastic range is shown in Figure 9.1(a). The ovalization induced to the cross section is shown in Figure 9.1(b). The ovalization causes a progressive reduction in bending rigidity, eventually resulting in a moment maximum (marked on the response with a caret "∧"). As discussed in Chapter 8, the moment maximum is the limit state of this structure. Consider the same bending test being conducted in the presence of external pressure of $P/P_o = 0.273$, as shown in the figures. The external pressure amplifies the ovalization induced by bending. The initial elastic part of the moment–curvature response remains relatively unaffected, as the ovalization is relatively small in this regime. Higher values of moment initiate inelastic action, which accelerates the growth of ovalization as illustrated in Figure 9.1(b). The

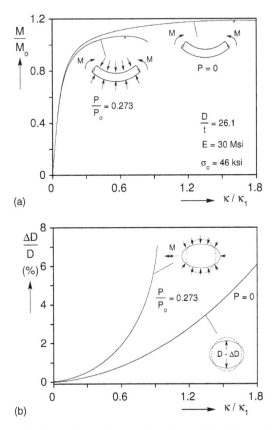

Figure 9.1   Responses of tubes subjected to pure bending and bending in the presence of external
pressure: (a) moment–curvature and (b) ovalization–curvature.

bending rigidity is reduced faster, thus lowering the maximum moment and causing it
to occur at a curvature much smaller than under pure bending. At higher pressures, this
degradation in moment and curvature capacity is more severe. These characteristics will
first be illustrated experimentally and then modeled with extensions of the formulations
in Chapters 4 and 8. Several factors that can alter this general behavior and influence the
associated critical states will be discussed at the end of the chapter.

As for pure bending, higher D/t pipes can develop wrinkling and local shell buckling
prior to the limit state induced by ovalization. It will be demonstrated that this can be
addressed adequately by a model that can predict the onset of wrinkling, similar to the
one outlined in Section 8.3.4.

9.2 COMBINED BENDING-EXTERNAL PRESSURE EXPERIMENTS

Experiments involving bending of pipes in the presence of external pressure have been
reported in Refs. [9.3–9.12] and by others. The experiments involved large diameter
UOE pipes, seamless pipes and small-scale tubes and pipes of steel and aluminum alloys

(a)

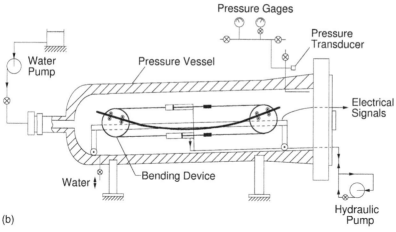

(b)

Figure 9.2   Combined bending-pressure test facility: (a) photograph and (b) schematic.

manufactured by several different methods. Both the material response as well as the manufacturing process can affect the response and collapse of pipes under this combined loading. For this reason, correlating the results of these studies is not attempted. Instead, we will primarily use results from controlled experiments on small-scale aluminum and steel tubes to illustrate the physics of the problem. Modern facilities available for full scale tests will also be introduced.

9.2.1 Test Facilities

The custom combined bending-pressure testing facility of Corona and Kyriakides [9.6] is shown in Figure 9.2(a). Its major components are labeled in the schematic shown in

Figure 9.2(b). The facility consists of a pure bending device that operates inside a pressure vessel. A data acquisition system is used to monitor the experiments. The four-point bending device, capable of applying bending and reverse bending, is a variation of the one shown in Figure 8.6. It consists of two heavy sprocket assemblies resting on two beams. Heavy chains run around the sprockets and are connected to two hydraulic cylinders and load cells, forming a closed loop. Each tube tested is fitted with solid rod extensions as shown in Figure 8.6(c). The rod-tube interface is sealed with a rubber hose. The test specimen assembly is engaged by the bending device through four rollers located on each sprocket, as shown in Figure 8.6(c). Bending is achieved by contracting either of the cylinders, causing rotation of the sprockets. The rolling contact between the test specimen and the device guarantees freedom of movement of the tube in the axial direction during bending.

The hydraulic system was designed so that the idle cylinder extends an amount equal to the contraction of the active one. This ensures that the periphery of the closed loop consisting of the chains, load cells and cylinders has constant length. Bending in the reverse sense is achieved by reversing the direction of flow in the hydraulic circuit. The hydraulic system is such that the device operates essentially as a *curvature-controlled* testing machine.

The curvature of the test specimen is directly proportional to the sum of the angles of rotation of the sprockets. The rotation of each sprocket is monitored via a rotary transducer (RVDT). The applied bending moment is directly proportional to the tension in the chain, which is monitored by the two load cells in the chain loop. The capacity of the bending device is 15,000 lb-in (1700 N-m). Tubes having maximum lengths of 40 in (1,000 mm) and diameters up to 1.5 in (38 mm) can be tested. The device was designed to be stiff relative to the test specimen. For the experiments reported, the energy stored in the device was less than 3% of the energy stored in the test specimen. Custom instruments like the ones shown in Figure 8.7 are mounted at the tube mid-span and are used to monitor changes in the major (and when necessary the minor) diameter of the tube cross section.

The pressure vessel is cylindrical in shape with an internal length of 72 in (1.8 m) and an internal diameter of 20 in (510 mm). A special closure system provides full opening access at one end of the vessel (see Figure 9.2(a)). The vessel has a working pressure of 5,000 psi (345 bar). The bending device rolls into the pressure vessel and is operated remotely. The pressure vessel is usually completely filled with water and pressurized in a nearly volume-controlled fashion using a hydraulic pump. Pressure gages and an electrical pressure transducer are used to monitor the pressure during the experiment. The applied moment (M), curvature (κ), ovalization ($\Delta D/D$) and pressure (P) are customarily monitored via the data acquisition system.

Before showing experimental results obtained in this facility, it is worthwhile to also describe a commercial bending-pressure testing facility used for full scale testing of pipes up to diameters of 28 inches. The C-FER Technologies facility is shown schematically in Figure 9.3(a), while Figure 9.3(b) shows a pipe being removed from the facility after having collapsed under combined loads (see also [9.11, 9.12]). The pressure vessel is 37.2 ft (11.35 m) long with an ID of 48 in (1,220 mm) and full ID openings at the ends. It has a working pressure of 8,000 psi (552 bar). The pipe is bent by extending two 1.2×10^6 lbf (5.34 MN) rams that form couple pairs (four-point bending) with cradle-type supports. The rams and the supports are internal to the vessel and their loads are reacted by the vessel as shown in the figure. The longest ram-to-ram distance that can be practically tested is 31.8 ft (9.7 m). The curvature range depends on the diameter of

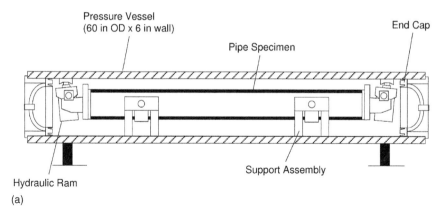

(a)

(b)

Figure 9.3   (a) Schematic of C-FER Technologies large-scale bending-pressure test facility. (b) 24-inch pipe collapsed under combined bending and pressure. (Courtesy C-FER Technologies)

the pipe tested and on the distance between the rams and the support assemblies chosen. The bending system operates as a curvature-controlled machine at a prescribed external pressure level. The pressurization medium is water. Pipes 16 inches in diameter or less are usually tested in the same pressure vessel, but are bent using an alternate bending mechanism. The pipe ends are attached to rigid disks, which are caused to rotate by actuator-applied forces as described in [9.11].

9.2.2   Experimental Results

The majority of the small scale tests were performed using seamless, stainless steel tubes with D/t values in the range of about 35 to 18. The lengths of the test specimens ranged between 18 and 24 tube diameters. The geometric and material properties of each tube tested were measured in accordance with the procedure outlined in Chapter 4. Average

Table 9.1   Mean geometric and material parameters of tubes referred to in this chapter.

Material	$\dfrac{D}{t}$	D in (mm)	t in (mm)	Δ_o %	E Msi (GPa)	σ_o ksi (MPa)	σ_y ksi (MPa)	n	S	$\dfrac{\sigma_R}{\sigma_o}$
SS-304	34.96	1.248 (31.70)	0.0357 (0.907)	0.04	26.9 (186)	37.6 (259)	32.5 (224)	9.7	0.94	–
SS-304	25.72	1.2475 (31.69)	0.0485 (1.232)	0.1	29.2 (201)	51.8 (357)	49.6 (342)	17	0.80	–
SS-304	19.20	1.248 (31.70)	0.065 (1.651)	0.05	29.5 (204)	45.6 (314)	41.1 (283)	10.9	0.987	–
SS-304	24.84[†]	1.252 (31.81)	0.0504 (1.651)	0.07	25.3 (174)	33.4 (230)	28.5 (197)	8.6	0.886	0.35
SS-304 W[††]	18.45	1.376 (39.95)	0.0746 (1.895)	0.04	29.3 (202)	42.2 (291)	38.0 (262)	11.5	0.97	0.235
Al-6061-T6	49.26	1.000 (25.4)	0.0203 (0.516)	0.15	10.2 (70.3)	44.8 (309)	44.5 (307)	29	0.90	–
X65 UOE	19.24	24.047 (610.8)	1.250 (31.75)	0.166	29.8 (206)	76.3 (526)	–	–	0.82	0.04
X65 UOE	16.01	25.980 (659.9)	1.6226 (41.21)	0.193	30.0 (207)	76.0 (524)	74.0 (510)	20	0.778	0.116

† Tubes used in residual stress study, †† Seam-welded

values of the key parameters of each set are listed in Table 9.1. The variables P, M, κ and ΔD are, respectively, normalized by the following parameters:

$$P_o = 2S\sigma_o \left(\frac{t}{D_o}\right), \quad M_o = \sigma_o D_o^2 t, \quad \kappa_1 = \frac{t}{D_o^2}, \quad \text{and } D. \tag{9.1}$$

Three loading paths were adopted in the experiments. In the first, the tube was first bent to a chosen value of curvature at zero pressure. The curvature was then fixed and pressure loading was applied until the tube collapsed (depicted by $\kappa \to P$). In the second loading path, pressure was applied first, followed by bending under constant pressure until collapse (depicted by $P \to \kappa$). In the third, the curvature and pressure were incremented in a way that kept them at a constant ratio (depicted as *radial* loading path). In all cases, loading was terminated when the tube collapsed.

For the first two loading paths, enough experiments were carried out to establish the configuration of the collapse envelopes for the full range of pressure and curvature values of interest. In the case of the radial path, a selected number of experiments were carried out in order to establish the trend of the results.

In the $P \to \kappa$ loading path, pressure (P) was applied first, amplifying to some degree small initial ovality present in the tube. The tube curvature was then gradually increased while the pressure was kept constant. Bending induced a further growth of ovalization of the tube cross section. It was further accelerated by the presence of pressure, as illustrated in Figure 9.1(b). A limit moment was eventually reached, which will be considered as the limit state of such structures. Figure 9.4(a) shows a set of moment–curvature responses for tubes with $D/t = 25.7$ bent at different pressures. The degrading effect of pressure on the responses is quite clear. In all cases, failure was catastrophic and occurred at the last point recorded in the response. At higher values of pressure, collapse followed immediately the attainment of the limit moment. At lower pressures, the tube could be bent beyond the limit moment, in some cases significantly. Bending beyond the limit moment results

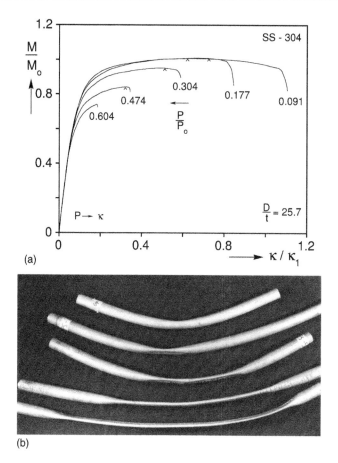

(a)

(b)

Figure 9.4   (a) Moment–curvature responses of tubes bent under several values of external pressure ($P \to \kappa$ loading path). (b) Tubes collapsed at increasing values of pressure (top to bottom).

in localized growth of ovalization and, as such, the recorded results cease to represent the moment–curvature response of the structure. Figure 9.4(b) shows the collapsed tubes corresponding to these tests. In this test setup, the extent of damage sustained by each tube during collapse depends on the value of pressure and the compliance of the pressure vessel. In a constant pressure environment, the tubes would be flattened by propagation of the collapse for all pressures higher than the propagation pressure of the tube.

In the $\kappa \to P$ loading path, the tube was first bent to a curvature $\bar{\kappa}_1$ with corresponding moment $\overline{M}_1$, as shown in Figure 9.5(a). The pressure was then increased gradually while the curvature was kept fixed (Figure 9.5(b)). This led to a drop in the moment, as illustrated in the figure and to a growth of ovalization. The tube collapsed when the ovalization grew to a value at which the applied pressure could no longer be supported. The collapse pressure (P_{CO}) corresponds to the limit pressure in the M–P response in Figure 9.5(b).

Interaction collapse envelopes were generated for each of these two paths for tubes of three different D/t s. The main features of the collapse envelopes and the strong effect of the loading history are illustrated in Figure 9.6 using the results from $D/t = 35.0$. In the case of the $P \to \kappa$ results, the lower curvature points (squares) represent the recorded limit

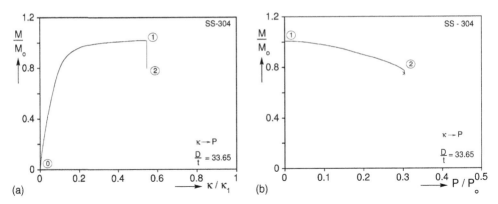

(a)   (b)

Figure 9.5   Response of a tube subjected to the $\kappa \to P$ loading path: (a) moment–curvature and (b) moment–pressure.

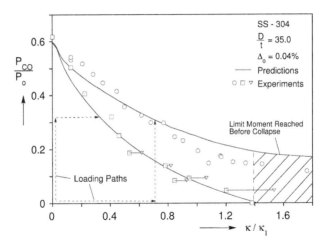

Figure 9.6   Experimental and predicted collapse envelopes for tubes with $D/t = 35.0$ for the $P \to \kappa$ and $\kappa \to P$ loading paths.

moment at a given pressure, and the points at the higher curvature (triangles) represent collapse. The $P \to \kappa$ collapse envelope is substantially lower than the $\kappa \to P$ envelope. The primary cause of this difference is the more severe plastic state induced in the areas of $\theta = 0$, $\pi/2$, π and $3\pi/2$ (see Figure 9.10(b)) on the tube cross section during loading. This enables an earlier onset of what is essentially an $n = 2$ collapse mode.

A select number of experiments were conducted for the radial path for $D/t = 35.0$. The established critical states are seen in Figure 9.7 to be somewhat lower than the $P \to \kappa$ path. This will be further discussed in Section 9.4 in the light of predictions.

Figures 9.8 and 9.9 show similar interaction collapse envelopes for tubes with D/t of 25.7 and 19.2. The trends of the results are similar for all three D/t values. A somewhat larger scatter is observed in the results for D/t of 25.7. This is attributed to the random orientation of the tube initial ovality to the plane of bending (see Section 9.5.4). (Numerical values of the data points in Figures 9.6–9.9 can be found in [9.4].)

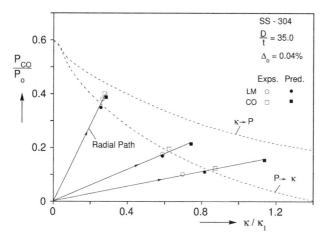

Figure 9.7 Experimental and predicted collapse results for the *radial* loading path and predictions for the $P \to \kappa$ and $\kappa \to P$ paths ($D/t = 35.0$).

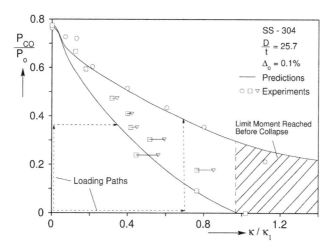

Figure 9.8 Experimental and predicted collapse envelopes for tubes with $D/t = 25.7$ for the $P \to \kappa$ and $\kappa \to P$ loading paths.

9.3 FORMULATION

The problem is once more assumed to be sufficiently represented by a long tube or pipe with uniform geometric and material properties along the length. In the most general case, the cross section is allowed to have an axially uniform, small initial geometric imperfection $\bar{w}(\theta)$, defined in Eq. (4.18), and the wall thickness is allowed to vary around the circumference in the sense of Eq. (4.19). The tube is bent in the x_1–x_3 plane to a uniform curvature κ under external pressure P (see Figure 9.10). It can thus be treated with the same basic kinematical equations given in Eqs. (8.2, 8.5, 8.6). In the most commonly used

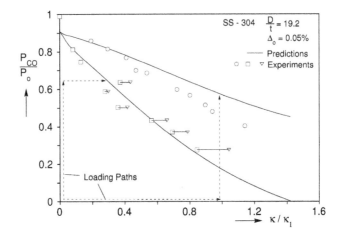

Figure 9.9   Experimental and predicted collapse envelopes for tubes with $D/t = 19.2$ for the $P \rightarrow \kappa$ and $\kappa \rightarrow P$ loading paths.

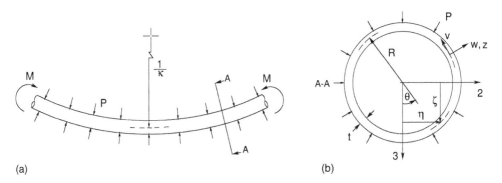

(a)   (b)

Figure 9.10   Problem parameters: (a) global view and (b) cross-sectional parameters.

version, the material is modeled as an elastic-plastic solid using the J_2 theory of plasticity along with isotropic hardening. The yield function includes Hill-type anisotropy, as represented by Eq. (13.11).

9.3.1   Principle of Virtual Work

The pipe is loaded incrementally by prescribing the curvature and the pressure. Under this loading, the Principle of Virtual Work can be expressed as

$$R \int_0^{2\pi} \int_{-t/2}^{t/2} (\hat{\sigma}_x \delta \dot{\varepsilon}_x + \hat{\sigma}_\theta \delta \dot{\varepsilon}_\theta) \left(1 + \frac{z}{R}\right) dz \, d\theta = \delta \dot{W}_e \tag{9.2a}$$

where

$$\delta \dot{W}_e = -\hat{P}R \int_0^{2\pi} \left[\delta \dot{w} + \frac{1}{2R}(2\hat{w}\delta\dot{w} + 2\hat{v}\delta\dot{v} + \hat{w}\delta\dot{v}' + \hat{v}'\delta\dot{w} - \hat{v}\delta\dot{w}' - \hat{w}'\delta\dot{v}) \right] d\theta, \tag{9.2b}$$

where ($\dot{\bullet}$) denotes a variable increment and ($\hat{\bullet}$) $\equiv$ ($\bullet + \dot{\bullet}$). The problem domain is discretized in the same fashion as in the previous problems by using the following series expansions for the displacements:

$$w \cong R \left[a_o + \sum_{n=1}^{N} (a_n \cos n\theta + b_n \sin n\theta) \right] \text{ and } v \cong R \sum_{n=2}^{N} (c_n \cos n\theta + d_n \sin n\theta). \quad \text{(a)}$$

9.3.2 Bifurcation Buckling Under Combined Bending and External Pressure

For thinner pipes bent at lower pressures, bifurcation buckling leading to wrinkling of the compressed side can precede the natural limit load discussed above. Since wrinkling becomes the limit state, it must be considered in a design process. The problem is addressed through an extension of the wrinkling bifurcation buckling analysis of Ju and Kyriakides [9.9] outlined in Section 8.3.4. The bifurcation criterion in Eq. (8.10) is extended to include the external pressure term as follows:

$$R \int_0^\lambda \int_0^{2\pi} \left\{ [\tilde{N}_{\alpha\beta,j}\tilde{\varepsilon}^o_{\alpha\beta,i} + \tilde{M}_{\alpha\beta,j}\tilde{\kappa}_{\alpha\beta,i} + N_{xxo}(\tilde{\phi}_{1,i}\tilde{\phi}_{1,j} + \tilde{\phi}_{,i}\,\tilde{\phi}_{,j}) + N_{\theta\theta o}(\tilde{\phi}_{2,i}\tilde{\phi}_{2,j} \right.$$
$$\left. + \tilde{\phi}_{,i}\,\tilde{\phi}_{,j})]\tilde{q}_i\tilde{q}_j \right\} dx\,d\theta + \frac{P}{2}\int_0^\lambda \int_0^{2\pi} \left\{ [(2\tilde{w}_{,i} + \tilde{v}'_{,i})\tilde{w}_{,j} + (2\tilde{v}_{,i} - \tilde{w}_{,i})\tilde{v}_{,j} \right.$$
$$\left. + \tilde{w}_{,i}\,\tilde{v}'_{,j} - \tilde{w}'_{,j}\,\tilde{v}_{,i}]\tilde{q}_i\tilde{q}_j \right\} dx\,d\theta = 0,$$

or

$$H_{ij}\tilde{q}_i\tilde{q}_j = 0, \quad i,j = 1, 2, \ldots (N_w + N_v + N_u). \quad \text{(9.3)}$$

All other aspects of this formulation remain the same. The bifurcation check algorithm works in conjunction with BEPTICO [9.13]. To determine the critical curvature, the check is applied at each converged curvature step as explained in Section 8.3.4.

9.4 PREDICTIONS

The solution procedure described in the previous section was used to simulate the experiments carried out using the mean geometric and material parameters of each set listed in Table 9.1. The main characteristics of the predicted responses for the $\kappa \to P$ loading path are summarized in a set of results for $D/t = 25.7$ presented in Figure 9.11. As in the experiments, the tube was incrementally loaded to the chosen value of curvature at zero pressure. Calculated moment–curvature (M–κ) and ovalization–curvature ($\Delta D/D$–κ) responses for this part of the path are shown in Figures 9.11(a) and 9.11(b), respectively. The curvature was then fixed (at the four values indicated by dotted lines in the figures), and the pressure was incremented. The ovalization developed during the bending parts of the path grew in the fashion shown in Figure 9.11(c). Eventually a limit pressure was reached that represents the collapse pressure. During the pressurization process the bending moment drops as shown in Figure 9.11(d). This causes material unloading at some points on the cross section, followed by reloading.

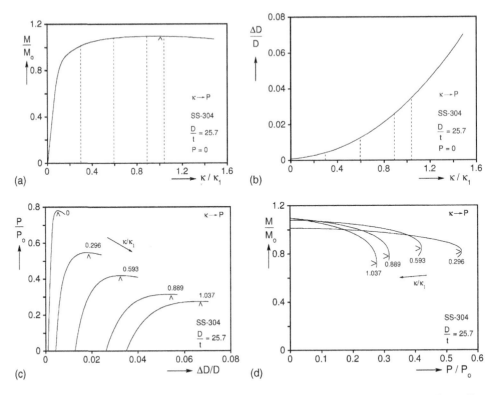

Figure 9.11   Calculated responses for the $\kappa \to P$ loading path: (a) moment–curvature, (b) ovalization–curvature, (c) pressure–ovalization and (d) moment–pressure.

Sample results for the $P \to \kappa$ path are presented in Figure 9.12. In this case, pressure was
applied first, and the calculated pressure–ovalization response is shown in Figure 9.12(a).
The pressure was then fixed (at the numerical values indicated in figures) and the curvature was incremented. The M–κ, $\Delta D/D$–M and $\Delta D/D$–κ responses corresponding
to four pressure levels are shown in Figures 9.12(b)–(d), respectively. At relatively low
values of pressure, the M–κ response exhibits a limit moment. In practice this is followed
by localized deformation and catastrophic collapse. Such localized deformation cannot
be simulated by this analysis. Instead, a good engineering practice is to consider the
attainment of a limit moment as the critical state. As was the case in the experiments,
collapse is more clearly identified at higher pressures. The critical curvature is identified
by a sharp decay in the bending moment with a small increase of curvature while the ovalization increases at a rapid rate to a value at which the external pressure can no longer be
supported.

Simulations of this type were conducted for each of the three main D/t values used
in the experiments. The predicted critical states for the $\kappa \to P$ and $P \to \kappa$ loading paths
are drawn with solid lines in Figures 9.6, 9.8 and 9.9. The generated collapse envelopes
follow the trend of the experimental points quite well for all three cases. The pure pressure
collapse data and the pure bending limit states are predicted with accuracy. The scatter can
be attributed to deviations of individual tube mechanical properties from the mean values
adopted in the calculations. For lower pressure levels, the calculated $\kappa \to P$ envelopes are

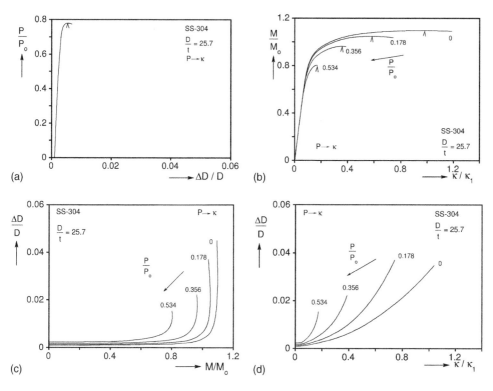

Figure 9.12 Calculated responses for the $P \to \kappa$ loading path: (a) pressure–ovalization, (b) moment–curvature, (c) ovalization–moment and (d) ovalization–curvature.

somewhat higher than the experiments. It will be illustrated below that the predictions can be improved by replacing isotropic hardening plasticity with a nonlinear kinematic hardening model.

The effect of the loading path on the collapse envelopes is clearly significant. The $P \to \kappa$ collapse envelopes are substantially lower than the $\kappa \to P$ envelopes for all D/t values considered. We reiterate that the collapse criteria adopted, despite being different for the two loading paths, are the ones deemed most appropriate for limit state design. The primary cause of the lower collapse envelopes for the $P \to \kappa$ loading history can be attributed to the more severe plastic state experienced in this path, as explained earlier.

Results for the radial path generated for tubes with $D/t = 35.0$ are shown in Figure 9.7 along with the three experimental data sets. For both the experiments and predictions distinction is made between points at which the limit moment (LM) and the actual collapse (CO) occurred. Once more, the limit state can be understood to be represented by the appearance of a limit moment in the response. Measured and predicted limit loads are seen to be in good agreement. By contrast, for the two lower pressure cases, the predicted collapse is delayed. This is to be expected as the BEPTICO uniform tube analysis precludes localization, which can delay the onset of collapse. It is also interesting to observe that the radial path yields critical values that are somewhat lower than the $P \to \kappa$ path. However, in most practical applications of interest, pipes are either bent in air or at a given water depth. As a result, the $P \rightarrow \kappa$ path is considered the most practically relevant way of testing pipe under this combined loading.

9.5 FACTORS THAT AFFECT COLLAPSE

9.5.1 Effect of Hardening Rule

As pointed out earlier, the $\kappa \rightarrow P$ loading path can involve unloading from a plastic state and reloading. It was found that nonlinear kinematic hardening captures the results from this loading history more closely than isotropic hardening. This is illustrated by substituting isotropic hardening plasticity with the Drucker-Palgen kinematic hardening model in the analysis (Section 13.4.1). The model was implemented using the Ziegler and the two-surface hardening rules (the two-surface version of the model uses the Mroz hardening direction as described in (Section 13.4.1).

The first observation is that ovalization induced by bending is sensitive to the hardening rule adopted in the constitutive model. This is illustrated in Figure 9.13, where the ovalization predicted by isotropic hardening and the two kinematic hardening models are compared. Isotropic hardening and the two-surface kinematic hardening models yield similar predictions, which are also seen to agree well with measured results. The ovalization predicted by the Ziegler hardening rule is substantially higher. The main difference between the two kinematic hardening models used is the direction of translation of the yield surface (see also [9.14, 9.15]).

The effect on bending followed by pressure is illustrated in Figure 9.14, where collapse envelopes from the three models are compared. The isotropic hardening envelope is somewhat higher than the experiments. The two-surface hardening rule yields the best overall comparison with the experiments, while the Ziegler hardening envelope underpredicts the experimental results significantly. The higher ovalization resulting from the latter model during the bending phase of the loading history leads to lower collapse pressures. Corona and Kyriakides [9.6] reported similar prediction trends for other D/t tubes. In summary,

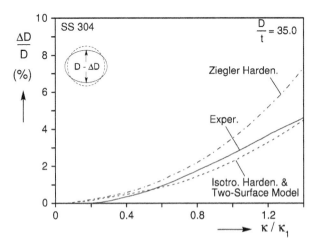

Figure 9.13 Measured ovalization in a pure bending experiment and predictions using three different hardening rules.

isotropic hardening plasticity is adequate for most bending-pressure loadings of interest. In cases where the pipe is bent plastically and then pressurized, a kinematic hardening model with a suitable hardening rule can yield better predictions of the critical states.

9.5.2 Bifurcation Buckling

For thinner pipes, bifurcation buckling in the form of wrinkling can precede the natural limit load induced by the Brazier effect under bending. This type of behavior was illustrated in a set of experiments conducted on Al-6061-T6 tubes with $D/t = 49.3$ by Ju and Kyriakides [9.9]. The major geometric and material parameters of this set of tubes are given in Table 9.1. The critical states measured for the $P \to \kappa$ and the $\kappa \to P$ loading paths are shown in the form of interaction envelopes in Figure 9.15. Bifurcation buckling was recorded for bending in the absence of external pressure and for bending under relatively low pressures. Figure 9.16 shows a set of collapsed tubes from this lower pressure regime. In the first four cases, the tubes collapsed by developing one sharp kink with a diamond mode of collapse. Tube kinking took place very soon after the appearance of wrinkles on the compressed side of the tube. As the pressure at which the tubes were bent was increased, the localized collapse spread over a larger section of the tube. In the case of the tube at the bottom of the photograph, the pressure was high enough to initiate a propagating collapse, which is seen to have spread over the RHS section of the tube (the propagation pressure of this tube was $P_P = 0.323 P_C$). All tubes tested at pressures equal to or higher than $P = 0.369 P_C$ developed similar propagating buckles.

Included in Figure 9.15 are predictions of the critical states for the two loading paths obtained by the solution procedures in Section 9.3. At lower pressures, a distinction is made between limit load and bifurcation buckling instabilities. At these pressures, bifurcation buckling precedes the limit moment instability as was observed in the experiments. At higher pressure levels, the calculated limit load instabilities agree very well with the measured collapse data.

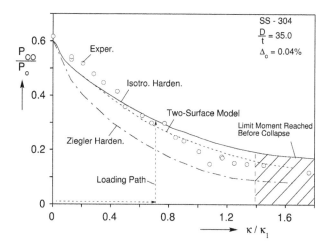

Figure 9.14 Experimental collapse envelopes and predictions using three hardening rules for tubes with $D/t = 35.0$ for the $\kappa \to P$ loading path.

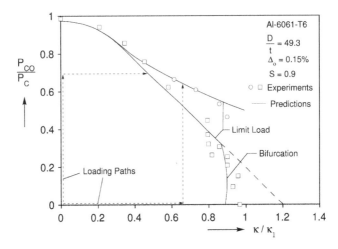

Figure 9.15 Experimental and predicted collapse envelopes for Al-6061-T6 tubes with $D/t = 49.3$ for the $P \to \kappa$ and $\kappa \to P$ loading paths exhibiting bifurcation instability at high curvatures.

Figure 9.16 Al-6061-T6 tubes with $D/t = 49.3$ collapsed by bending and pressure. Pressure increases from top to bottom.

In summary, for relatively high D/t pipes, bifurcation buckling can replace the limit load instability as the limit state and must be considered in the analysis of the structures.

9.5.3 Effect of Residual Stresses

The minor axis of the oval-shaped cross section of a tube under bending usually lies in the plane of bending. External pressure enhances the growth of this ovalization and leads to the eventual collapse of the tube. In the bending-pressure facility shown in Figure 9.2, tubes are bent in a vertical plane, and as a result the collapse is normal to this plane, as shown in Figure 9.17(a). In the course of the experimental program reported in [9.5, 9.6], several tubes collapsed in the plane of bending as shown in Figure 9.17(b), that is, in a direction orthogonal to the one expected.

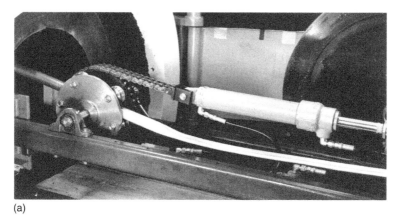

(a)

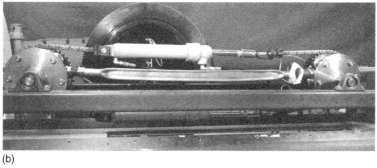

(b)

Figure 9.17 (a) Expected collapse mode and (b) reverse collapse mode under combined bending and pressure.

Pressure-curvature interaction collapse data, for the $\kappa \to P$ loading path, obtained from tubes that exhibited this behavior, are shown in Figure 9.18(a) (mean geometric and material parameters are listed in Table 9.1). The mode of collapse is identified by the line drawn through the experimental points. The unusual behavior occurred for bending curvatures $\kappa/\kappa_1 < 0.3$. In this regime, the collapse pressure measured under increasing prescribed values of curvature first decreased, then increased, and then continued to decrease in the general fashion reported earlier. This behavior was found to be repeatable in a second group of experiments carried out in this curvature regime, shown in Figure 9.18(b) [9.5]. Measurements revealed that in the curvature regime in question, the tubes exhibited reverse ovalization during bending, i.e., the major axis of the deformed cross section was in the plane of bending. For higher values of curvature, the induced ovalization returned to the normal type.

Careful consideration of all the problem parameters revealed that bending-induced reverse ovalization, similar to that observed in the experiments, can be caused by the presence of a circumferentially uniform, bending-type residual stress. The sense of the residual bending stress required is such that if the tube is cut along a generator, it would spring open as shown in Figure 9.19(a). The amplitude of reverse ovalization and the range of curvature in which it appears depend on the strength of the residual stress field (i.e., σ_R/σ_o). Experiments carried out on the tubes used in the collapse tests verified that the major component of the residual stress field was of this type.

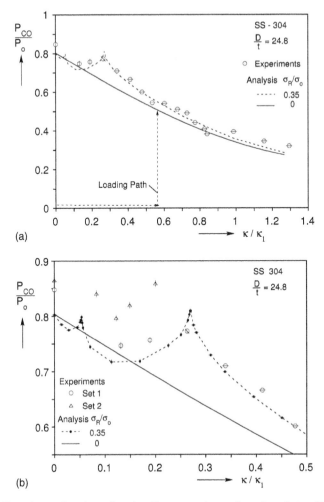

Figure 9.18 Experimental and predicted collapse envelopes for tubes that exhibited the reverse collapse mode ("|" indicates the mode of collapse): (a) complete range and (b) close-up of the loading regime where the reverse mode occurred.

The effect this type of residual stress has on the ovalization induced by bending is illustrated in Figure 9.19(b) for $\sigma_R/\sigma_o = 0$ and 0.35. Plotted are the changes of the diameters of the tube in the plane of bending ($\Delta D_1/D$) and normal to it ($\Delta D_2/D$), as a function of curvature. In this case, κ is normalized by the curvature when the tube first yields (κ_o). In the absence of the residual stress field, the ovalization follows the expected trend. When the residual stress is added, the ovalization initially grows normally, as the material is elastic. However, as the tube is plasticized by bending, the reverse ovalization starts to develop. The plastic deformation is more pronounced at the extremities of the cross section ($\theta = 0$ and π). Here the material stiffness is reduced, which allows the residual stress field to induce reverse ovalization. The ovality reaches a maximum value for $\kappa \approx 3\kappa_o$ and reverts back to the usual mode for $\kappa > 5.12\kappa_o$ because the Brazier effects negates the effect of the residual stress.

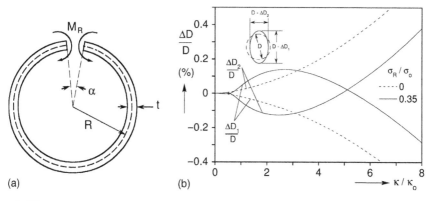

(a)   (b)

Figure 9.19   (a) Bending-type residual stress field. (b) Predicted ovalization as a function of curvature with and without residual stress.

Figure 9.18(a) shows predictions of the collapse envelopes of the tubes tested for $\sigma_R = 0$ and $0.35\sigma_0$. Figure 9.18(b) shows the same results plotted in a view zoomed for clarity in the low curvature regime where the phenomenon is exhibited. In the presence of the residual stress field, the collapse envelope exhibits two cusps in this region. The predicted collapse mode for curvature values between the two cusps was found to be of the unusual type, as indicated by the vertical lines drawn through the solid bullets of the predictions. All experimental points that exhibited the reverse collapse mode are also seen to lie within the two cusps. Outside this regime, the usual mode of collapse was predicted. Overall, the agreement between the experimental and predicted results is good for all values of curvatures. In the absence of residual stress, the unusual features are not present.

9.5.4 Asymmetric Modes of Collapse

Recent applications involving pipelines for water depths in the range of 5,000–10,000 ft (1,500–3,000 m) have motivated several experimental programs aimed at establishing the bending capacity of large-diameter pipes in the presence of external pressure. The experiments involved UOE pipes with diameters ranging between 20 and 28 inches tested following the $P \rightarrow \kappa$ loading path. In the course of these experiments, several pipes collapsed with the major axis of the heavily ovalized cross section lying at an angle to the axis of bending. In such cases, the measured critical curvature exceeded expected values. Such anomalous buckling modes were also observed in experiments on small scale tests on stainless steel 304 tubes. An investigation of this phenomenon was reported in [9.10].

The small scale tests involved SS-304-W tubes with $D/t = 18.5$. These particular tubes were formed from plate, welded, and then finished by drawing through a die. As a result, they had very uniform wall thickness and very small initial ovalities. The mean geometric and material parameters of the tubes are listed in Table 9.1. Included are the amplitude of the bending residual stress field ($\bar{\sigma}_R = \sigma_R/\sigma_0$) and the small anisotropy found in the tubes. Seventeen tubes were tested using the $P \rightarrow \kappa$ loading path for pressures $P/P_o > 0.5$. The results are summarized in the interaction collapse envelope in Figure 9.20 (numerical values of the critical states appear in Table 9.1 of Ref. [9.10]). Ten of the tubes, tested at pressures higher than $0.7P_o$, collapsed such that the flattening direction was at an angle

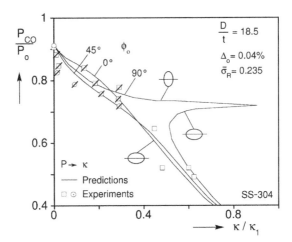

Figure 9.20   Calculated $P \rightarrow \kappa$ collapse envelopes including residual stress for three initial ovality
orientations and experimental results.

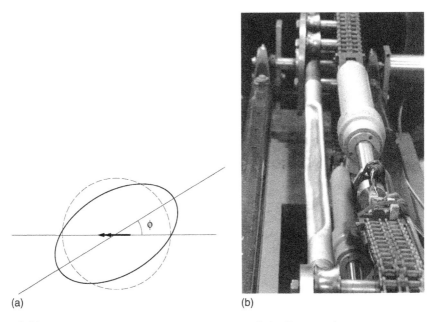

(a)   (b)

Figure 9.21   (a) Ovality orientation and (b) tube with angled collapse mode.

(ϕ_C) to the bending axis, as shown in the photograph in Figure 9.21 (data points marked
with slash in Figure 9.20). The angle ϕ_C varied between 20° and 45°.

The version of the formulation that allows for a general imperfection of the cross section
to be considered (Sections 8.3, 9.3), was used to examine the effect of the orientation of
initial ovality on the critical bending curvature. Collapse envelopes calculated with the
properties in Table 9.1 for initial ovality orientations (ϕ_o) of 90°, 85°, 70°, 45° and 0°
are shown in Figure 9.22(a). For $\phi_o = 0°$, the collapse envelope follows the usual trend

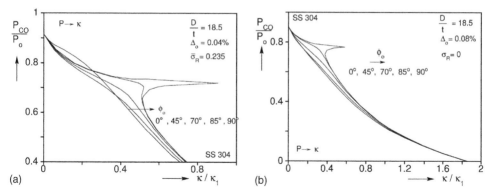

Figure 9.22   Calculated $P \to \kappa$ collapse envelopes for various initial ovality orientations: (a) with residual stress and (b) without residual stress.

although for $P/P_o > 0.7$ the residual stresses distort it to some degree. When $\phi_o = 90°$, the ovality is orthogonal to the orientation usually induced by bending. On the application of pressure its amplitude grows further. Bending initially makes the tube cross section more circular and, as a result, at these pressures collapse occurs at higher values of curvature. This results in the formation of a cusp in the collapse envelope. For the upper part of the cusp, bending does not negate the initial ovality and the tube buckles with $\phi_C = 90°$. For the lower part of the cusp, bending overcomes the initial ovality and the tube buckles with $\phi_C = 0°$. We note that the perfectly aligned and axially uniform initial ovality assumed is an idealization, which is partly responsible for the sharp cusp. In reality, initial imperfections are more complex than the $w_o = a_2 \cos 2\theta$ shape assumed here, and their amplitude and orientation may vary to some degree along the length of the pipe tested. Such effects will tend to round and reduce the extent of the cusp.

The envelope for $\phi_o = 85°$ is seen to follow that of $\phi_o = 90°$, but the small angular deviation results in a truncation and smoothing of the region around the cusp. Reduction of ϕ_o to 70° and 45° results in a further shift of the envelope to the left. In fact, for this particular set of parameters and for high pressures, $\phi_o = 45°$ is close to the lower bound (this is mainly due to the influence of the residual stress field). The effect of the orientation of the initial ovality is reduced significantly at lower pressures, and the collapse envelopes for all cases approach each other. This is illustrated in Figure 9.22(b), where the complete collapse envelopes are shown for the same values of ϕ_o (note that $\Delta_o = 0.08\%$).

The very small value of initial imperfections in the tubes tested, and variations along the length of each specimen, prevented us from establishing initial orientations of the imperfections. For this reason, in Figure 9.20 the experimental results are compared with three calculated envelopes for $\phi_o = 0°$, 45° and 90°, which approximately represent bounds of the predictions in Figure 9.22(a). With the exception of four points in the high pressure regime, the experimental results are seen to be mainly contained within the three envelopes. This trend, coupled with the observed angled collapse modes obtained in the experiments, indicates that the scatter in the experimental results is mainly due to the orientation of the initial imperfections.

The effect of the amplitude of the residual stress field $\bar{\sigma}_R$ (acting as in Figure 9.19(a)) on the results is illustrated in Figure 9.23(a). The collapse envelopes for $\phi_o = 0°$ and 90° are compared for $\bar{\sigma}_R = 0$, 0.235 and 0.355. In the absence of residual stresses, the

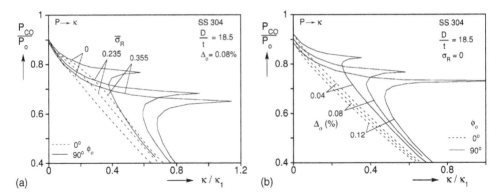

Figure 9.23 Calculated $P \to \kappa$ collapse envelopes for $\phi_o = 0$ and 90°: (a) effect of residual stress and (b) effect of initial ovality.

two envelopes do not intersect. Increasing the residual stress causes the two envelopes to cross at high pressures, as the envelope for $\phi_o = 0°$ is distorted and the one for $\phi_o = 90°$ is lowered. In addition, the cusp in the outer envelope extends to higher curvatures when the residual stress is increased.

Figure 9.23(b) shows how the two extreme ovality orientation envelopes are affected by the magnitude of the initial ovality for $\sigma_R = 0$. Results for $\Delta_o = 0.04\%$, 0.08% and 0.12% are compared. Increasing the initial ovality lowers both envelopes, while at the same time the cusp in the outer envelope extends to higher values of curvature.

Angled collapse modes, such as the ones encountered in the small scale tests, have also been observed in full scale tests on large diameter UOE pipe [9.16]. This test program involved several pure external pressure collapse tests (EPx in Table 5.2) and eight tests in which pipes were bent to collapse under relatively high values of external pressure (Table 9.2, AR). In pipes manufactured by this process, the plane passing through the seam and the pipe axis is usually a plane of symmetry. In four of the tests, the seam was placed in the plane of bending and in the other four at 90° to it. All specimens collapsed such that the major axis of collapse was at an angle ϕ_C that varied from 5° to 20°. Furthermore, in all cases the bending curvature was generally higher than expected from calculations in which the ovality orientation was placed in the unfavorable direction ($\phi_o = 0°$). The experimental results are shown in the form of $P \to \kappa$ interaction envelope in Figure 9.24. Included are predictions using the mean geometric and material properties of this set of pipes listed in Table 9.1. Predictions for the two extreme ovality orientations of $\phi_o = 0°$ and 90° are shown. All but two of the experimental points are within the two envelopes, most closer to the one corresponding to $\phi_o = 90°$. (It is noted that individual predictions of the critical states were generally closer to the measured values because of variations in geometric and material parameters.)

In summary, the trend of the full scale results is generally similar to that of the small scale tests in Figure 9.20. Both reinforce the fact that for pipes bent under high pressure, ovality orientation and residual stresses can influence the critical curvature. A conservative design approach is to orient the ovality in the most favorable orientation in the analysis ($\phi_o = 0°$) and the residual stress in the sense shown in Figure 9.19(a). More details on the collapse of UOE pipe will be given in Section 9.6.

Table 9.2 Pipe parameters and collapse loads for 24-inch pipes bent to collapse under constant external pressure [9.16].

No.	13P	14P	15P	16P	17P	18P	19P	20P	3P	6P
D in	24.047	24.024	24.028	24.019	24.016	24.019	24.037	24.029	24.014	24.026
t in	1.250	1.2516	1.2503	1.249	1.251	1.249	1.250	1.248	1.247	1.250
D/t	19.24	19.19	19.22	19.23	19.25	19.23	19.23	19.29	19.26	19.22
L/D	12.84	12.86	12.85	12.85	12.86	12.86	12.85	12.86	12.86	12.84
Δ_o (%)	0.144	0.237	0.172	0.139	0.181	0.156	0.139	0.205	0.188	0.188
Condition	AR	AR	AR	AR	AR	AR	AR	AR	HT	HT
σ_{oO} ksi	−55.9	−61.1	−55.4	−56.8	−59.5	−58.9	−58.7	−61.6	−70.1	−71.9
σ_{ol} ksi	−63.2	−69.2	−62.9	−64.0	−67.9	−67.7	−66.9	−68.2	−76.0	−78.6
σ_{ox} ksi	74.6	78.9	74.3	75.6	77.3	76.5	76.1	76.4	77.9	80.5
S	0.822	0.827	0.799	0.796	0.831	0.817	0.822	0.847	0.931	0.941
$|\sigma_R|$ ksi	1.43	0.84	5.16	4.54	1.70	6.95	0.74	2.93	2.57	6.1
P psi	4,400	4,200	3,800	4,200	4,800	4,500	4,700	4,000	5,300	5,100
ε_{bL} %	0.665	0.918	0.912	0.786	0.471	0.469	0.571	0.876	0.478	0.749

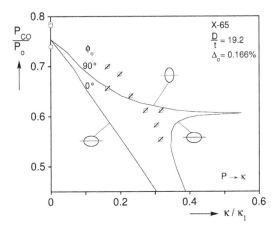

Figure 9.24 Comparison of predicted collapse envelopes for $\phi_o = 0$ and $90°$ to results of full-scale tests.

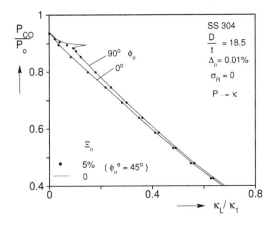

Figure 9.25 Comparison of collapse envelopes for $\phi_o = 0$ and $90°$ with and without wall eccentricity.

9.5.5 Effect of WallThickness Variations

Seamless pipe usually exhibits some thickness variation around the circumference introduced by the manufacturing process (see Chapter 2). A common type of wall thickness variation is eccentricity between the centers of the inner and outer surfaces of the pipe, as shown in Figure 4.11(b). The eccentricity can be described by Eq. (4.29), where $\Xi_o = \xi_o/t$ (4.26). We examine how this type of imperfection affects the interaction between bending and pressure loads and how it influences buckle orientation. In order to reduce the dominating effect of initial ovality, Δ_o will be reduced to 0.01%, but a wall eccentricity of $\Xi_o = 5\%$ will be added. The remaining variables are taken from Table 9.1 ($D/t = 18.5$).

Four interaction collapse envelopes calculated for this set of parameters are shown in Figure 9.25. The lines represent cases with uniform wall thickness and ovality orientations

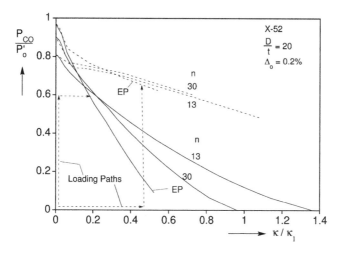

Figure 9.26   Effect of hardening on the predicted collapse envelopes for X52 tubes with $D/t = 20$.

of $\phi_o = 0°$ and $90°$. The solid dots represent corresponding results for a tube with the same ovality value and orientations, but with the addition of wall eccentricity of $\Xi_o = 5\%$. Furthermore, in an attempt to increase the effect of this eccentricity its axis of symmetry is oriented at $45°$ to the axis of bending (ϕ_o^e).

The collapse pressure of the tube with uniform wall thickness is $P_{CO} = 0.941P_o$. When the wall eccentricity is added, the collapse pressure is reduced by less than 1% to $0.935P_o$. Thus, consistent with the results in Section 4.4.5, the effect of this level of eccentricity on P_{CO} is small.

For the tube with uniform wall thickness, the $P \rightarrow \kappa$ interaction collapse envelopes are similar to those presented earlier, except that the difference between $\phi_o = 0°$ and $90°$ is smaller, and the extent of the cusp is less pronounced and limited to higher pressure levels. The wall eccentricity is seen to have a negligible influence on the lower envelope. Its main effect on the envelope for $\phi_o = 90°$ is a rounding of the cusp region. The most interesting consequence of wall eccentricity is that for pressures higher than approximately $0.85P_o$, the buckles were angled. Accordingly, although the effect of this imperfection on the interaction collapse envelopes is relatively small, at high pressures it can be another source of angled collapse modes.

9.5.6   Effect of Material Stress–Strain Response

As was the case for collapse under pure external pressure, the shape of the stress–strain response influences the bending–pressure interaction collapse envelopes. This is demonstrated for pipes with $D/t = 20$ and the three X52 stress–strain responses shown in Figure C.2. Figure 9.26 shows calculated pressure–curvature interaction collapse envelopes for the three material responses for both the $P \rightarrow \kappa$ and $\kappa \rightarrow P$ loading paths. In this figure, the applied pressure is normalized by the yield pressure (P_o') based on the common value of σ_o' (52 ksi). There are no primes in the Figures 9.26 and 9.27. Collapse under pure external pressure is as discussed in Section 4.4.6. The elastic-perfectly plastic response yields the highest collapse pressures, and the one with $n = 13$ the lowest. As bending increases, the $\kappa \rightarrow P$ envelopes are relatively insensitive to the hardening exponent. By

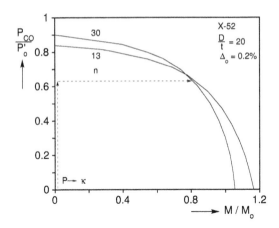

Figure 9.27   Effect of hardening on $\kappa \rightarrow P$ moment–pressure collapse envelopes for X52 tubes with $D/t = 20$.

contrast, the $P \rightarrow \kappa$ envelopes are significantly affected by material hardening. Away from the very high pressure regime, the limit moment instability occurs at lower curvatures as the hardening decreases, which is consistent with the pure bending results in Figure 8.21. Note that the EP material response assumed is an idealization of hot-finished line grade steels, where yielding is followed by Lüders banding behavior. Neglecting this effect is permissible for higher pressure/low bending loadings. For bending under low pressure levels, prediction of collapse requires that the Lüders bands be accounted for (e.g., see [9.17]).

In some load-controlled applications the moment–pressure collapse envelopes may replace the commonly used curvature–pressure envelopes discussed this far. Figure 9.27 shows such envelopes calculated for the $P \rightarrow \kappa$ loading path for the same X52 pipe as in Figure 9.26, for $n = 13$ and 30. Collapse again represents limit pressures and limit moments. The interaction here is of the classical quadratic type, which can be approximated by

$$\left(\frac{P}{P_{CO}}\right)^2 + \left(\frac{M}{M_L}\right)^2 = 1, \tag{a}$$

where P_{CO} and M_L are the critical quantities for the corresponding pure external pressure and bending loadings.

9.5.7   Effect of Anisotropic Yielding

The simplest type of anisotropy that can affect the bending–pressure interaction collapse envelope is the type represented by

$$S_r = S_\theta = S = \frac{\sigma_{o\theta}}{\sigma_{ox}}. \tag{b}$$

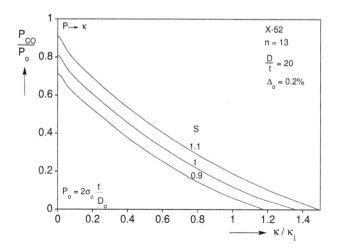

Figure 9.28   Effect of yield anisotropy parameter S on the $P \rightarrow \kappa$ collapse envelopes for X52 tubes with $D/t = 20$.

The yield function appropriate for the present problem reduces to

$$ f = \left[\sigma_x^2 - \sigma_x \sigma_\theta + \frac{1}{S^2}\sigma_\theta^2 \right]^{\frac{1}{2}} = \sigma_{e\,\text{max}}. \qquad (c)$$

A difference in yield stress between the axial and circumferential directions affects the collapse pressure as well as the bending-induced ovalization. Consequently, the interaction collapse envelopes are also affected. Figure 9.28 shows the effect of parameter S on the $P \rightarrow \kappa$ collapse envelope of an X52 pipe with $D/t = 20$. In order to best demonstrate the effect of S on the results, pressure is normalized by $2\sigma_o(t/D_o)$. $S = 1$ represents isotropic yielding. When $S > 1$, both the collapse pressure under pure external pressure and the limit curvature under pure bending increase, and the collapse envelope shifts outwards. When $S < 1$, the collapse pressure and the limit curvature under pure bending both decrease, and the collapse envelope shifts inwards.

9.6   COLLAPSE OF UOE PIPE BENT UNDER EXTERNAL PRESSURE

In Chapter 5 it was demonstrated that the UOE pipe manufacturing process introduces changes to the material properties that tend to reduce the collapse pressure of the pipe. The property changes also affect the response and stability of such pipe under combined bending and external pressure. Because of the importance of this loading to deep water applications, several full-scale experimental programs have been performed during the last decade aimed at understanding differences in performance vis-à-vis seamless pipe. Results from two such experimental programs are reviewed. Included is also an outline of a procedure for calculating the extent to which such pipe can be bent under external pressure.

9.6.1 Experiments

A set of eight 26-inch pipes were tested in 1994 under combined bending and external pressure in an experimental program undertaken in support of the Oman-India pipeline project [9.18]. The tests were performed in the combined loading testing facility first developed for this project by C-FER Technologies of Edmonton, Calgary (see Section 9.2.1 and Figure 9.3). The pipes were manufactured by British Steel in their pipe mill in Hartlepool (presently Corus Tubes). They were nominally X65 grade, with a wall thickness of 1.625 in (41.3 mm). 13.3D long pipe specimens were loaded by four-point bending inside the 48 in (1219 mm) internal diameter pressure chamber shown schematically in Figure 9.3(a). In this setup, the length of pipe under pure bending was 10D. The loading followed $P \to \kappa$ paths. The chamber was filled with water and pressurized to the desired level. The pressure was kept fixed as the specimen was gradually bent by controlling the rotation of the ends. The end rotations were monitored by transducers. In addition, the bending strain was measured by strain gages placed on the pipe. Displacement transducers placed along the length monitored the vertical displacement of the top and bottom surfaces of the deforming pipe.

The geometry of each specimen was measured in the manner described in Section 5.2. The average diameter, wall thickness and maximum ovality of each specimen is listed in Table 9.3. As is typical for UOE pipe, the wall thickness was quite uniform. The maximum ovality varied between 0.18% and 0.235%, confirming that the pipe can be manufactured to low ovality even for thicknesses above 40 mm. Tensile and compressive stress–strain properties were measured using specimens extracted from the axial and circumferential directions. Four specimens came from locations close to the inner surface of the pipe and four as close to the outer surface as possible (see Figure A2(b) for circumferential specimens). Figure 5.1 shows two typical circumferential compression responses. The cold work, and in particular the final expansion, caused significant rounding of the response and lowering of the yield stress (see Chapter 5 for causes). By contrast, the corresponding tensile responses, shown in Figure 5.2, exhibit some strengthening. In the axial direction, the difference between tension and compression was not so significant. Figure 9.29 shows the responses measured from specimens extracted from a location close to the inside wall (A-T-I) and close to the outer wall (A-T-O, see inset). Both responses exhibit some rounding, with the outer response having a somewhat lower yield stress. At the same time, the yield stresses are significantly higher than those of the two circumferential compression specimens in Figure 5.1. This general pattern was observed for all pipes in the set.

The circumferential compression yield stresses from the inside and outside specimens are listed in Table 9.3 under σ_{oI} and σ_{oO}, respectively. σ_{oO} varies between 47.9 ksi (330 MPa) and 63.8 ksi (439 MPa), with the average being 56 ksi (386 MPa), a value significantly lower than the nominal pipe yield stress of 65 ksi (448 MPa). σ_{oI} varies between 51.4 ksi (354 MPa) and 74.6 ksi (514 MPa) with an average of 63.3 ksi (437 MPa). In the axial direction, the average yield stress of the four measured responses is listed under $\bar{\sigma}_{ox}$. It varies between 73.3 ksi and 81.5 ksi (506–562 MPa) with the average being 77 ksi (530 MPa). In addition, the axial responses had a much higher elastic limit and a sharper knee at yield.

Two of the pipes were collapsed under pure external pressure (LP loading for ZFV5). The collapse pressures were 7,144 psi (492.7 bar) and 7,117 psi (490.1 bar). The collapse pressure of pipes with the same geometry but assigned the properties of the plate

Table 9.3 Pipe parameters and collapse loads for 26-inch pipes bent to collapse under constant external pressure [9.18].

No.	ZFV 14	ZFV 5	ZFV 18	ZFV 20	ZFV 21	ZFV 22	ZFV 23	ZFV 7	ZFV 16	ZFV 19
D in	25.971	25.970	25.989	25.977	25.961	25.993	25.962	25.992	25.982	25.980
t in	1.6232	1.6328	1.6180	1.6210	1.6234	1.6220	1.6277	1.6160	1.6195	1.6200
D/t	16.03	15.91	16.06	16.03	16.00	16.03	15.96	16.08	16.04	16.04
L/D	13.3	14.5	13.3	13.3	13.3	13.3	13.3	13.3	13.3	13.3
Δ_o (%)	0.20	0.18	0.204	0.216	0.189	0.194	0.196	0.162	0.193	0.235
Condition	AR	AR	AR	AR	AR	AR	AR	AR	AR	HT
$|\sigma_R|$ ksi	10.0	11.1	9.5	10.8	8.35	8.8	6.0	8.6	7.67	0
σ_{oO} ksi	−54.5	−54.4	−47.9	−63.0	−57.0	−55.7	−49.9	−58.0	−63.8	−69.0
σ_{ol} ksi	−71.9	−61.6	−51.4	−60.9	−68.6	−61.5	−60.9	−58.0	−74.6	−78.0
$\bar{\sigma}_{ox}$ ksi	73.3	–	75.3	78.2	80.2	76.75	75.75	75.0	81.55	75.78
S	0.867	–	0.659	0.793	0.782	0.764	0.731	0.773	0.849	0.97
P psi	7,144	7,117†	6,300	5,000	6,800	5,500	5,000	5,400	6,700	6,500
ε_{bL} %	0	0	0.38	1.06	0.22	0.80	1.04	0.85	0.26	0.87

† Lateral Pressure Loading

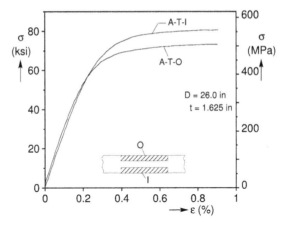

Figure 9.29   Axial tensile stress–strain responses of an X65 UOE pipe.

($\sigma_o = 70$ ksi) were calculated to be 9,568 psi (659.9 bar) and 8,751 psi (603.5 bar – see Section 5.2), respectively, illustrating the degrading effect of the cold forming of the pipe. Seven of the pipes were bent to collapse at fixed pressures ranging between 5,000 psi and 6,800 psi (344.8–469 bar). The pipes bent plastically, reached a limit moment and failed by local diffuse ovalization as shown in Figure 9.3(b). The average of the strains recorded for each pipe at the top and bottom surfaces of the pipe at the maximum moment is listed in Table 9.3 under ε_{bL}. The strains at collapse are plotted against the pressure in Figure 9.30. Despite the relatively large variation in yield stresses between the nine specimens, the results follow a nearly linear trajectory. This indicates that the average mechanical properties of the pipes varied less than what is indicated in Table 9.3. As mentioned in Section 3.2, plate properties vary to some degree, more so at the ends of the plates. The uniaxial test specimens were extracted from rings cut off the ends of the pipes. It is believed that this may be contributing to a relatively large property variation in the measured yield stresses.

Pipe ZFV19 was heat-treated before testing, and as a result σ_{oI} and σ_{oO} increased significantly. Because of this recovery in yield stress, the pipe could be bent to a larger strain at a higher pressure, as reflected in Figure 9.30.

A second set of full scale bending under external pressure tests on UOE pipe was conducted in 1999 in support of the Blue Stream pipeline project in the Black Sea [9.16]. The project had maximum depth of 2,150 m (7,050 ft). The pipes tested were nominally X65 grade and were manufactured by Europipe. They had diameters of 24 in and wall thicknesses of 1.252 in (31.8 mm) ($D/t = 19.2$). These tests were also performed in the C-FER testing facility using the same general testing procedures as those described above. Similar material property measurements were also performed. The results of this project are summarized in [9.16], while the major test parameters are also listed in Table 9.2 (see Table 5.2 for pure external pressure collapse tests). The maximum ovalities ranged between 0.14% and 0.24%. The degradation in compressive properties is similar to that reported in Chapter 5 and in the results in Table 9.3.

In these tests, the orientation of the pipe cross section with respect to the plane of bending was varied, breaking the symmetry about the plane of bending (in several specimens the

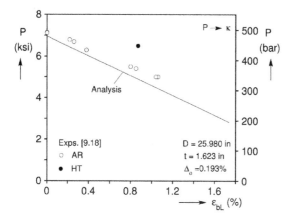

Figure 9.30 Measured and predicted $P \rightarrow \kappa$ collapse envelopes for 26-inch UOE pipe (experimental data from [9.18]).

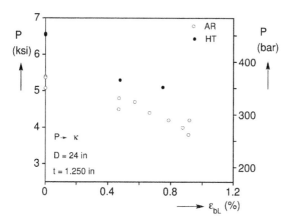

Figure 9.31 Measured $P \rightarrow \kappa$ collapse envelopes for 24-inch UOE pipe (from [9.16]). Included are as-received and heat treated pipe results.

seam was placed at 90° to the plane of bending). In addition, the manufacturing process introduced some mild asymmetries to the cross section. As a result, all pipes collapsed so that the major axis of the cross section was at an angle to the plane of bending. The consequences of this are discussed in Section 9.5.4.

Several pipes were heat-treated using the low-temperature process described in Section 5.3 (see also [9.16]). As evidenced by the results in Table 9.2, the heat treatment resulted in significant recovery in compressive strength. The pipes were subsequently collapsed under external pressure and under combined external pressure and bending. The heat-treated (HT) results are compared to those from the as-received pipes in Figure 9.31. As demonstrated in the figure, the improvement in strength increased the pressure and bending capacity of the pipes, therefore causing an upward shift of the data. Thus, this type of heat treatment has the potential of alleviating at least part of the reduction in pipe performance caused by the UOE process. It is once more pointed out that the degree to which the mechanical properties recover by the strain aging process depends on the temperature level, the duration and on the alloy content of the material. It is thus imperative that the heat treatment cycle be custom designed to the needs of each project.

9.6.2 Analysis

The formulation and solution procedure outlined in Section 9.3 can be used to predict the extent to which such pipe can be bent under external pressure. The model includes circumferential-type residual stresses and allows for anisotropic yielding, but otherwise the material is assumed to be in a virgin state. We have seen that the UOE forming process introduces significant changes in the stress–strain responses. Such changes can only be captured by first simulating the forming and locking in the induced stress states and histories at each integration point, as is done in the 2-D forming model presented in Section 5.5. With these in place, the pipe can then be pressurized and bent to collapse. Such simulations require knowledge of the plate properties, the pipe mill forming parameters, and other details that are not available to a design engineer. This prompted the development of the following approximate scheme for establishing the collapse loads of as-received UOE pipe under combined bending and pressure.

The pipe geometry is measured as described in Section 5.2. D, t and Δ_o are evaluated from these measurements. The compressive stress–strain responses of the pipe in the circumferential direction are measured as described in Appendix A. The tensile response of the pipe material in the axial direction is measured using either a full thickness specimen or two specimens close to the inner and outer surfaces of the pipe. The circumferential compressive responses are averaged and compared to the axial one (averaged if more than one). Such a comparison is made in Figure 9.32, which shows the average of the two tensile axial stress–strain responses for ZFV14 and the average of the compressive circumferential responses. Clearly, not only is the yield stress in the hoop direction lower, but the shapes of the two responses are also different. We choose the axial response to represent the fundamental response of the material. The softer behavior in the circumferential direction is introduced by an anisotropy of the type

$$S = S_\theta = \frac{\sigma_{o\theta}}{\sigma_{ox}},$$

where σ_{ox} and $\sigma_{o\theta}$ are the stresses at strain offsets of 0.2% of the average axial and average circumferential responses of the pipe, respectively. For the results in Figure 9.32, $S = 0.867$. This procedure was used to generate the anisotropy parameters for the pipes listed in Tables 9.2 and 9.3 under the variable S.

The adequacy of this scheme was evaluated by individually simulating the experiments listed in Table 9.3. In general, the predictions exhibited some scatter, while the limit strains were somewhat lower than the measured values. The scatter was caused by the significant variation in the material properties of the pipes in the set mentioned earlier. The predicted limit strains were lower than the measured ones primarily because the ovality used was the maximum measured in the pipe and was oriented in the worst sense. More representative predictions were made by averaging all the material and geometric parameters of the eight pipes in the set. This included generating an average stress–strain response from the eight

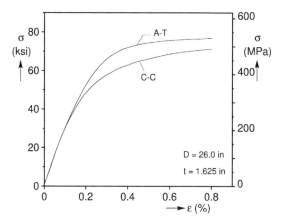

Figure 9.32   Comparison of the average axial tensile and compressive circumferential responses for a 26-inch UOE pipe.

axial responses of the combined loading specimens. The response was found to be well represented by a Ramberg–Osgood fit with the following parameters

$$E = 30 \times 10^3 \, \text{ksi}, \sigma_y = 74.0 \, \text{ksi} \quad \text{and} \quad n = 20.$$

The average anisotropy was found to be 0.778. The rest of the averaged parameters appear in the last row of Table 9.1.

These parameters were used to generate a $P \rightarrow \kappa$ collapse envelope in the pressure regime of the experiments that is included in Figure 9.30. The calculated envelope is somewhat lower than the experiments, but follows their linear trend closely. The same scheme was used to generate the average properties of the 24-inch pipes in Table 9.2. The averaged parameters listed in Table 9.1 were used to generate the two bounding collapse envelopes included in Figure 9.24. Overall, the averaging scheme can be declared successful and dependable.

9.7   CONCLUSIONS AND RECOMMENDATIONS

In Chapter 4 it was demonstrated that initial ovality can significantly reduce the collapse pressure of a long tube, even at relatively small values. In Chapter 8 it was shown that bending a tube ovalizes its cross section and induces a limit load instability. Hence, when the two loads are applied together, they interact strongly through ovalization. Consequently, bending a tube under pressure results in a lower collapse curvature. Correspondingly, pressurizing externally a bent tube results in a lower collapse pressure. Although wrinkling bifurcation remains possible for relatively high D/t tubes bent under relatively low external pressure, for commonly used pipeline geometries limit load instabilities dominate the behavior. These can be evaluated by a uniform ovality model such as the one presented in Section 9.3, or by a corresponding FE model. The following additional points derive from the work outlined in the chapter:

- The limit state depends on the loading path followed. For example, bending under constant pressure is more severe than pressurizing a pre-bent pipe.

- Initial ovality can lower the interaction collapse envelope. For design calculations, ovality should be oriented so that the minor axis is in the plane of bending (worst orientation).
- The pipe yield stress in the circumferential direction can differ (often is lower) from that in the axial direction. Such anisotropy influences the interaction collapse envelope and must be included in the analysis.
- Wall thickness variations do not significantly influence the interaction collapse envelopes. Eccentricity (Ξ_o) of less than 10% is desirable.
- Circumferential bending-type residual stress fields can influence the collapse performance at higher pressures.
- The shape of the stress–strain response can influence the collapse envelope. Pipes with more material hardening can be bent to a larger curvature before collapse.
- UOE pipe has lower collapse pressure and curvature than seamless pipe of the same D/t and steel grade. This is caused by the degradation in the circumferential yield stress in compression resulting from the UOE cold forming process. This degradation in yield stress is introduced as a yield anisotropy in numerical simulations.

REFERENCES

9.1. Murphey, C.E. and Langner, C.G. (1985). Ultimate pipe strength under bending, collapse and fatigue. *Proc. 4th International Conference on Offshore Mechanics and Arctic Engineering*, Vol. 1, Houston, TX, 467–477.

9.2. Kyriakides, S., Corona, E., Madhavan, R. and Babcock, C.D. (1989). Pipe collapse under combined pressure, bending and tension loads. *Proc. Offshore Technology Conference* 1, OTC6104, 541–550.

9.3. Johns, T.G., Mesloh, R.E., Winegardner, R. and Sorenson, J.E. (1975). Inelastic buckling of pipelines under combined loads. *Proc. Offshore Technology Conference* II, OTC2209, 635–646.

9.4. Kyriakides, S., Corona, E., Madhavan, R. and Babcock, C.D. (1987). *Factors Affecting Pipe Collapse–Phase II*. Final Report to the American Gas Association PR-106-512, Engineering Mechanics Research Laboratory Report No. 87/8, December 1987.

9.5. Corona, E. and Kyriakides, S. (1987). An unusual mode of collapse of tubes under combined bending and pressure. *ASME J. Pressure Vessel Technol.* 109, 302–304.

9.6. Corona, E. and Kyriakides, S. (1988). On the collapse of inelastic tubes under combined bending and pressure. *Int. J. Solid. Struct.* 24, 505–535.

9.7. Corona, E. and Kyriakides, S. (1988). Collapse of pipelines under combined bending and pressure. *Proc. 5th International Conference on the Behaviour of Offshore Structures*, Vol. 3, Trondheim, Norway, June 1988, 953–964.

9.8. Fowler, J.R., Hormberg, B. and Katsounas, A. (1990). *Large Scale Collapse Testing*. Final Report to the American Gas Association PR-201-818, Stress Engineering Services, Inc., PN3667, June 1990.

9.9. Ju, G.-T. and Kyriakides, S. (1991). Bifurcation versus limit load instabilities of elastic–plastic tubes under bending and pressure. *ASME J. Offshore Mech. Arctic Eng.* 113, 43–52.

9.10. Corona, E. and Kyriakides, S. (2000). Asymmetric collapse modes of pipes under combined bending and external pressure. *ASCE J. Eng. Mech.* 126, 1232–1239.

9.11. Toscano, R.G., Dvorkin, E.N., Timms, C.M. and DeGeer, D.D. (2003). Determination of the collapse and propagation pressure of ultra-deepwater pipelines. *Proc. 22nd International Conference on Offshore Mechanics and Arctic Engineering*, June 8–13, Cancun, Mexico, ASME ISBN:0-7918-3672-X.

9.12. DeGeer, D., Marewski, U., Hillenbrand, H.-G., Weber, B. and Crawford, M. (2004). Collapse testing of thermally treated pipe for ultra-deepwater applications. *Proc. 23rd International Conference on Offshore Mechanics and Arctic Engineering*, June 20–25, Vancouver, Canada, ASME ISBN:0-7918-3672-X.

9.13. Kyriakides, S., Dyau, J.-Y. and Corona, E., (1994). *Pipe Collapse Under Bending, Tension and External Pressure (BEPTICO)*. Computer Program Manual, Engineering Mechanics Research Laboratory Report No. 94/4, January 1994.

9.14. Shaw, P.-K. and Kyriakides, S. (1985). Inelastic analysis of thin-walled tubes under cyclic bending. *Int. J. Solid. Struct.* **21**, 1073–1100.

9.15. Kyriakides, S., Corona, E. and Miller, J.E. (2004). Effect of yield surface evolution on bending induced cross sectional deformation of thin-walled sections. *Int. J. Plast.* **20**, 607–618.

9.16. DeGeer, D., Timms, C. and Lobanov, V. (2005). Blue Stream collapse test program. *Proc. 24th International Conference Offshore Mechanics and Arctic Engineering*, June 12–17, Halkidiki, Greece, Paper OMAE2005-67260.

9.17. Aguirre, F., Kyriakides, S. and Yun, H.D. (2004). Bending of steel pipes with Lüders bands. *Int. J. Plast.* **20**, 1199–1225.

9.18. Stark, P.R. and McKeehan, D.S. (1995). Hydrostatic collapse research in support of the Oman-India gas pipeline. *Proc. Offshore Technology Conference* **2**, OTC7705, 105–120.

10
Inelastic Response Under Combined Bending and Tension

Bending in the presence of tension is experienced by pipelines mainly during their installation, and in some special cases during their operation. For example, in the S-lay installation method the pipe is bent over the stinger while it is simultaneously subjected to significant tension (Figure 10.1). In conventional applications, the pipe is installed empty, reducing the required tension, while the angle at which the line leaves the stinger (*departure angle*) is relatively small. Under these conditions the pipe remains elastic, which is desirable for the stability of the line once it reaches the sea floor. In a few deep-water applications, pipelines have been installed flooded, increasing the required tension. In addition, in order to reduce the horizontal component of the force required from the lay vessel, the pipe departure angle is much larger and the stinger much longer, as shown in Figure 2.18 [10.1]. The combined effect of the relatively high tension and stinger curvature can plastically deform the line while on the stinger [10.2]. Once off the stinger, the line is under combined bending and tension, which is required in order to control its configuration, in particular the curvature of the sagbend.

In the reeling installation method shown in Figure 2.4, some tension is applied as the pipe is wound onto the reel (see Section 2.2.3). During unreeling, most of the tension is applied by the tensioner mounted on the ramp; in other words, after the pipe is straightened.

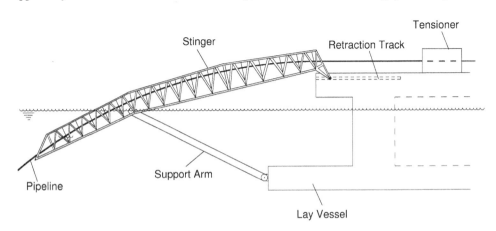

Figure 10.1 Conventional S-lay vessel stinger.

In deeper waters, however, the tension required may exceed the capacity of the tensioner, in which case part of the tension is reacted by the reel. In this case the pipe is straightened, bent again over the ramp, and straightened once more while under tension. On the way to the sea floor, tension is again used to control the curvature of the sagbend and remains high closer to the vessel. The particular installation process shown in Figure 2.4 includes pulling the end of the line through a J-tube to the top of a fixed platform [10.3]. In this case, the pipe is again plastically deformed by combined tension and bending while also under external pressure.

Bending ovalizes the pipe cross section, which in turn reduces its resistance to external pressure. The ovalization can be significantly increased by tension if the pipe is bent plastically over a stiff surface like a reel or a stinger. This motivates a more detailed study of this interaction. In the case of a suspended line, tension interacts with bending and pressure, primarily through its effect on yielding. This chapter deals with the mechanics of inelastic bending in the presence of tension. The distinction between the two types of combined tension and bending is addressed in detail.

10.1 FEATURES OF TUBE BENDING UNDER TENSION

Tension applied to a curved pipeline is balanced by a transverse distributed force. In the suspended section of an installation process, the transverse forces result from the weight of the line and the buoyancy. Their resultant is normal to the local axis of the pipe. In the case of a pipe in contact with a curved surface, the applied tension is reacted by a distributed force acting on the surface of the pipe. For simplicity, we will assume a section of tube bent to a uniform curvature κ^* by a bending moment M^*. When a tensile force T is also applied, a normal reaction force $T\kappa^*$ develops and the moment changes to M, as shown in Figure 10.2. When the tube is in contact with a rigid circular surface (Figure 10.2(a)) the reaction is on the tube generator that contacts the surface. This contact force tends to increase the ovalization induced by bending. When the tube is suspended (Figure 10.2(b)), the reaction force is best assumed to act on the axis of the tube [10.4] and does not significantly affect the ovalization. Both of these types of combined loading will be considered in an extension of the long tube formulation developed in previous chapters. The experimental study that follows was aimed at the more severe case of bending over a curved rigid surface in the presence of tension.

10.2 COMBINED BENDING–TENSION EXPERIMENTS

10.2.1 Test Facility

The problem was studied using the custom testing facility shown in Figure 10.3(a). Its major components are identified in the schematic shown in Figure 10.3(b) [10.5, 10.6]. It consists of a stiff structure that supports a smooth steel mandrel of circular shape (typical radii ρ^*: 30.3 in (770 mm). 50 in (1,270 mm)). The tube is bent over the mandrel by applying axial and shear loads at the ends, as shown in the figure. The loads are applied by two hydraulic actuators pinned to the support structure and to the ends of the tube. Simultaneous contraction of the two actuators causes the tube to bend and conform to the geometry of the mandrel. Full contact of the tube with the mandrel can be achieved at

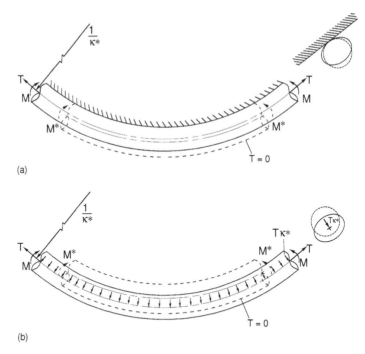

Figure 10.2 Combined bending–tension loadings: (a) pipe bent over a rigid surface and (b) model of suspended section of pipe.

various values of axial load by selecting the test facility configuration (i.e., the vertical position of the mandrel and the horizontal position at which the actuators engage the support structure). Once the tube is in full contact with the mandrel, the applied axial load can be further increased to the desired value.

The actuators are operated by a hydraulic system that loads the tube symmetrically about its mid-span. The load and displacement of each actuator are monitored by load cells and displacement transducers (LVDTs). The rotation of the fixed ends of the cylinders is monitored by rotary transducers (RVDTs). The position of the ends of the tube during the loading history is evaluated from the axial contraction and rotation of the cylinders.

A special resistance paper with regularly spaced, continuous, conductive silver lines deposited on one side is placed between the tube and the mandrel and is used to monitor the length of tube in contact with the mandrel. The resistance paper and the tube are connected into two constant current circuits in the manner shown in Figure 10.4. The circuits are calibrated prior to each experiment. The contact lengths, s_1 and s_2, are monitored by registering changes in voltage in each of the two circuits.

The measured values of the applied loads F_α ($\alpha = 1, 2$), the lengths of tube in contact with the mandrel s_α, and the positions of the tips of the tube are used to calculate the axial tension, T, applied to the tube at the points of separation A (see Figure 10.5) from the mandrel as follows:

Let $(x, y)_A$ be the coordinates of point A and $(x, y)_B$ those of the tip of the tube. They are related to the setup geometric variables defined in Figure 10.5 through

$$x_A = \rho^* \sin \phi, \quad y_A = \rho^*(1 - \cos \phi), \quad \phi = \frac{s}{\rho^*}, \quad x_B = b - l \cos \beta, \quad y_B = h - l \sin \beta. \quad \text{(a)}$$

(a)

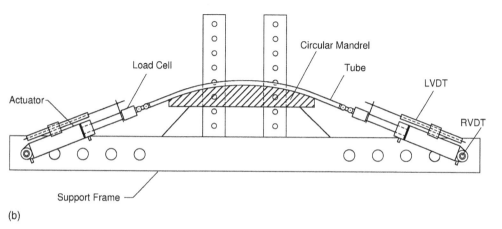

(b)

Figure 10.3 Combined bending–tension test facility: (a) photograph and (b) scaled schematic.

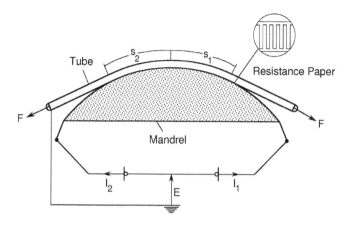

Figure 10.4 Contact of tube with mandrel monitored by using resistance paper.

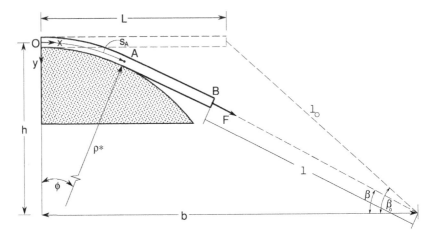

Figure 10.5   Geometric variables of test setup shown in Figure 10.3(a).

The initial values of β and l are given by

$$\beta_o = \tan^{-1}\left(\frac{h}{b-L}\right) \quad \text{and} \quad l_o = \sqrt{h^2 + (b-L)^2}, \tag{b}$$

and the current values are evaluated from the measurements of the LVDTs and RVDTs. The axial and shear forces at A are then given by

$$T = F\sin(\beta - \phi) \quad \text{and} \quad Q = F\cos(\beta - \phi). \tag{c}$$

The ovalization of the tube cross section is an important measure of the structural degradation sustained by the tube during the loading. A custom instrument, shown in Figure 10.6, was used to monitor the change in the minor axis of the cross section of the tube ($\Delta D/D \equiv$ ovalization). The instrument consists of a frame structure that engages the mandrel with four wheels. A traversing beam which can freely move in the vertical direction contacts the top of the tube with a roller. The motion of the traversing beam is monitored by a miniature LVDT.

During an experiment, the instrument monitors the ovalization at mid-span of the test specimen. The loading is periodically interrupted and the ovality transducer is rolled along the length of the tube to record variations in ovality. The position along the length is monitored by an encoder consisting of a photo-diode (transmitter/receiver) that looks at a strip of graduated light/dark markings (Figure 10.6(a)). In the experiments described here, the axial resolution of the encoder was 0.2 in (5 mm).

10.2.2   Experimental Procedure and Results

The experiments conducted involved seamless carbon steel 1020 tubes with nominal diameters of 1.250 in (31.8 mm) and various wall thicknesses. The test specimens were typically 30D long. The geometric and material parameters of tubes that will be analyzed in this chapter are listed in Table 10.1. Tubes are referred to by the number they are

(a)

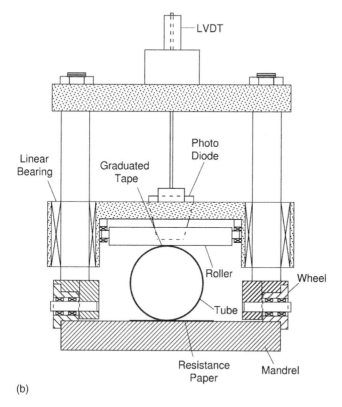

(b)

Figure 10.6   Transducer used to scan ovalization along the length of the test specimen: (a) photograph and (b) schematic.

Table 10.1 Test specimen geometric and material parameters.

Tube No.	D in (mm)	$\dfrac{D}{t}$	E Msi (GPa)	σ_y ksi (MPa)	n	σ_o ksi (MPa)	S
1	1.254 (31.84)	35.2	30.8 (212)	83.0 (572)	15.0	87.5 (603)	1.0
2	1.255 (31.88)	25.3	28.0 (193)	68.0 (469)	9.5	73.7 (508)	1.0
3	1.251 (31.78)	36.1	27.6 (190)	63.0 (434)	10.0	68.0 (469)	1.0

Table 10.2 Test facility configurations.

Configuration No.	ρ^* in (mm)	b in (mm)	h in (mm)	s_o [†] in (mm)
I	50 (1270)	40 (1020)	12.06 (306)	13.13 (334)
II	50 (1270)	40 (1020)	10.06 (256)	13.13 (334)
III	30.3 (770)	40 (1020)	15.63 (397)	13.50 (343)

[†] $2s_o$ = mandrel arc length

assigned in the table. Special end-plugs welded to each test specimen pin-connected the ends to the actuators.

The test facility geometry can be changed according to the loading history require-ments of the experiment. The geometric parameters varied are the mandrel radius ρ^*, the mandrel height h and the horizontal positions of the fixed ends of the actuators b (see Figure 10.5). Three different configurations, used in the experiments discussed here, are listed in Table 10.2 (configurations identified by Roman numerals).

The appropriate facility geometry was decided by numerically simulating the experi-ment through a nonlinear beam analysis (Appendix B in [10.6]). The analysis is based on large deflection beam kinematics and incorporates the measured elastic-plastic material properties of the tube. Given the radius of the mandrel, the analysis was used to select b and h values that yield the load conditions desired when the tube is in full contact with the mandrel, and to estimate the value of the maximum achievable axial load. Once the geometry of the test facility was decided, the beam analysis was used to extract the curvature–tension ($\kappa - T$) loading history followed by different points along the length of the tube.

The loading was applied in a quasi-static fashion in a manner that allowed the two actu-ators to keep the loading close to symmetric. Figure 10.7(a) shows plots of the measured forces F_1 and F_2 versus the length of tube in contact with the mandrel (s) for experiment 1/I (Tube 1/Configuration I). Corresponding results from the nonlinear beam analysis are included for comparison. The predicted load is higher than the measured values, but the general trend of the predictions is close to that of the measurements.

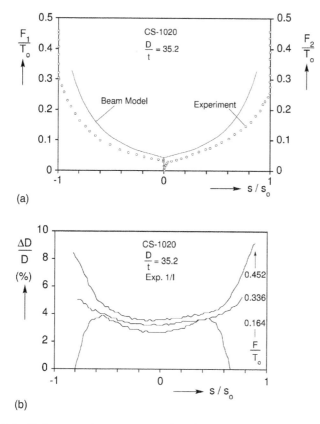

(a)

(b)

Figure 10.7 (a) Applied actuator force vs. contact length. (b) Measured ovalization along tube length.

Figure 10.7(b) shows a set of three scans of ovality along the tube length, taken during the same experiment. The scans are identified by the values of axial load at which they were taken. During the first scan, taken at $F/T_o = 0.164$, the tube was only partially in contact with the mandrel ($T_o = \pi\sigma_o D_o t$ is the yield tension). The ovalization recorded is seen to be reasonably uniform along a length of approximately 8 tube diameters. The maximum values of ovalization are seen to occur at the lift-off points, beyond which the measurement of ovality with the specific instrument used is not meaningful. At the higher loads of $F/T_o = 0.336$ and 0.452, the tube was in full contact with the mandrel. The scans show the ovality to be nearly uniform over the central section of the tube, approximately 12 to 16 tube diameters long. This is a crucial observation that will allow us to compare the results from an axially uniform analysis to the ovalization measured at the mid-span of the test specimens in the experiments. The ovalization is larger at the ends of the mandrel due to the presence of concentrated reaction forces (line of action of load not along the tangent at the end of mandrel for this test facility configuration).

Figure 10.8(a) shows the ovalization measured at the mid-span of the tube in the same experiment as a function of the tensile force T acting along the axis of the section of the tube in contact with the mandrel (OA in Figure 10.5). Included in the figure is a plot of the length of tube in contact with the mandrel. As the axial force is increased, the

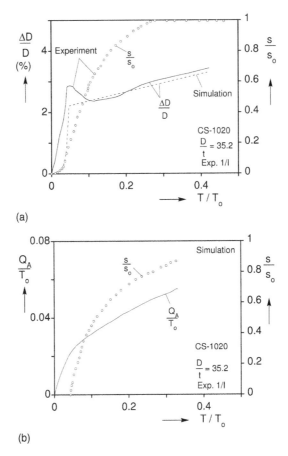

(a)

(b)

Figure 10.8   (a) Ovalization measured at mid-span as a function of axial tension. (b) Shear force Q_A as a function of axial tension.

tube gradually bends and starts conforming to the geometry of the circular mandrel. The induced bending causes the cross section of the tube to ovalize. Additional ovalization results from the transverse reaction force along the generator of the tube in contact with the mandrel. The ovalization at the mid-span is seen to reach a local maximum, followed by a local minimum. Once approximately one half of the tube is in contact with the mandrel, the ovalization increases monotonically with T.

The initial transient in the measured ovalization clearly demonstrates that the problem is three-dimensional. Under the idealized conditions used in the nonlinear beam model of the set-up, the section of beam in contact with the mandrel sees a uniform reaction given by $T\kappa^*$ ($\kappa^* = 1/(\rho^* + D/2)$) per unit length. However, additional point reactions, Q_A, act on the tube at the points of separation with the mandrel. Thus, at any given value of tension, the biggest ovalization occurs at the lift-off points (e.g., see ovalization distribution for $T/T_o = 0.164$ in Figure 10.7(b)). As a result, a stationary observer will record a small decrease in ovalization as the point of separation with the mandrel moves away from him. This explains the local maximum developed in the ovalization recorded at mid-span in Figure 10.8(a). In the experiments conducted, the length of tube affected

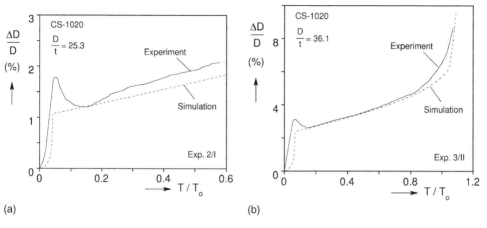

Figure 10.9 Comparison of ovalization measured at mid-span with predictions. (a) $D/t = 25.3$ and (b) $D/t = 36.1$.

by the discontinuity of the reaction force at the lift-off point was on the order of 3 to 5 diameters long.

Some understanding of the magnitude of the concentrated reaction force Q_A can be gained from Figure 10.8(b), where the calculated values of Q_A and contact length s for experiment 1/I are plotted against T. The tube is initially in contact with the mandrel only at one point. During initial loading, Q_A grows sharply with T. As more of the tube comes into contact with the mandrel, the rate of growth of Q_A decreases. In general, the value of Q_A was relatively small for the test facility configurations used. (Note that the point reaction force is a direct result of the assumptions of Bernoulli–Euler beam theory. In practice, the reaction can be expected to be distributed over a small section of the tube. However, this does not alter any of the conclusions drawn above.)

In spite of the three dimensional effects described above, once a significant length of the tube is in contact with the mandrel, the central part of it can be viewed as being under uniform loading for the reasons given. In the case shown in Figure 10.8, this condition was reached when $s/s_o \simeq 0.5$. Results like those presented in this figure will be used to verify the formulation and solution procedures developed in the next section for analyzing tubes under axially uniform bending and tension loads.

Figure 10.9(a) shows a plot of the ovalization recorded at mid-span vs. T for a tube with $D/t = 25.3$ (Exp. 2/I). The trend of the ovalization is similar to the one from Exp. 1/I. In this case the tube is thicker, and as a result the net value of ovalization recorded for comparable tension values is smaller. Figure 10.9(b) shows a similar set of results obtained from a tube with $D/t = 36.1$ (Exp. 3/II). This tube had a lower yield stress than tube 1 (see Table 10.1). As a result, it was possible to load it to the yield tension. The ovalization–tension response recorded is initially similar to the one shown earlier in Figure 10.8(a). Following the initial transient behavior, the ovalization at mid-span grows linearly with tension. When the tension reaches a value of approximately $0.8T_o$, the ovalization starts growing at a highly accelerated rate with T. When the tension reaches T_o, the ovality grows to approximately $0.1D$, and the slope of the response is very steep. The tube is thus very close to a condition of catastrophic collapse.

The experiments presented here were conducted in the related study of the problem reported in [10.6]. Subsequently, a more versatile bend-stretch facility suitable for the conduct of such experiments was developed. This facility is shown in Figure 10.10 and described in detail in [10.7]. The tension actuators operate under load control and rotate freely so that they can follow the motion of the ends of the tube and remain approximately tangential to it as it conforms to the circular mandrel. The mandrel is mounted on a pair of actuators which move upwards in displacement control, bringing it into contact with the pretensioned tube. The tube gradually conforms to the shape of the mandrel at a prescribed tension. Once in full contact with the mandrel, the tube can be post-tensioned to the required level. The kinematics of the mandrel and tensioners, as well as the ability to prescribe the tension history, enable a wider choice of tension–bending histories than those available in the facility in Figure 10.3.

10.3 FORMULATION

We once more focus on the 2-D version of the problem involving a long tube or pipe with uniform geometric and material properties along the length. The tube is bent to a uniform curvature κ^* while simultaneously being under a tensile force T. The tensile force is reacted by a lateral distributed force $T\kappa^*$ per unit length, as shown in Figure 10.11 (note that the small difference between the curvature of the tube and that of the rigid surface is neglected in this section for easier reading). In this problem, we will limit consideration to tube cross sections that are symmetric about the plane of bending (1–3). The problem can be treated by the same basic formulation developed in Section 8.3 by imposing this symmetry and adding the contribution of tension to the work term in the PVW (8.7a) as follows:

$$2R \int_0^\pi \int_{-t/2}^{t/2} (\hat{\sigma}_x \delta\dot{\varepsilon}_x + \hat{\sigma}_\theta \delta\dot{\varepsilon}_\theta) \left(1 + \frac{z}{R}\right) dz \, d\theta = \delta\dot{W}_e. \tag{10.1}$$

Because of the imposed symmetry, the displacement functions (8.7b) reduce to

$$w \cong R \left[a_o + \sum_{n=1}^N a_n \cos n\theta \right] \quad \text{and} \quad v \cong R \sum_{n=2}^N d_n \sin n\theta, \tag{10.2}$$

where a_n and d_n are unknown degrees of freedom of the problem. The material is modeled with the same J_2-type flow rule outlined in Section 4.3.2. The tensile force is reacted by a distributed force acting either on the axis of the tube or on a surface, as shown in Figure 10.2. It is treated in the following manner.

Transverse Loading on Tube Surface

This type of loading is for bending a pipe over a rigid surface to a curvature κ^*. Contact is assumed to occur only along the generator of the pipe at $\theta = \pi$ (see Figure 10.11(b)), and

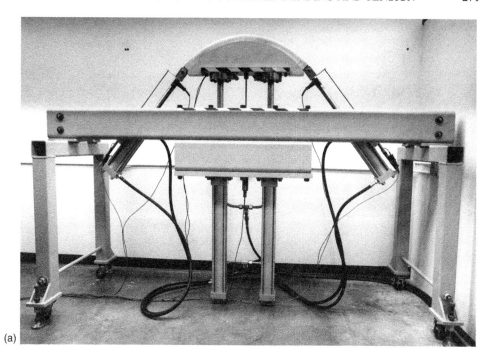

(a)

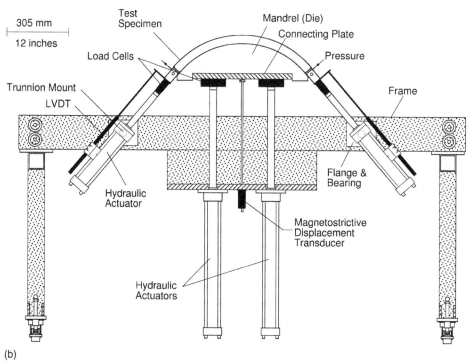

(b)

Figure 10.10   Bend-stretch testing facility suitable for establishing the bending–tension response of tubes under various loading histories [10.7]: (a) photograph and (b) schematic.

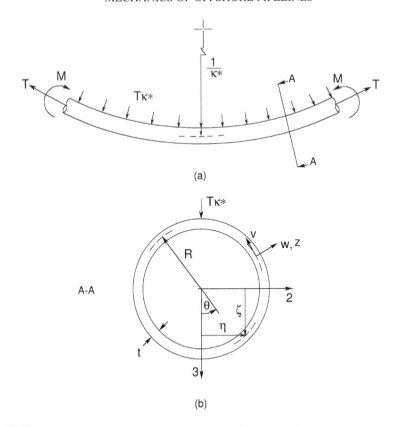

Figure 10.11   Problem parameters: (a) global view and (b) cross-sectional parameters.

the surface remains in contact with the pipe. The zero displacement condition at $\theta = \pi$ is introduced through a constraint as follows:

$$\Lambda \delta \dot{w}(\pi) = \Lambda[\delta \dot{a}_o - \delta \dot{a}_1 + \delta \dot{a}_2 - \delta \dot{a}_3 + \cdots + (-1)^n \delta \dot{a}_n],$$

where Λ is a Lagrange multiplier. The RHS of Eq. (10.1) is then given by

$$\delta \dot{W}_e = \hat{T} \delta \dot{\varepsilon}^o - \Lambda \delta \dot{w}(\pi). \tag{10.3a}$$

Transverse Loading on Tube Axis

This type of loading is used for suspended sections of pipe. In this case, the RHS of Eq. (10.1) is simply

$$\delta \dot{W}_e = \hat{T} \delta \dot{\varepsilon}^o. \tag{10.3b}$$

The numerical solution procedure follows the steps described in Section 8.3.2, with T being an additional variable prescribed incrementally. The bending moment is evaluated from a converged solution from

$$M = 2R \int_0^\pi \int_{-t/2}^{t/2} \sigma_x \zeta \left(1 + \frac{z}{R}\right) dz \, d\theta. \tag{10.4}$$

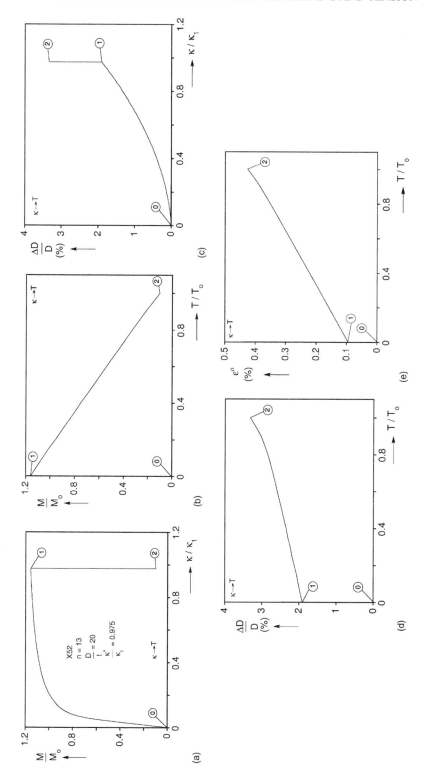

Figure 10.12 Calculated responses for the $\kappa \rightarrow T$ loading path: (a) moment–curvature, (b) moment–tension (c) ovalization–curvature, (d) ovaliza-tion–tension and (e) axial strain–tension.

If external pressure is also present, then the pressure work term in (4.22b) is also added to the RHS of (10.1).

10.4 PREDICTIONS

The formulation was incorporated in the computer code BEPTICO [10.8]. The main features of the pipe response under this combined loading are illustrated in Figure 10.12 for an X52 pipe ($n = 13$, Figure C2) with $D/t = 20$. The pipe is first bent to a curvature κ^*. It is then tensioned incrementally to a value $T = T_o$, with the lateral reaction coming from a rigid surface of the appropriate curvature ($\kappa \to T$ loading path). Shown in the figure are the calculated moment–curvature, moment–tension, ovalization–curvature, ovalization–tension and the axial strain–tension responses. In the ⓪–① phase of the loading history, the pipe is bent by prescribing κ incrementally with the tension kept at zero. In other words, the pipe is under pure bending. The ovalization is seen to grow to a value of approximately $\Delta D/D \simeq 0.02$. During the subsequent phase of the loading history (①–②), the curvature is kept constant at $\kappa^* = 0.975\kappa_1$ ($\kappa_1 = t/D_o^2$) while the axial tension is increased. The circumferential and axial stresses and deformations interact due to inelastic action while, simultaneously, the transverse reaction from the rigid surface causes the ovalization to grow. The combined effects lead to a drop in the moment required for equilibrium of the pipe. The ovalization of the pipe cross section experiences a nearly linear growth with tension up to $T \approx 0.8T_o$, and subsequently the rate of growth accelerates. By the end of the process, the ovalization reaches a value of nearly 3.4%. The axial strain grows nearly linearly with tension in the ①–② part of the loading history, reaching a value of about 0.425% by the highest value of tension.

10.4.1 Simulation of Experiments

The 2-D model was used to simulate several of the bending–tension experiments reported in [10.6] as follows: The nonlinear beam model of the test set-up (Appendix B of [10.6]) was used to evaluate the $\kappa - T$ loading history experienced at the mid-span. This history was used in the 2-D formulation, assuming that the tube is in continuous contact with a rigid surface. Initially, the surface curvature changes with T. When the pipe is bent to the curvature of the mandrel, the curvature is fixed, while the tension continues to increase until it matches the final test value. Figure 10.8(a) includes the calculated $\Delta D - T$ response for Exp. 1/I. As expected, the initial part of the response is governed by 3-D effects, and clearly cannot be reproduced by a 2-D treatment. However, as pointed out in the experimental section, after a significant part of the test specimen comes into contact with the mandrel, its central portion, approximately 10–12 diameters long, is essentially under 2-D loading. In this particular experiment, an axially uniform section of sufficient length was deemed to have developed for $T \geq 0.1T_o$. As a result, the ovalization calculated for $T \geq 0.1T_o$ is seen to be in very good agreement with the ovalization measured at $s = 0$.

Similar comparisons of measured $\Delta D - T$ responses with predictions from the 2-D model are included in Figure 10.9. In Figure 10.9(a), the predictions for the thicker tube from Exp. 2/I are seen to be of similarly good quality. In the results in Figure 10.9(b), the specimen was loaded to a tensile load that was somewhat higher than T_o. The ovalization predicted after the initial transient is again seen to be in good agreement with the

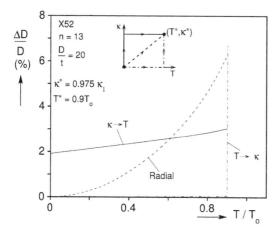

Figure 10.13   Ovalization–tension responses predicted for three different tension-bending loading paths.

experimental values for the whole range of axial loading applied, including the nearly collapsed state at the highest tension. Additional results of this type reported in [10.6] are of similarly high quality.

In summary, we conclude that the 2-D formulation is sound and suitable for this type of combined loading. It can be used to calculate the expected degradation of the pipe from such loadings in situations in which a nearly 2-D state of loading is applicable. This, for instance, can be done for plastically winding and unwinding pipelines onto reels in the presence of tension. This subject will be discussed in a dedicated section in the third volume of this book series.

The model is now used to examine the effect of some of the problem parameters on the induced ovalization for an X52 pipe ($n = 13$) with $D/t = 20$.

10.5   PARAMETRIC STUDY

10.5.1   Effect of Loading Path

The experimental facility in Figure 10.3 induces to the tube a loading history unique to the set-up. Because in pipeline applications the loading will be different, it is worth investigating the effect of the loading path on the induced ovalization. $\Delta D - T$ responses corresponding to the $\kappa \to T$, $T \to \kappa$, and *Radial* paths are compared in Figure 10.13. In all cases, the pipe is assumed to be bent over a rigid surface. In the $\kappa \to T$ loading path, discussed earlier in Figure 10.12, the pipe is bent to a required curvature ($\kappa^* = 0.975\kappa_1$) and then tensioned to $0.9T_o$ with the curvature kept fixed. In the $T \to \kappa$ loading path, the pipe is first tensioned to $T^* = 0.9T_o$ and then bent to κ^* with the tension kept fixed. In the Radial path, the curvature and tension are increased proportionately until the values of $\kappa = 0.975\kappa_1$ and $T = 0.9T_o$ are reached (see inset in Figure 10.13). Interestingly, the ovalization produced by the $T \to \kappa$ path is nearly double the amount produced by the $\kappa \to T$ path. This increase is a direct result of the presence of the transverse force $T^*\kappa^*$ during bending. The Radial path also results in a significant increase in final ovality for the same reason. The results demonstrate that plastically bending a pipe over a rigid surface

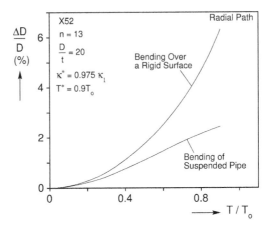

Figure 10.14   Ovalization–tension responses for an X52 pipe bent over a rigid surface and one suspended.

like a reel or a stinger under tension can result in significant ovality, with corresponding consequences on the pipe collapse capacity. If external pressure is added to the process, as is the situation when pulling a pipe through a J-tube (Figure 2.4), the ovality growth is even more severe and the possibility of collapse greater.

10.5.2 Transverse Force on Axis of Pipe

The transverse force $T\kappa^*$ has been modeled in two different ways, representing (a) bending under tension over a rigid surface, and (b) combined bending and tension experienced in a suspended section of a pipeline. The effect of this difference on the induced ovalization is illustrated in Figure 10.14. $\Delta D - T$ responses are compared for the same X52 pipe bent under tension to $\kappa^* = 0.975\kappa_1$ and $T^* = 0.9T_o$, following the radial path. The ovalization for the pipe bent over a rigid surface reaches a value exceeding 6.3%, while for the suspended section it only reaches 2.4% (the possibility of instability other than uniform collapse of the tube was not considered in these simulations). By comparison, a pipe bent to the same curvature in the absence of tension develops an ovality of 1.92%. Thus, when the transverse force is reacted on the axis of the pipe the interaction between bending and tension is mainly through inelastic action, which has only a mild effect on ovality.

10.5.3 Effect of Curvature

The values of curvature and tension to which the pipe is loaded are, of course, the main factors that decide the ovalization induced on the pipe. Their combined effect is illustrated in Figure 10.15 for the same X52 pipe loaded under the $\kappa \to T$ loading path. The value of κ^* was varied from $0.65\kappa_1$ to $1.137\kappa_1$ and T^* from 0 to T_o (the possibility of instability other than uniform collapse of the tube was not considered in these simulations). As expected, as κ^* is increased, the ovalization in the pure bending phase of the loading path increases. The rate of growth of ovalization during the axial loading phase of the history is also seen to increase with κ^*. In addition, for higher values of κ^*, the value of axial force at the transition when the ovalization starts to grow precipitously decreases.

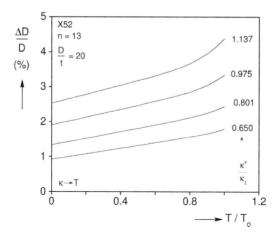

Figure 10.15   Ovalization–tension responses for an X52 pipe bent over surfaces of different curvatures through the $\kappa \to T$ loading path.

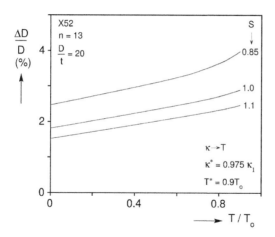

Figure 10.16   Effect of anisotropy on ovalization–tension responses of an X52 pipe bent over a rigid surface through the $\kappa \to T$ loading path.

10.5.4   Effect of Yield Anisotropy and Residual Stresses

Yield anisotropy is another factor that can influence the ovalization induced by this combined loading. Its effect is illustrated in Figure 10.16 for the same X52 pipe loaded under the $\kappa \to T$ loading path to $\kappa^* = 0.975\kappa_1$ and $T^* = 0.9T_o$. $\Delta D - T$ responses for isotropic and anisotropic ($S = 0.85$ and 1.1) materials are compared. $S = 0.85$ reduces the yield stress in the circumferential direction, resulting in an increase in the ovalization induced by both the pure bending and the tensioning parts of the loading history. By contrast, $S = 1.1$ increases the circumferential yield stress, which results in a reduction of the induced ovalization.

Another consequence of some manufacturing processes is residual stresses in the pipe. As was the case for pure bending and combined bending and pressure, the most severe residual stress field is a bending-type stress in the circumferential direction. The effect

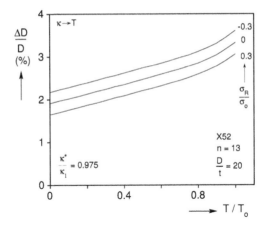

Figure 10.17   Effect of bending-type residual stress on ovalization–tension responses of an X52 pipe
bent over a rigid surface through the $\kappa \to T$ loading path.

of such a stress field on the induced ovalization under the combined loading considered
here is illustrated in Figure 10.17. This was done for the same X52 pipe loaded under the
$\kappa \to T$ loading path to $\kappa^* = 0.975\kappa_1$ and $T^* = 0.9T_o$. $\Delta D - T$ responses for $\sigma_R/\sigma_o = 0$,
0.3 and −0.3 are compared. Positive σ_R is the type which results in opening in the split-
ring test, as shown in Figure A8. Positive σ_R reduces the induced ovalization, while a
negative value increases it.

10.6   CONCLUSIONS AND RECOMMENDATIONS

Bending a circular tube ovalizes its cross section and leads to a limit load instability.
Bending in the presence of tension can aggravate the growth of ovalization. Two types
of combined bending–tension loading have been identified. The first is encountered in a
suspended line, where the tension is used to control the shape of the suspended section.
In this case, the tension is reacted by a distributed force normal to the axis of the tube
resulting from the weight of the pipe and the buoyancy force. The transverse force is best
assumed to act at the axis of the tube and thus does not influence ovality. In this type
of loading, the tension interacts with bending (and possibly pressure) through inelastic
action by aggravating the state of stress. In the second type of combined bending–tension
loading, the pipe is bent over a rigid surface, in which case the tension is reacted by a
distributed force acting on the surface of the bent tube. This type of loading is encountered,
for example, by a pipeline passing over a curved stinger, when winding or unwinding a
line on a reel, etc. Such a transverse force has been shown to increase the ovalization. The
extent of ovalization depends on the magnitude of the tension and on the curvature of the
surface.

   In most pipeline applications, the variation of forces and deformations is over lengths
exceeding 10 pipe diameters. In view of this, the formulation developed in previous
chapters, in which the tube remains uniform along its length, has been extended to include
bending–tension interaction. When in contact with a rigid surface, the tube is bent to the
curvature of the surface by gradually increasing the radius of the contact surface. The

formulation has been shown to capture the induced ovalization in experiments in which the tubes came into contact with rigid mandrels by a fully three-dimensional process. This type of model has been shown to effectively predict the ovality induced to pipelines in applications such as a pipe passing over a stinger, or in the more complicated reeling and unreeling process in the presence of tension (not reported here). Implementation of this model in these applications requires that a separate (usually beam) analysis of the process be conducted, from which the $T - \kappa$ history is extracted (see [10.8] for examples). The following additional comments should be helpful to engineers modeling such loadings:

- The induced ovalization is strongly influenced by the $T - \kappa$ history experienced by the pipe.
- Yield anisotropy and residual stresses can influence the induced ovality and should be included in the modeling when known.
- In the case of reeling, tension applied during winding and unwinding should be minimized to the extent possible. This will minimize the ovality induced by the process to the pipe paid into the sea.
- In the case of passing a pipe over a shaped stinger, it is most desirable that the loads are such that the pipe remains elastic. For some deepwater stingers, this may have to be violated. In this case, the ovality induced to the pipe should be evaluated and proved to be acceptable for the water depth of the application at the design stage.

REFERENCES

10.1. Solitaire (2006). *Description of D.P. Pipelay Vessel Solitaire*. Allseas Brochure.

10.2. Yun, H.D., Peek, R.R., Paslay, P.R. and Kopp, F.F. (2004). Loading history effects for deep-water S-lay pipelines. *ASME J. Offshore Mech. Arctic Eng.* **126**, 156–163.

10.3. Walker, A.C. and Davies, P. (1983). A design basis for the J-tube of riser installation. *ASME J. Energy Resour. Technol.* **105**, 263–270.

10.4. Fabian, O. (1977). Collapse of cylindrical, elastic tubes under combined bending, pressure and axial loads. *Intl J. Solid. Struct.* **13**, 1257–1270.

10.5. Tischler, J.C. (1988). *An Experimental Study on the Response of Inelastic Thin Walled Tubes Under Combined Bending and Tension*. M.S. Thesis, Department of Aerospace Engineering and Engineering Mechanics, The University of Texas at Austin, EMRL Report No. 88/2.

10.6. Dyau, J.-Y. and Kyriakides, S. (1992). On the response of elastic–plastic tubes under combined bending and tension. *ASME J. Offshore Mech. Arctic Eng.* **114**, 50–62.

10.7. Miller, J.E., Kyriakides, S. and Bastard, A.H. (2001). On bend-stretch forming of aluminum extruded tubes: Part I experiments. *Intl J. Mech. Sci.* **43**, 1283–1317.

10.8. Kyriakides, S., Dyau, J.-Y. and Corona, E. (1994). *Pipe Collapse Under Bending, Tension and External Pressure (BEPTICO)*. Computer Program Manual, University of Texas, Engineering Mechanics Research Laboratory Report No. 94/4.

10.9. Kyriakides, S., Corona, E., Mahadavan, R. and Babcock, C.D. (1989). Pipe collapse under combined pressure, bending, and tension load. *Proc. Offshore Technology Conference* **1**, OTC6104, 541–550.

11
Plastic Buckling and Collapse Under Axial Compression

Long cylindrical tubes and pipes under axial compression will usually bend, behaving as beam columns. Shell-type localized buckling occurs mainly when the structure is restrained from lateral movement. This, for example, is the case for a pipeline buried in a trench (Figure 11.1) or resting on a deformable foundation. Compression can be caused by the passage of hot hydrocarbons carried from the well to a central gathering point by buried flowlines in offshore operations [11.1]. Foundation motion caused by fault movement, landslides, ground subsidence, permafrost melting, or soil liquefaction can also result in severe compression of the lines [11.2–11.6]. Both loading scenarios can impose compressive strains high enough to result in shell-type buckling. In most onshore

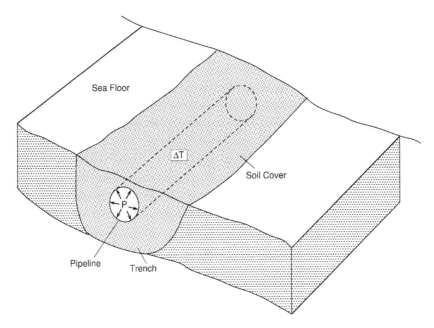

Figure 11.1 Schematic of a buried pipeline that develops compression due to change in temperature and internal pressure.

and offshore pipeline operations, diameter-to-thickness ratios (D/t) and steel grades are such that buckling occurs in the plastic range. In this chapter, the main features of plastic buckling under axial compression are first illustrated experimentally. The formulation for predicting the onset of plastic wrinkling is then developed, followed by a study of how wrinkles grow, localize and lead to collapse.

11.1 FEATURES OF AXIAL PLASTIC BUCKLING

Unlike elastic shell buckling in which collapse is sudden and catastrophic, plastic buckling failure is preceded by a cascade of events, where the first instability and collapse can be separated by average strains of 1–5%. The behavior is summarized schematically in the axial stress-shortening response of a long tube shown in Figure 11.2. Initially, the tube deforms uniformly (OA). At some strain level indicated by "↓" on the response, axisymmetric wrinkling becomes preferred. The wrinkles, initially small in amplitude, gradually grow (AB) to visible levels, as illustrated in Figure 11.3. In the process, the axial rigidity of the tube is reduced. For thicker shells, this eventually leads to a limit load instability (indicated by "∧") that can be considered the limit state of the structure. Under displacement-controlled loading, deformation localizes with the load dropping (BC). The localized deformation can be in the form of one axisymmetric lobe that grows until its folded walls come into contact. The process can subsequently be repeated, resulting in concertina folding, as illustrated in Figure 11.4(a) (e.g., [11.7]). Alternatively, a non-axisymmetric mode with 2, 3 or more circumferential waves develops in the zone of localization. Under persistent compression, this can again be repeated, as illustrated in Figures 11.4(b) and (c).

For thinner shells, the non-axisymmetric mode develops before the limit load associated with the purely axisymmetric deformation (e.g., point B'). This results in additional

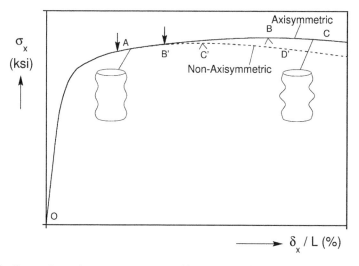

Figure 11.2 Stress-shortening responses expected in a compression test of an inelastic circular cylinder. Shown are the onset of wrinkling (A) followed by axisymmetric collapse (B) or non-axisymmetric collapse (C').

softening of the response (dashed line) (B′ C′), causing an earlier limit load (C′). Beyond this limit load, the non-axisymmetric deformation localizes (C′ D′), followed by folding similar to what was described previously. The material stress–strain response and D/t determine which of the two paths a given cylinder takes.

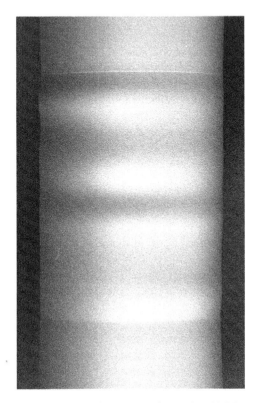

Figure 11.3   Specimen wrinkled under axial compression ($D/t = 28.97$).

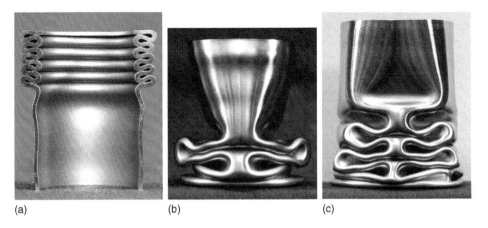

(a)   (b)   (c)

Figure 11.4   (a) Carbon steel tube that developed axisymmetric concertina folding; (b) mode 2 folding and (c) mode 3 folding of stainless steel tubes.

Axial plastic buckling is influenced by geometric imperfections, as well as by stress concentrations such as those generated by clamped or other end constraints. In addition, a full development of the progression of events described above requires that the specimen be of sufficient length. For example, in the experiments of Lee [11.8] on a soft aluminum (Al-3003-0), clamped edges resulted in the early development of edge bulges. Such bulges mask the onset of axisymmetric wrinkling. As a result, the main results reported are the stresses associated with the onset of collapse. Batterman [11.9] tested Al-2024-T4 shells with D/t values between about 20 and 180. The specimens were compressed between lubricated rigid platens that provided less constraint to the ends. The shells had length-to-diameter ratios ranging between 0.18 and 1.5. Such relatively short lengths are thought to have influenced the development of both the axisymmetric wrinkling and the subsequent non-axisymmetric modes. Following are results from a more systematic experimental study (Bardi and Kyriakides [11.10]), in which care was taken to limit the effect of both edge effects and shell size. As a result, they can be considered to be representative of the behavior expected in a pipeline.

11.2 AXIAL BUCKLING EXPERIMENTS

11.2.1 Experimental Setup

The test specimens were machined out of cold-finished seamless tube stock of SAF 2507 super-duplex, a type of stainless steel. The tube stock came in lengths of about 20 ft (6 m), and had a nominal diameter of 2.375 in (60.3 mm) and a nominal wall thickness of 0.154 in (3.91 mm). The test specimens were designed and fabricated in a way that minimized the effects of both end constraints and initial geometric imperfections. They were first machined and reamed on the inside to the desired dimensions. A 3 in (76.2 mm) long test section was then turned down on the outside to a wall thickness t in a way that ensured that the inner and outer surfaces of the test section were concentric (within a tolerance of about 0.001 in – 0.025 mm). Linear tapers connected the test section to thicker end sections that were left at the as-received diameter. The length of the tapers was selected through FE simulations of the test setup so as to minimize the effect of the thickness discontinuities on the axial stress. In this manner, the onset and growth of wrinkling was approaching that of a long uniform pipe. The overall length of the test specimens was 11 in (280 mm).

The tests were performed in a 225 kip (1 MN) servohydraulic testing machine equipped with custom axisymmetric grips. A photograph of the grips and the rest of the experimental setup used is shown in Figure 11.5(a). The main components of the setup are identified in the scaled schematic shown in Figure 11.5(b). Solid inserts were placed inside the thicker ends of the specimens for support. Gripping was achieved by the application of circumferential pressure using *Ringfeder* locking assemblies. The local circumferential and axial strains in the test section were measured by strain gages. A custom extensometer that spanned the test section was mounted on the test specimen as shown in the figures. The extensometer was used to measure the shortening of the test section.

A scanning device was used to periodically scan the surface of the test section during the tests. The device consists of an LVDT displacement transducer mounted on a linear encoder that allows monitoring of the axial position of the transducer. The encoder, in turn, is mounted on a ring that smoothly rotates concentrically to the test specimen.

(a)

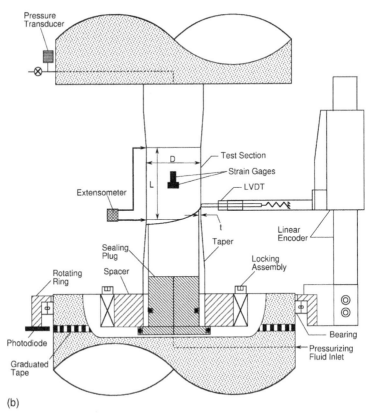

(b)

Figure 11.5   (a) Photograph of axial buckling test specimen in the custom test set-up used and
(b) scaled schematic of the experimental set-up.

The angular position is decided by a polar encoder, consisting of a photodiode and a graduated black–white tape mounted on the lower grip, as shown in Figure 11.5(b). The tube was first scanned axially at zero axial load, and then at regular strain intervals as compression progressed. A limited number of circumferential scans were performed once non-axisymmetric buckling modes initiated.

The tests were conducted under constant displacement rate, which for the homogeneous strain part of the test resulted in a strain rate of approximately $2 \times 10^{-5} \text{ s}^{-1}$. The tests were monitored via a computer-operated data acquisition system, which recorded signals from the extensometer, the testing machine displacement, the strain gages and the load cell on a common time base. Compression was periodically interrupted for a few seconds in order to perform an axial scan. The scans were recorded on a second data acquisition system. Compression eventually led to a limit load indicating the onset of localization. Under the displacement-controlled loading used, it was possible to follow the response past the limit load. In most experiments, the test was terminated before a well-defined fold developed.

11.2.2 Experimental Results

Fifteen specimens with various D/t values in the range of 23–52 were compressed to failure. The diameter and wall thickness were measured along the length and around the circumference of the test section prior to each test. Mean values of the measurements are listed under D and t, respectively, in Table 11.1. The specimens came from three different SAF 2507 tube stock. The mechanical properties of each mother tube stock are listed in Table 11.2.

Figure 11.6(a) shows a typical nominal axial stress-shortening ($\sigma_x - \delta_x/L$) response measured in such an experiment, on a tube with $D/t = 26.3$ (Exp. CW16). Figure 11.6(b) shows a set of axial scans taken at the positions identified on the response by solid bullets. Initially, the test section deforms uniformly and the stress-displacement response closely follows the compressive stress–strain response of the material. At some stage, axisymmetric wrinkles become discernible. Initially, they grow rather slowly (slower for lower D/t tubes). Because the axial scans were conducted at discrete strains, a range of strain values is quoted in Table 11.1 that bounds the strain at the onset of wrinkling. The width of this range depends on the frequency at which the scans were conducted. In this case, the first signs of wrinkling occurred between strains of 1.08% and 1.18%, marked with "↓" on the response in Figure 11.6(a) and also listed under ε_C in Table 11.1. Corresponding stress values are listed under σ_C. These values will be associated with the critical "bifurcation" values of the tubes. The wavelengths of the wrinkles were measured and found to also vary within a range. For this experiment $0.285 < \lambda_C/R < 0.377$. The onset of wrinkling is influenced by small initial imperfections specific to each specimen, and as a result the wrinkling pattern, the recorded range of wavelengths, and to some degree the "bifurcation" stresses and strains were not exactly repeatable.

As the specimen was further compressed, the amplitude of the waves grew, and in the process its axial rigidity was gradually reduced. For CW16, the growth of the wave, approximately at the mid-span of the test section, accelerated, indicating localization of deformation. A load maximum developed at an average strain ($\bar{\varepsilon}_L$) of approximately 4.75% and a stress of $\sigma_L = 111.2$ ksi (767 MPa), marked by a caret "∧" on the response. Following the limit load, deformation in the central wave accelerated significantly. At some stage in the process, possibly before the limit load, deformation in the neighborhood

Table 11.1 Specimen parameters and critical variables measured in axial buckling experiments.

Exp. No.	Mat.	D in (mm)	t in (mm)	$\frac{D}{t}$	$t_{min}-t_{max}$ in (mm)	σ_C ksi (MPa)	% ε_C	$\frac{\lambda_C}{R}$	σ_L ksi (MPa)	% $\bar{\varepsilon}_L$	m
CW20	SAF3	2.2975 (58.36)	0.1005 (2.55)	22.86	0.100–0.1009 (2.54–2.56)	98.2–100.3 (677–692)	1.19–1.44	0.30–0.374	117.7 (812)	6.35	2
CW16	SAF3	2.2676 (57.60)	0.0863 (2.19)	26.28	0.0850–0.0874 (2.16–2.22)	95.4–96.2 (658–663)	1.08–1.18	0.285–0.377	111.2 (767)	4.75	2
CW9	SAF3	2.2533 (57.23)	0.0791 (2.01)	28.49	0.0779–0.0800 (1.98–2.03)	96.6–98.4 (666–679)	1.2–1.4	0.276–0.313	–	–	–
CW3	SAF3	2.2530 (57.23)	0.0790 (2.01)	28.52	0.0781–0.0796 (1.98–2.02)	96.0–98.1 (662–677)	1.00–1.25	0.248–0.294	108.9 (751)	3.96	2
CW8	SAF3	2.2509 (57.17)	0.0777 (1.97)	28.97	0.0760–0.0791 (1.93–2.01)	91.9–94.3 (634–650)	0.80–1.01	0.304–0.433	–	–	–
CW19	SAF3	2.2333 (56.73)	0.0685 (1.74)	32.60	0.0681–0.0689 (1.73–1.75)	98.2–100.3 (677–692)	1.14–1.38	0.23–0.25	108.0 (745)	3.12	2
CW14	SAF3	2.2175 (56.32)	0.0608 (1.54)	36.47	0.0593–0.0625 (1.51–1.59)	93.9–96.7 (648–667)	0.89–1.15	0.287–0.343	103.3 (712)	2.31	3
CW15	SAF3	2.1978 (55.82)	0.0508 (1.29)	43.26	0.0498–0.0518 (1.26–1.32)	92.2–97.8 (636–674)	0.69–1.08	0.214–0.252	101.0 (697)	1.60	3
CW24	SAF3	2.1938 (55.72)	0.0486 (1.23)	45.14	0.0479–0.0489 (1.22–1.24)	94.8–97.4 (654–672)	0.78–0.92	0.243–0.265	103.7 (715)	1.74	3
CW17	SAF3	2.1826 (55.44)	0.0431 (1.10)	50.64	0.0425–0.0438 (1.08–1.11)	98.5–100.4 (679–692)	0.68–0.76	0.196–0.280	103.0 (710)	1.09	3
CW18	SAF3	2.1809 (55.39)	0.0420 (1.07)	51.93	0.0417–0.0426 (1.06–1.08)	98.5–100.7 (679–694)	0.80–1.02	0.187–0.224	101.1 (697)	1.13	3
CW23*	SAF5	2.2879 (58.11)	0.0936 (2.38)	24.43	0.0931–0.0942 (2.37–2.39)	100.4–101.5 (692–700)	1.29–1.50	0.221–0.273	116.7 (805)	5.22	2
IPC5*	SAF4	2.2555 (57.29)	0.0802 (2.04)	28.12	0.0798–0.0807 (2.03–2.05)	96.7–98.5 (667–678)	1.21–1.41	0.313–0.335	109.6 (756)	3.71	2
CW21*	SAF4	2.2189 (56.36)	0.0611 (1.55)	36.33	0.0606–0.0615 (1.54–1.56)	95.5–98.1 (659–677)	0.97–1.17	0.222–0.250	105.2 (726)	2.46	2
IPC11*	SAF5	2.2112 (56.16)	0.0555 (1.41)	39.84	0.0546–0.0567 (1.39–1.44)	98.2–99.6 (677–687)	0.86–1.01	0.260–0.295	104.8 (723)	1.89	3

*L = 5 in (127 mm).

Table 11.2 Mechanical properties of SAF 2507 tubes used in the experiments.

Mat.	E Msi (GPa)	σ_o ksi (MPa)	S_θ	S_r	$S_{r\theta}$
SAF3	28.2 (194.5)	85.75 (591.4)	1.15	0.85	1.05
SAF4	28.5 (196.5)	81.34 (561.0)	1.15	0.85	–
SAF5	28.7 (197.9)	90.19 (622.0)	1.11	0.87	–

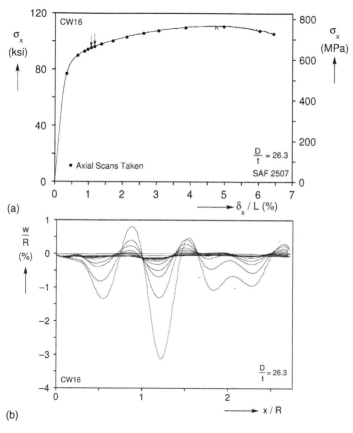

(a)

(b)

Figure 11.6 Typical results from a pure compression test: (a) recorded axial stress-shortening response and (b) axial scans showing evolution of wrinkles in test section.

of this wave reverted to a non-axisymmetric mode with two lobes ($m = 2$). The test was terminated at a net shortening of 6.5%. Continued compression would lead to progressive concertina folding, with two lobes for each fold. Figure 11.7 shows a photograph of the test section of this specimen after unloading. The axial wrinkles are clearly discernible. The central wave is seen to have grown significantly more than the others, and the non-axisymmetric buckling mode has set in.

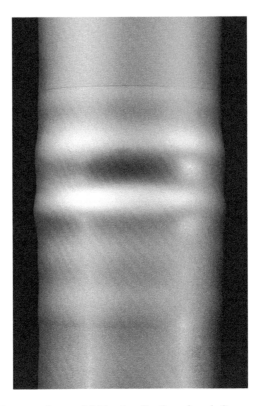

Figure 11.7 Buckled test specimen exhibiting localization of mode 2 near mid-span.

Eleven experiments were conducted on tubes from the mother tube stock SAF3. These tests had a test section with a length (L) of 3.0 in (72.2 mm). The two bounds for the critical strain, stress and wavelength ranges are listed in Table 11.1 and are plotted against D/t, respectively, in Figures 11.8–11.10 (depicted as ●, O). Despite the careful machining of the specimens, imperfections were unavoidable. The imperfections include small wall thickness variations around the circumference and along the length, small eccentricities in the applied load, and some surface hardening and surface marking from machining. In addition, in the perfect case, wrinkling is a nearly tangential bifurcation, making it very difficult to pinpoint its onset experimentally. Thus, the results in Figures 11.8–11.10 have a clear trend, but also exhibit some scatter.

The wrinkle patterns also varied from experiment to experiment. Figure 11.11 shows a second set of axial scans from specimen CW8 with a $D/t = 28.97$ (see Table 11.1). The development and growth of wrinkles in the test section is apparent. In this experiment, compression was terminated before the limit load was reached, and thus localization had not yet commenced. The photograph in Figure 11.3 shows the axisymmetric wrinkles at the end of the test.

In order to examine the influence of the test section length on the results, four experiments were run with longer test sections ($L = 5.0$ in–127 mm). The results are listed in Table 11.1, where they are identified with (CWX*). They are also included in Figures 11.8–11.10, where they are depicted with symbols (■, □). These specimens came from

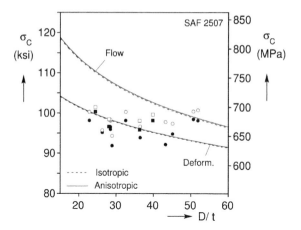

Figure 11.8   Critical stress (onset of wrinkling) vs. specimen D/t from 15 experiments.

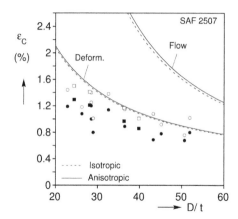

Figure 11.9   Critical strain (onset of wrinkling) vs. specimen D/t from 15 experiments.

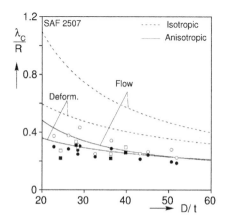

Figure 11.10   Wrinkle half-wavelength vs. specimen D/t from 15 experiments.

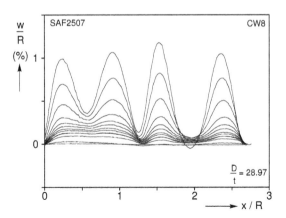

Figure 11.11　Axial scans showing evolution of wrinkles in the test section of Spec. CW8 $(D/t = 28.97)$.

mother tubes with somewhat different mechanical properties (see Table 11.2). As a result, the critical stresses are somewhat different from the rest (Figure 11.8). However, the critical strains and wavelengths follow the same trend as the other 11 data sets. It was thus concluded that a test section length of 3.0 in (76 mm $\sim$3R) was sufficient for the range of parameters considered.

　　When the tubes were compressed far enough, a second bifurcation always took place. The onset of the second bifurcation was even more difficult to determine than the first. The non-axisymmetric deformation grows slowly, which made pinpointing its onset difficult. The circumferential wave number m determined from observations at the end of the tests is reported in Table 11.1: for lower D/t tubes m was 2, and for higher values m was 3. The two modes at well-developed stages are illustrated in Figures 11.12 and 11.13 in two cases where loading was continued until a well-defined lobe developed. In Figure 11.12 the most deformed cross section is ovalized ($m = 2$), whereas in Figure 11.13 it has three circumferential waves.

　　The mean strain at the limit load ($\bar{\varepsilon}_L$) constitutes the critical limit state of the problem. Results from 13 experiments are plotted on log–log scales against D/t in Figure 11.14. Distinction is made between results from tubes with the shorter test section (circular symbols) and those with longer (square symbols). The results are seen to fall on a linear trajectory, indicating a powerlaw relationship between the two variables with an exponent of -2.081. Experimentally, the limit strain is a more well-defined variable, and as a result the scatter in the results is relatively small. Included in the same figure are the measured bifurcation strains (ε_C), which are seen to be significantly lower than $\bar{\varepsilon}_L$. The two coincide for elastic buckling, but their difference increases for tubes with lower D/t values, where instability is governed by inelastic action. The stress difference between the two depends on the hardening of the material and on the tube D/t. For the material used in this study, σ_L/σ_C varied from about 1.19 for $D/t = 22.9$ to about 1.02 for $D/t = 51.9$. In a stress-controlled application, the bifurcation stress can serve as a conservative design criterion. By contrast, in a deformation-controlled problem like a buried pipe compressed by thermal loads, the pipe can maintain its structural integrity well after the onset of wrinkling, and thus $\bar{\varepsilon}_L$ is a more appropriate design variable.

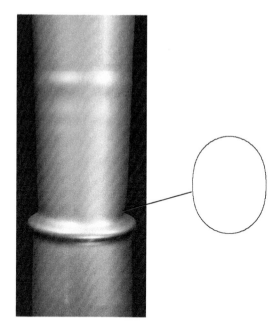

Figure 11.12   Localized mode 2 in a stainless steel tube $(D/t = 28.77)$.

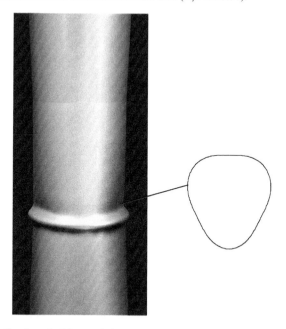

Figure 11.13   Localized mode 3 in a stainless steel tube $(D/t = 45.14)$.

A set of five additional experiments was conducted in a parallel study of the crushing behavior of tubes of the same material under axial loading. In these experiments, the tubes were uniform, had L/R ratios of about 6.6, and were compressed between rigid platens under displacement control at approximately the same strain rate as the experiments in

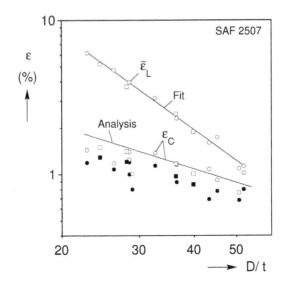

Figure 11.14   Critical and average limit strains vs. D/t from 15 experiments.

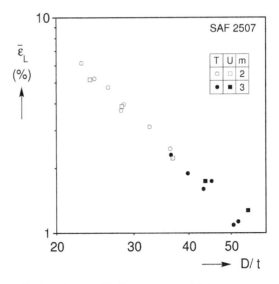

Figure 11.15   Average limit strains vs. D/t from two sets of data.

Table 11.1. The tubes had D/t values ranging between about 24 and 55 (Table 11.3 [11.10]). The results are similar to those described above, except that the development of edge bulges masks the onset of axisymmetric wrinkling. In Figure 11.15, $\bar{\varepsilon}_L$ from both sets of results are plotted vs. D/t. Data from the tapered tubes are depicted as circles (T $\equiv$ "Tapered Ends") and from the uniform tubes as squares (U $\equiv$ "Uniform"). Empty symbols represent tubes that developed $m = 2$ buckling modes, and solid symbols $m = 3$ modes. The two sets of results correlate very well for $D/t < 37$. The data for $D/t > 37$ exhibit more scatter. Clearly, higher D/t tubes exhibit the $m = 3$ mode and lower D/t the

$m = 2$ mode. Furthermore, this expanded data set points to the possibility that the single exponent fit of $\bar{\varepsilon}_L$ in Figure 11.14 may be inappropriate. This issue will be revisited in the light of numerical results.

11.3   ONSET OF AXISYMMETRIC WRINKLING

11.3.1   Formulation

The experiments have shown that the first instability encountered when a circular cylinder is compressed plastically is axisymmetric wrinkling. Following is the formulation leading to the classical result for this critical state [11.9, 11.11]. Consider a long circular cylindrical shell with mid-surface radius R and wall thickness t under axial compression. The nonlinear equilibrium equations for axisymmetric deformations of such a shell are given by

$$N'_{xx} = 0, \tag{11.1}$$

$$M''_{xx} - \frac{N_{\theta\theta}}{R} = -N_{xx}w'',$$

where $(\bullet)' \equiv (\bullet)_{,x}$. (Eq. (11.1) can be deduced from Sanders' shell equations in Appendix D). The corresponding kinematical relationships are:

$$\varepsilon = \varepsilon^o + z\kappa = \begin{Bmatrix} \varepsilon^o_{xx} \\ \varepsilon^o_{\theta\theta} \end{Bmatrix} + z \begin{Bmatrix} \kappa_{xx} \\ \kappa_{\theta\theta} \end{Bmatrix} \tag{a}$$

with

$$\varepsilon^o_{xx} = u' + \frac{1}{2}(w')^2, \quad \varepsilon^o_{\theta\theta} = \frac{w}{R}, \tag{11.2}$$

$$\kappa_{xx} = -w'', \quad \kappa_{\theta\theta} = 0.$$

For thicker shells, buckling occurs in the plastic range, and the appropriate incremental linearized buckling equations are:

$$\dot{N}'_{xx} = 0, \tag{b}$$

$$\dot{M}''_{xx} - \frac{\dot{N}_{\theta\theta}}{R} = -N_{xxo}\dot{w}'' = \sigma t \dot{w}''$$

where σ is the applied axial stress. The corresponding kinematical relations become

$$\dot{\varepsilon}^o_{xx} = \dot{u}', \quad \dot{\varepsilon}^o_{\theta\theta} = \frac{\dot{w}}{R}, \quad \dot{\kappa}_{xx} = -\dot{w}'' \quad \text{and} \quad \dot{\kappa}_{\theta\theta} = 0. \tag{c}$$

The instantaneous stress–strain relations are given by

$$\begin{Bmatrix} \dot{\sigma}_x \\ \dot{\sigma}_\theta \end{Bmatrix} = \begin{bmatrix} C_{11} & C_{12} \\ C_{12} & C_{22} \end{bmatrix} \begin{Bmatrix} \dot{\varepsilon}_x \\ \dot{\varepsilon}_\theta \end{Bmatrix} \tag{d}$$

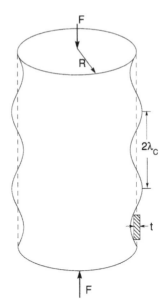

Figure 11.16 Axisymmetric buckling mode of a cylindrical shell under axial compression.

where $[C_{\alpha\beta}]$ are the incremental deformation theory moduli (inversion of (13.26)). The instantaneous force and moment intensities are given by

$$
\dot{N} = \int_{-t/2}^{t/2} \dot{\sigma}\, dz = t\, \boldsymbol{C}\dot{\boldsymbol{\varepsilon}}^{o} \quad \text{and} \quad \dot{M} = \int_{-t/2}^{t/2} \dot{\sigma}\, z dz = \frac{t^{3}}{12}\boldsymbol{C}\dot{\boldsymbol{k}}. \tag{e}
$$

The buckling equations can be expressed in terms of the displacements using (b) through (e). It can then be easily recognized that the following buckling mode (see Figure 11.16) satisfies the resulting differential equations

$$
\tilde{w} = a \cos\frac{\pi x}{\lambda} \quad \text{and} \quad \tilde{u} = b \sin\frac{\pi x}{\lambda}. \tag{11.3}
$$

Substituting (11.3) in the equations above results in the following eigenvalue problem:

$$
\begin{bmatrix} \dfrac{C_{12}}{R}\left(\dfrac{\pi}{\lambda}\right) & C_{11}\left(\dfrac{\pi}{\lambda}\right)^{2} \\[2ex] -\dfrac{t^{2}C_{11}}{12}\left(\dfrac{\pi}{\lambda}\right)^{4} - \dfrac{C_{22}}{R^{2}} + \sigma\left(\dfrac{\pi}{\lambda}\right)^{2} & -\dfrac{C_{12}}{R}\left(\dfrac{\pi}{\lambda}\right) \end{bmatrix} \begin{Bmatrix} a \\ b \end{Bmatrix} = 0. \tag{f}
$$

For non-trivial solutions

$$
\sigma = \frac{1}{\pi^{2}}\left(\frac{C_{11}C_{22} - C_{12}^{2}}{C_{11}}\right)\left(\frac{\lambda}{R}\right)^{2} + \frac{\pi^{2}}{12}C_{11}\left(\frac{t}{R}\right)^{2}\left(\frac{R}{\lambda}\right)^{2}. \tag{g}
$$

Minimizing σ with respect to λ results in

$$\lambda_C = \pi \left[\frac{C_{11}^2}{12(C_{11}C_{22} - C_{12}^2)} \right]^{1/4} (Rt)^{1/2} \tag{11.4a}$$

and

$$\sigma_C = \left[\frac{C_{11}C_{22} - C_{12}^2}{3} \right]^{1/2} \left(\frac{t}{R} \right). \tag{11.4b}$$

The critical bifurcation strain ε_C is the value that corresponds to an axial stress σ_C. If the material is linearly elastic, Eqs. (11.4) reduce to the classical elastic results

$$\lambda_{Ce} = \pi \left[\frac{1}{12(1 - v^2)} \right]^{1/4} (Rt)^{1/2} \tag{11.5a}$$

and

$$\sigma_{Ce} = \frac{E}{\sqrt{3(1 - v^2)}} \left(\frac{t}{R} \right). \tag{11.5b}$$

It is also instructive to solve the same problem through the principle of virtual work (*PVW*). In this case, equilibrium is represented by

$$\int_A \{N_{\alpha\beta}\delta\varepsilon_{\alpha\beta}^o + M_{\alpha\beta}\delta\kappa_{\alpha\beta}\}dA = 0, \quad (\alpha, \beta) = (1, 2) \equiv (x, \theta) \tag{11.6}$$

where the integral is taken over the mid-surface of the cylinder. The prebuckling solution is $N_{xxo} = -\sigma t$, with all other force and moment intensities zero. In the incremental solution procedure, we assume that at a given stage of deformation two solutions become possible, denoted by $\dot{u}^1$ and $\dot{u}^2$. Their difference $\tilde{u} = \dot{u}^2 - \dot{u}^1$ is associated with the buckling mode given by Eq. (11.3). The corresponding strains are

$$\tilde{\varepsilon}_{xx}^o = \tilde{u}_{,x} + w_{o,x}\tilde{w}_{,x}, \quad \tilde{\varepsilon}_{\theta\theta}^o = \frac{\tilde{w}}{R}, \tag{h}$$

$$\tilde{\kappa}_{xx} = -\tilde{w}_{,xx}$$

with all other components zero. Here $(\bullet)_o$ represents the uniform prebuckling solution. The perturbed *PVW*, with terms corresponding to the prebuckling equilibrium state eliminated, becomes [11.12]

$$\int_A \{\tilde{N}_{\alpha\beta}\tilde{\varepsilon}_{\alpha\beta} + \tilde{M}_{\alpha\beta}\tilde{\kappa}_{\alpha\beta} + N_{\alpha\beta o}\tilde{w}_{,\alpha} \, \tilde{w}_{,\beta}\}dA = 0. \tag{11.7}$$

We identify a vector of the coefficients of the buckling mode (11.3) as

$$\tilde{q} = \begin{Bmatrix} a \\ b \end{Bmatrix}. \tag{i}$$

Then Eq. (11.7) can be written as

$$\int_0^\lambda 2\pi\{\tilde{N}_{\alpha\beta,j}\tilde{\varepsilon}_{\alpha\beta,i} + \tilde{M}_{\alpha\beta,j}\tilde{\kappa}_{\alpha\beta,i} + N_{\alpha\beta o}\tilde{w}_{,\alpha i}\,\tilde{w}_{,\beta j}\}Rdx\tilde{q}_i\tilde{q}_j$$

$$\equiv H_{ij}\tilde{q}_i\tilde{q}_j = 0, \quad \text{where } (\bullet)_{,j} \equiv (\bullet)_{,qj}. \tag{11.8}$$

For nontrivial solutions, $\det |\boldsymbol{H}| = 0$. The following constitutive equations hold:

$$\tilde{\boldsymbol{N}} = t\boldsymbol{C}\tilde{\boldsymbol{\varepsilon}}^o \quad \text{and} \quad \tilde{\boldsymbol{M}} = \frac{t^3}{12}\boldsymbol{C}\tilde{\boldsymbol{\kappa}}.$$

The force and moment intensity derivatives relevant to this problem are:

$$\tilde{N}_{xx,1} = tC_{12}\tilde{\varepsilon}^o_{\theta\theta,1}, \quad \tilde{N}_{xx,2} = tC_{11}\tilde{\varepsilon}^o_{xx,2}, \quad \tilde{N}_{\theta\theta,1} = tC_{22}\tilde{\varepsilon}^o_{\theta\theta,1}, \quad \tilde{N}_{\theta\theta,2} = tC_{12}\tilde{\varepsilon}^o_{xx,2}, \tag{j}$$

$$\tilde{M}_{xx,1} = \frac{t^3}{12}C_{11}\tilde{\kappa}_{xx,1},$$

where $\boldsymbol{C}$ are the deformation theory instantaneous moduli (d). The following derivatives of the strain components (h) survive in (11.8), with all others zero:

$$\tilde{\varepsilon}^o_{xx,2} = \frac{\pi}{\lambda}\cos\frac{\pi x}{\lambda}, \tag{k}$$

$$\tilde{\varepsilon}^o_{\theta\theta,1} = \frac{1}{R}\cos\frac{\pi x}{\lambda},$$

$$\tilde{\kappa}_{xx,1} = \left(\frac{\pi}{\lambda}\right)^2\cos\frac{\pi x}{\lambda}.$$

Substituting (j) and (k) into (11.8) and performing the integrations results in Eq. (f); taking the determinant of the matrix leads to the critical stress and wavelength given in Eqs. (11.4).

11.3.2 Predictions

In the course of this investigation, it was established that the particular tubular stock used in the tests exhibited anisotropic yielding. Yield anisotropy can affect many aspects of axial compression, and therefore must be incorporated in modeling. The anisotropy was adequately modeled through Hill's anisotropic yield criterion (13.21a). The constants S_r and S_θ were evaluated experimentally as described in Appendix B, and their values appear in Table 11.2. The anisotropy was of the type shown in Figure B.3; that is, $S_\theta > 1$ and $S_r < 1$. The cause of this type of anisotropy is the cold finishing of the mother tubes during their manufacture. The problem variables affected include aspects of the onset of wrinkling. For this reason, the anisotropic deformation theory (13.28) will also be used in the predictions.

Predictions of the critical state $\{\sigma_C, \varepsilon_C, \lambda_C\}$ corresponding to (11.4) using both the isotropic and anisotropic deformation and flow theories, along with the mechanical properties of SAF3 in Table 11.2, are included in Figures 11.8–11.10. As expected, the J_2 flow theory significantly overpredicts both the critical stress and strain. By contrast, the J_2 deformation theory is seen to yield predictions that are quite close to the experimental results for both. The critical wavelength is also grossly overpredicted by the flow rule. Interestingly, J_2 deformation theory is closer to the experimental results, but still overpredicts them by nearly a factor of two for the whole range of D/ts considered. Since λ_C is an essential starting point for postbuckling analyses, this discrepancy is debilitating.

When the measured anisotropy is added to the flow and deformation theories, σ_C and ε_C are hardly affected. On the other hand, the effect on λ_C is quite significant. Both anisotropic models predict wavelengths that pass through the data, with the deformation theory yielding slightly better agreement.

The effect of anisotropy on $\{\sigma_C, \varepsilon_C, \lambda_C\}$ was examined in more detail in [11.13] by varying S_θ and S_r. The calculated ε_C and λ_C for a tube with $D/t = 26.3$ are plotted in Figure 11.17. Interestingly, the anisotropy has the opposite effect on ε_C (and σ_C) to that on λ_C. Thus, when $S_\theta > 1$ and $S_r < 1$, as was the case for the tubes tested, the effect on λ_C is large and the effect on ε_C is small. A similar effect is seen when $S_\theta < 1$ and $S_r > 1$, but now the predicted wavelength is longer than that of the isotropic case. On the other hand, when both S_θ and S_r are either >1 or <1, the effect on ε_C is large and on λ_C is small.

11.4 EVOLUTION OF WRINKLING

As the experiments have shown, the onset of axial buckling and failure are distinctly different events, separated by a significant deformation. In deformation-controlled applications, it is important to distinguish between the two by establishing the conditions at the onset of failure. This can be achieved by considering somewhat more complex models that involve cylinders with small initial imperfections [11.2, 11.14, 11.15]. Often, purely axisymmetric imperfections suffice. In some cases, consideration of more complex non-axisymmetric imperfections is required.

We consider a thin-walled circular cylindrical shell with mid-surface radius R and wall thickness t. Sanders' shell kinematics are adopted based on the assumptions of small strains and moderately small rotations (Appendix D).

11.4.1 Kinematics

The strains at any point on the shell are given by

$$\varepsilon_{\alpha\beta} = (\varepsilon^o_{\alpha\beta} + z\kappa_{\alpha\beta})/(A_\alpha A_\beta)^{1/2} \quad \text{where} \quad A_1 \cong 1, \quad A_2 \cong 1 + \frac{z}{R}. \tag{11.9a}$$

For the imperfect structure, the strains are given by

$$\varepsilon_{\alpha\beta} = \varepsilon_{\alpha\beta}(u, v, w + \bar{w}) - \varepsilon_{\alpha\beta}(0, 0, \bar{w}). \tag{11.9b}$$

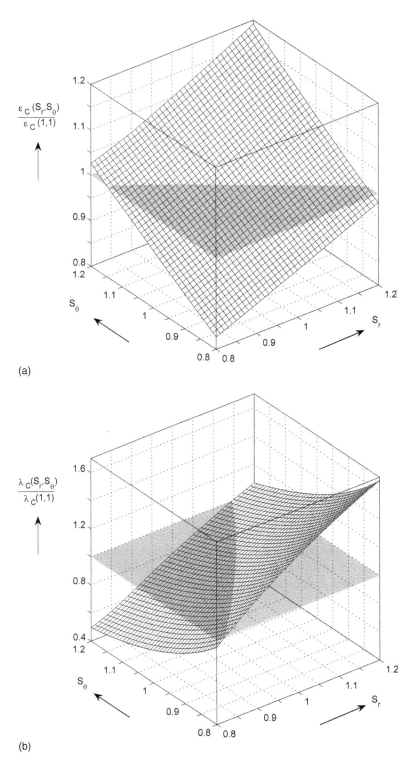

Figure 11.17 (a) Critical strain and (b) half-wavelength as functions of anisotropy variables $(D/t = 26.3)$.

The membrane and bending strains in terms of the displacements (u, v, w) are:

$$\varepsilon_{xx}^o = u_{,x} + \frac{1}{2}w_{,x}^2 + \frac{1}{2}\phi^2,$$

$$\varepsilon_{\theta\theta}^o = \frac{v_{,\theta}}{R} + \frac{w}{R} + \frac{1}{2R^2}(v - w_{,\theta})^2 + \frac{1}{2}\phi^2,$$

$$\varepsilon_{x\theta}^o = \frac{1}{2}\left(\frac{u_{,\theta}}{R} + v_{,x}\right) + \frac{1}{2R}w_{,x}(w_{,\theta} - v),$$

$$\kappa_{xx} = -w_{,xx},$$

$$\kappa_{\theta\theta} = \frac{-1}{R^2}(w_{,\theta\theta} - v_{,\theta}),$$

$$\kappa_{x\theta} = -\frac{1}{2R}(2w_{,x\theta} - v_{,x} - \phi), \quad \text{where} \quad \phi = \frac{1}{2}\left(v_{,x} - \frac{u_{,\theta}}{R}\right). \tag{11.10}$$

11.4.2 Principle of Virtual Work

Equilibrium will be satisfied through the principle of virtual work (*PVW*), which can be written in general terms as

$$\int_A \{N_{\alpha\beta}\delta\varepsilon_{\alpha\beta}^o + M_{\alpha\beta}\delta\kappa_{\alpha\beta}\}dA = \delta W, \tag{11.11a}$$

where the membrane and bending moment intensities are given by

$$N_{\alpha\beta} = \int_{-t/2}^{t/2} \frac{A_1 A_2}{(A_\alpha A_\beta)^{1/2}}\sigma_{\alpha\beta}\,dz, \quad M_{\alpha\beta} = \int_{-t/2}^{t/2} \frac{A_1 A_2}{(A_\alpha A_\beta)^{1/2}}\sigma_{\alpha\beta}\,z\,dz, \quad (\alpha, \beta) \text{ not summed.} \tag{11.11b}$$

The RHS of (11.11a) represents the virtual external work, which for displacement-controlled axial compression is zero.

11.4.3 Constitutive Equations

The material will be modeled through the flow theory of plasticity based on isotropic hardening (Eqs. (13.18)–(13.21)). The incremental form of the corresponding deformation theory will be used in the bifurcation check for non-axisymmetric buckling modes (Eqs. (13.25)–(13.28)).

11.4.4 Axisymmetric Solution

The postbuckling response of the wrinkled cylinder can be followed by introducing a small initial imperfection corresponding to the critical buckling mode (11.3) as follows [11.2, 11.14–11.18]:

$$\bar{w} = -t\omega_o \cos\left(\frac{\pi x}{\lambda_C}\right), \tag{11.12}$$

where λ_C is given in (11.4a) (Figure 11.16). In such a model, the domain of interest can be limited to just one half of the axial wavelength. The model will predict the evolution of the axial response, and will also allow us to check for the possible onset of a non-axisymmetric bifurcation.

For axisymmetric deformations, v and $(\bullet)_{,\theta}$ in Eqs. (11.10) are zero, and as a result (11.11a) reduces to

$$\int_0^{\lambda_C} \{N_{xx}\delta\varepsilon_{xx}^o + N_{\theta\theta}\delta\varepsilon_{\theta\theta}^o + M_{xx}\delta\kappa_{xx}\}2\pi R\, dx = 0. \tag{11.13}$$

This is solved by adopting the following admissible expansions for the displacements

$$w = a_0 + \sum_{i=1}^{N_w} a_i \cos\left(\frac{i\pi x}{\lambda_C}\right) \quad \text{and} \quad u = b_0 x + \sum_{i=1}^{N_u} b_i \sin\left(\frac{i\pi x}{\lambda_C}\right). \tag{11.14}$$

Substituting (11.14) into (11.13), the PVW can be restated as follows:

$$\int_0^{\lambda_C} [N_{xx}\varepsilon_{xx,i}^o + N_{\theta\theta}\varepsilon_{\theta\theta,i}^o + M_{xx}\kappa_{xx,i}]2\pi R\, dx\delta q_i = 0, \quad i = 1, 2, \ldots, N_w + N_u + 1 \quad \text{(a)}$$

where

$$(\bullet)_{,i} \equiv \frac{\partial(\bullet)}{\partial q_i} \quad \text{and} \quad q = [a_0, a_1, \ldots, a_{N_w}, b_1, b_2, \ldots b_{N_u}]^T$$

and $b_0 \, (=\delta_x/\lambda_C)$ is the average axial strain that is prescribed incrementally. In view of the arbitrariness of δq_i, the following algebraic equations represent equilibrium

$$G_i(q^* + \dot{q}) = \int_0^{\lambda_C} [N_{xx}\varepsilon_{xx,i}^o + N_{\theta\theta}\varepsilon_{\theta\theta,i}^o + M_{xx}\kappa_{xx,i}]2\pi R\, dx = 0, \quad i = 1, 2, \ldots, N_w + N_u + 1. \tag{b}$$

In the incremental solution procedure followed, q^* represents the previous converged solution, and $\dot{q}$ is the increment of q required for the current solution. In addition, $N_{\alpha\beta} = N_{\alpha\beta}^* + \dot{N}_{\alpha\beta} \ldots$, etc. The instantaneous constitutive equations are given by

$$\begin{Bmatrix} \dot{N}_{xx} \\ \dot{N}_{\theta\theta} \\ \dot{M}_{xx} \\ \dot{M}_{\theta\theta} \end{Bmatrix} = \int_{-t/2}^{t/2} \begin{bmatrix} A & zA \\ zA & z^2A \end{bmatrix} dz \begin{Bmatrix} \dot{\varepsilon}_{xx}^o \\ \dot{\varepsilon}_{\theta\theta}^o \\ \dot{\kappa}_{xx} \\ \dot{\kappa}_{\theta\theta} \end{Bmatrix}, \tag{c}$$

$$A = \begin{bmatrix} A_2/A_1 C_{11} & C_{12} \\ C_{12} & A_1/A_2 C_{22} \end{bmatrix}$$

where $[C_{\alpha\beta}]$ $\alpha, \beta = 1, 2$ come from the inverse of the constitutive matrix in (13.21b) with $\sigma_{x\theta} = d\sigma_{x\theta} = 0$. Gauss' integration rule with I_x points along the axial direction, and I_z points through the thickness that was used in the integrations. Typical values for these, arrived at through convergence studies, are $I_x = 16$ and $I_z = 5$. Through convergence studies, it was also shown that $N_u = N_w = 4$ in the displacement series (11.14) is sufficient.

A set of typical stress–displacement responses calculated with this formulation for a tube with $D/t = 26.3$ is shown in Figure 11.18 ($L = \lambda_C$). Because of the larger strains associated with axisymmetric collapse, the stress–strain response used is the *compressive*

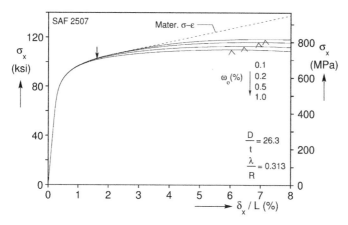

Figure 11.18   Effect of imperfection amplitude on axisymmetric wrinkling stress–displacement response.

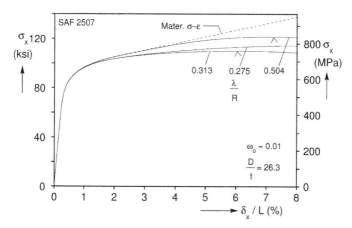

Figure 11.19   Effect of imperfection wavelength on axisymmetric wrinkling stress–displacement response.

version of the uniaxial response of SAF3 given in Table 11.2 (see [11.6]) along with the measured yield anisotropy parameters. Responses for imperfection amplitudes of $\omega_o = (0.1,\ 0.2,\ 0.5,\ 1.0)\%$ are shown in the figure. Initially, the calculated responses follow the material stress–strain response, drawn in the figure with a dashed line. The imperfect shell responses start deviating from the trivial response at strains between 1% and 2%; in other words, in the neighborhood of the bifurcation point (marked on the trivial response by "↓"). Subsequently, with continued compression, the amplitude of the wrinkles grows, progressively reducing the stiffness of the structure and eventually leading to a load maximum. The results in Figure 11.18 illustrate how the response and the limit state are affected by the amplitude of the imperfection. The order of magnitude of ω_o used is considered to be representative of the imperfections present in the tubes.

An essential starting point of such calculations is an accurate value of the wavelength of the axisymmetric wrinkles. The sensitivity of the calculated response to λ is illustrated in Figure 11.19. λ_C was evaluated by the bifurcation check to be $0.313R$. Increasing λ to

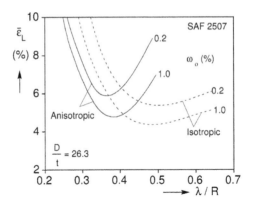

Figure 11.20 Average limit strain as a function of imperfection wavelength for two imperfection amplitudes and two material models.

$0.504R$ or decreasing it to $0.275R$ results in a stiffer response with a larger average limit strain. The sensitivity of the response to λ is further illustrated in Figure 11.20, where the calculated average limit strain $\bar{\varepsilon}_L$ (δ_L/L) is plotted against λ for two imperfection amplitudes $\omega_o = 0.2\%$ and 1.0%. For the smaller imperfection, the minimum value of $\bar{\varepsilon}_L$ occurs quite closely to λ_C, and for the larger imperfection at a somewhat higher value. Similar calculations were performed for the corresponding isotropic material (i.e., $S_r = S_\theta = 1$). In this case, $\lambda_C = 0.510R$, which corresponds quite well with the minimum of $\bar{\varepsilon}_L$ for the smaller imperfection. Interestingly, for $\omega_o = 1.0\%$ the minimum $\bar{\varepsilon}_L$ occurs at a somewhat smaller value of λ. In addition, for the isotropic material the variation of $\bar{\varepsilon}_L$ with λ is seen to be quite different from that of the anisotropic materials.

Anisotropy plays a significant role in the calculated response and the resultant limit load. Its effect stems from its strong influence on λ_C reported in [11.13]. Altering λ_C has the effect demonstrated in Figure 11.19. In addition, anisotropy directly affects the material response. The influence of anisotropy is demonstrated for four cases involving a tube with $D/t = 26.3$ of the same basic material assigned the different anisotropy values $\{S_\theta, S_r\}$ listed in Table 11.3. The calculated responses are designated numbers 1–4 in Figure 11.21. Case 1 has the anisotropy of SAF3. In Case 2, the material is isotropic. In Case 3, $S_\theta = 0.85$ and $S_r = 1$, and in Case 4 $S_\theta = S_r = 0.85$. Going from Case 1 to 4, the critical bifurcation strain ε_C progressively decreases. The corresponding value of λ_C increases for Cases 2 and 3 and decreases for Case 4 (see Figure 11.17). Figure 11.21 demonstrates that the four responses develop limit loads at progressively smaller strains (also listed in Table 11.3).

Because of the relatively high hardening exhibited by the material stress–strain response of SAF 2507 super duplex stainless steel, all tubes tested developed non-axisymmetric buckling modes. Despite this, it is worth examining how $\bar{\varepsilon}_L$ based on the axisymmetric solution compares with the experimental values. Calculations were performed for D/t values in the same range as the experiments. The material properties are those of SAF3, including the measured anisotropies. The imperfections considered had λ_C values appropriate for each D/t and   amplitudes of $\omega_o = (0.1, 1.0, 5.0)\%$. The predicted $\bar{\varepsilon}_L$ are plotted vs. D/t in log–log scales in Figure 11.22, together with the experimental results. The material response does not follow a powerlaw. Thus for fixed ω_o, the results

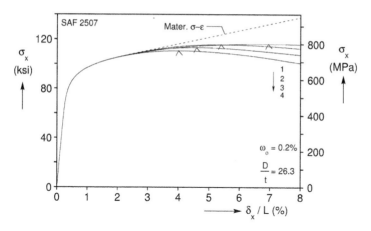

Figure 11.21   Effect of anisotropy on axisymmetric wrinkling stress–displacement response ($\{S_r, S_\theta\}$ given in Table 11.3).

Table 11.3   Anisotropy variables and critical and limit strains for axisymmetric calculations for a tube with $D/t = 26.3$.

Case	S_θ	S_r	$\dfrac{\lambda_C}{R}$	ε_C %	$\bar{\varepsilon}_L$ %
1	1.15	0.85	0.313	1.63	6.93
2	1	1	0.512	1.59	5.38
3	0.85	1	0.633	1.48	4.58
4	0.85	0.85	0.493	1.40	4.00

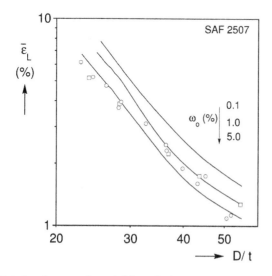

Figure 11.22   Predicted axisymmetric wrinkling limit strains vs. D/t for three imperfection amplitudes and experimental values.

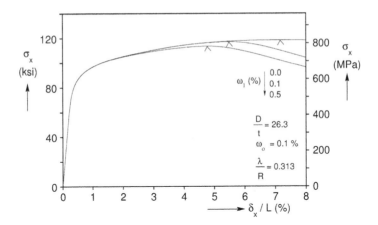

Figure 11.23   Effect of imperfection amplitude ω_1 on axisymmetric wrinkling stress-displacement response.

trace a somewhat nonlinear trajectory. The predicted $\bar{\varepsilon}_L$ for $\omega_o = 0.1\%$ are higher than the measured values. For $\omega_o = 1\%$, $\bar{\varepsilon}_L$ is closer to the experimental results for $D/t > 30$. For the extreme value of $\omega_o = 5\%$, the predictions are closer to the results for $D/t < 30$. The experimentally observed buckling modes coupled with this generally unfavorable comparison between experimental and predicted values of $\bar{\varepsilon}_L$ clearly indicate that consideration of non-axisymmetric buckling modes is necessary for this combination of material and tube geometries.

11.4.5   Localization of Axisymmetric Wrinkling

Structures that exhibit limit load instabilities tend to develop localized buckling patterns beyond the limit load [11.19–11.21]. This possibility is examined with the axisymmetric formulation, but the length of shell domain analyzed is increased to $L = N\lambda_C$. N is an odd integer and the imperfection is modulated by a second sinusoid as follows:

$$\bar{w} = -t\left[\omega_o + \omega_1 \cos\left(\frac{\pi x}{N\lambda_C}\right)\right] \cos\left(\frac{\pi x}{\lambda_C}\right). \tag{11.15}$$

This imperfection introduces an amplitude bias towards the valley at $x = 0$. The upper limit of integration in the PVW becomes $N\lambda_C$, λ_C in the denominator of the displacement functions (11.14) is replaced by $N\lambda_C$, and $b_0 = \delta_x/N\lambda_C$. In the cases discussed here $N = 7$. For this value of N, 81 Gauss integration points were used in the axial direction and $N_u = N_w = 21$.

A set of results from this model applied to the same tube as in the previous figures appears in Figure 11.23 ($L = 7\lambda_C$). The basic imperfection amplitude ω_o is 0.1%, while the biasing imperfection ω_1 takes amplitudes of 0.1% and 0.5%. Included in the figure is the uniform wrinkling case ($\omega_1 = 0$). The average limit strain is governed essentially by the amplitude of the valley at $x = 0$ given by ($\omega_o + \omega_1$). Thus, increasing the amplitude of ω_1 results in a lower average limit strain, as illustrated in the figure.

The larger domain considered, coupled with the imperfection bias, facilitate localization. The growth of the wrinkles at different average strain values is illustrated in

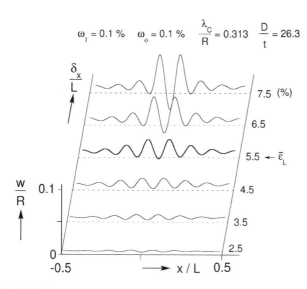

Figure 11.24 Radial displacement axial profiles at different axial displacements illustrating localization of axial wrinkles.

Figure 11.24 (here the full length is shown for clarity). Initially, the wrinkles grow in a controlled manner. In the neighborhood of the limit load ($\delta_x/L = 5.5\%$) the growth of the central wrinkles accelerates. After the limit load, parts of the structure away from the localizing zone start to unload, and the central wrinkles grow very fast. Eventually, this will lead to local folding. Three-dimensional renderings of the localization of axisymmetric wrinkling, calculated for the same parameters as those of Figure 11.24, are shown in Figure 11.25. The first configuration is before the limit point, the second corresponds to it, and the third is at a higher displacement. The localization results in the more precipitous drop in the axial stress seen in Figure 11.23. The steepness of the descending part of the response following the limit load depends on the overall size of the structure. In a larger structure, the localization can occur in an uncontrolled manner, leading to the rapid formation of a fold of the type shown in Figure 11.4(a).

11.4.6 Bifurcation into Non-Axisymmetric Buckling Modes

As illustrated in the experiments in Section 11.2, for the particular combination of tube geometries and material properties used, consideration of non-axisymmetric buckling modes becomes necessary. The first step in such an endeavor is to test the structure for possible bifurcations from axisymmetric wrinkling to non-axisymmetric modes. This prospect is checked by a special algorithm based on the following [11.2, 11.15]. It is again assumed that at a particular equilibrium state two possible incremental solutions exist. We denote their difference by ($\sim$) and identify it with the eigenmode given by

$$\tilde{w} = \cos m\theta \sum_{i=1}^{M_w} c_i \cos\left(i - \frac{1}{2}\right)\frac{\pi x}{\lambda_C},$$

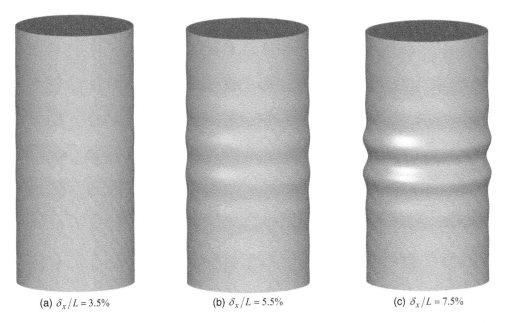

(a) $\delta_x/L = 3.5\%$ (b) $\delta_x/L = 5.5\%$ (c) $\delta_x/L = 7.5\%$

Figure 11.25 Deformed configurations of wrinkled tubes (a) before, (b) at and (c) after the limit load.

$$\tilde{u} = \cos m\theta \sum_{i=1}^{M_u} d_i \sin\left(i - \frac{1}{2}\right) \frac{\pi x}{\lambda_C},$$

(11.16)

$$\tilde{v} = \sin m\theta \sum_{i=1}^{M_v} e_i \cos\left(i - \frac{1}{2}\right) \frac{\pi x}{\lambda_C}$$

(i.e., it is assumed that the eigenmode will have an axial wavelength which is twice that of the prebuckling solution [11.16]). The corresponding strains take the form

$$\tilde{\varepsilon}_{xx}^o = \tilde{u}_{,x} + w_{o,x}\tilde{w}_{,x} + \bar{w}_{,x}\,\tilde{w}_{,x}\,,\ \text{etc.}\quad\text{and}\quad \tilde{\kappa}_{xx} = -\tilde{w}_{,xx}\,,\ \text{etc.}$$

(f)

We identify (11.16) with the vector of unknown coefficients

$$\tilde{q} = [c_1, c_2 \ldots, c_{M_w}, d_1, d_2 \ldots, d_{M_u}, e_1, e_2 \ldots, e_{M_v}]^T.$$

(g)

Due to its construction, $\{\sim\}$ must also satisfy the *PVW*. Thus,

$$\left\{\int_0^{2\lambda_C} \int_0^{2\pi} [\tilde{N}_{\alpha\beta,j}\tilde{\varepsilon}_{\alpha\beta,i}^o + \tilde{M}_{\alpha\beta,j}\tilde{\kappa}_{\alpha\beta,i} + N_{xxo}(\tilde{\phi}_{x,i}\tilde{\phi}_{x,j} + \tilde{\phi}_{,i}\,\tilde{\phi}_{,j})\right.$$

$$\left. + N_{\theta\theta o}(\tilde{\phi}_{\theta,i}\tilde{\phi}_{\theta,j} + \tilde{\phi}_{,i}\,\tilde{\phi}_{,j})]Rdxd\theta\right\} = H_{ij}\tilde{q}_i\tilde{q}_j = 0,$$

$$i, j = 1, 2, \ldots, (M_w + M_u + M_v),\quad (11.17)$$

where $(\bullet)_{,i} \equiv (\bullet)_{,\bar{q}_i}$. If the solution is unique, the LHS of (11.17) is positive for arbitrary $\tilde{q}$. It becomes negative for some $\tilde{q}$ as we pass the bifurcation point. As a result, bifurcation is identified by checking the sign of the determinant of the matrix H. Note that a full wavelength is used in the bifurcation test. In the case of the longer domain ($N\lambda_C$) the integration is performed over the wave with the largest amplitude. The following incremental constitutive relations are adopted in evaluating (11.17)

$$
\begin{Bmatrix}
\tilde{N}_{xx} \\
\tilde{N}_{\theta\theta} \\
\tilde{N}_{x\theta} \\
\tilde{M}_{xx} \\
\tilde{M}_{\theta\theta} \\
\tilde{M}_{x\theta}
\end{Bmatrix}
= \int_{-t/2}^{t/2}
\begin{bmatrix}
A & zA \\
zA & z^2A
\end{bmatrix} dz
\begin{Bmatrix}
\tilde{\varepsilon}^o_{xx} \\
\tilde{\varepsilon}^o_{\theta\theta} \\
\tilde{\varepsilon}^o_{x\theta} \\
\tilde{\kappa}_{xx} \\
\tilde{\kappa}_{\theta\theta} \\
\tilde{\kappa}_{x\theta}
\end{Bmatrix}
\tag{h}
$$

where

$$
A =
\begin{bmatrix}
A_2/A_1 C_{11} & C_{12} & 0 \\
C_{12} & A_1/A_2 C_{22} & 0 \\
0 & 0 & C_{33}
\end{bmatrix}.
$$

$[C_{ij}]$ are the incremental deformation theory moduli (13.26) evaluated by adopting the concept of Hill's *Comparison Solid* and the state of stress from the axisymmetric solution. If the solid is anisotropic, $[C_{\alpha\beta}] \, \alpha, \beta = 1, 2$ are the inverse of the matrix in (13.28) with $\sigma_{x\theta} = 0$ and

$$
C_{33} = \frac{\gamma}{2} + \left(1 + v - \frac{\gamma}{2}\right)\frac{E_s(\sigma_e)}{E}, \quad \gamma = \frac{1}{S_{r\theta}^2}. \tag{i}
$$

A typical set of results for the parameters of a tube with $D/t = 26.3$ appear in Figure 11.26. Shown are the calculated axisymmetric responses for ω_o of 0.2% and 1.0%. As pointed out earlier, as the imperfection amplitude increases, the response becomes somewhat softer and the limit load occurs at a smaller mean strain. Bifurcation points established through the bifurcation check described above for $m = 2, 3$ and 4 are marked on each response with arrows ($\downarrow$). They occur at increasingly higher strains. $m = 2$ occurs at the lowest average strain and it will thus be designated as the critical value. The corresponding average strain will be designated as the second bifurcation critical strain ($\bar{\varepsilon}_{C2}$). The results show that the second bifurcation strains are influenced by the amplitude of ω_o.

The last point is further illustrated in Figure 11.27, where $\bar{\varepsilon}_{C2}$ corresponding to $m = 2, 3$ and 4 are plotted together with $\bar{\varepsilon}_L$ against D/t. Results for $\omega_o = 0.2\%$ and 1.0% are depicted in Figure 11.27(a) and (b), respectively (SAF3 material). The first point that can be made is that for all cases $\bar{\varepsilon}_{C2} < \bar{\varepsilon}_L$, which corresponds to the experimental observations. For lower D/t values, $\bar{\varepsilon}_{C2}$ (2) is the lowest of the bifurcation strains, while for higher values, $\bar{\varepsilon}_{C2}$ (3) becomes lower. The switch occurs around $D/t \approx 47$ for the larger imperfection and around 55 for the smaller imperfection. The switch from $m = 2$ to 3 was also observed in the experiments, where it was found to occur at the somewhat smaller value of $D/t \approx 37$. The results illustrate the strong influence of the axisymmetric imperfection on the solution of the problem.

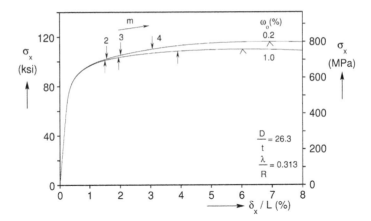

Figure   11.26   Axisymmetric   wrinkling   stress–displacement   responses   and   positions   of non-axisymmetric bifurcations.

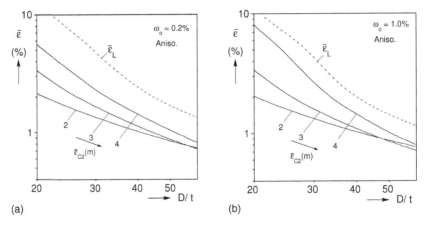

Figure 11.27   Axisymmetric limit strain and non-axisymmetric bifurcation strains vs. D/t. (a) $\omega_o = 0.2\%$ and (b) $\omega_o = 1.0\%$.

11.5   NON-AXISYMMETRIC BUCKLING AND COLLAPSE

A model for following the response of shells which switch to non-axisymmetric buckling modes has been developed in [11.15] (see also [11.18, 11.22, 11.23]). In this case, the shell is assigned an initial geometric imperfection of the type

$$\bar{w} = -t\left[\omega_o \cos\left(\frac{\pi x}{\lambda_C}\right) + \omega_m \cos\left(\frac{\pi x}{2\lambda_C}\right)\cos m\theta\right] \qquad (11.18)$$

where ω_o is the amplitude of the axisymmetric imperfection with wavelength $2\lambda_C$ and ω_m is the amplitude of the non-axisymmetric imperfection with circumferential wave number

m and axial wavelength $4\lambda_C \cdot \lambda_C$ comes from the first bifurcation calculation and m from the second.

All membrane and bending stress intensities and strains as listed in (11.10) are now included in the *PVW*, which becomes

$$\int_0^{2\lambda_C} \int_0^{2\pi} \{N_{\alpha\beta}\delta\varepsilon_{\alpha\beta}^o + M_{\alpha\beta}\delta\kappa_{\alpha\beta}\}Rd\theta dx = 0. \tag{11.19}$$

The problem is now discretized with the following displacement series:

$$w = a_0 + \sum_{i=1}^{I} a_i \cos\left(\frac{i\pi x}{\lambda_C}\right) + \sum_{j=1}^{J}\sum_{k=1}^{K} a_{jk} \cos\left(\frac{(2j-1)\pi x}{2\lambda_C}\right)\cos(km\theta), \tag{11.20}$$

$$u = b_0 x + \sum_{i=1}^{I} b_i \sin\left(\frac{i\pi x}{\lambda_C}\right) + \sum_{j=1}^{J}\sum_{k=1}^{K} b_{jk} \sin\left(\frac{(2j-1)\pi x}{2\lambda_C}\right)\cos(km\theta), \quad \left(b_0 = \frac{\delta_x}{2\lambda_C}\right),$$

$$v = \sum_{j=1}^{J}\sum_{k=1}^{K} c_{jk} \cos\left(\frac{(2j-1)\pi x}{2\lambda_C}\right)\sin(km\theta).$$

The *PVW* now becomes

$$\int_0^{2\lambda_C} \int_0^{2\pi} [N_{xx}\varepsilon_{xx,i}^o + N_{\theta\theta}\varepsilon_{\theta\theta,i}^o + 2N_{x\theta}\varepsilon_{x\theta,i}^o + M_{xx}\kappa_{xx,i} + M_{\theta\theta}\kappa_{\theta\theta,i} + 2M_{x\theta}\kappa_{x\theta,i}]Rd\theta dx\delta q_i = 0$$

$$i = 1, 2, \ldots, (2I + 3JK), \quad (\bullet)_{,i} \equiv \frac{\partial(\bullet)}{\partial q_i} \tag{a}$$

and

$$q = [a_o, a_2, \ldots, a_I, b_2, \ldots, b_I, a_{11}, a_{12}, \ldots, a_{JK}, b_{11}, b_{12}, \ldots, b_{JK}, c_{11}, c_{12}, \ldots, c_{JK}]^T.$$

Typically $I = J = K = 5$; 16 Gauss integration points are typically used in the axial direction, 24 in the circumferential direction and 5 through the thickness.

The problem can also be solved with a nonlinear finite element code by considering a periodic domain of length $2\lambda_C$, which is assigned an initial imperfection of the type of (11.18). In such an effort, λ_C and m need to be first determined from bifurcation analyses, like the ones presented in Sections 11.3.1 and 11.4.6.

11.5.1 Results

A typical set of results corresponding to a tube with $D/t = 28.5$ (Exp. CW3) is shown in Figure 11.28. The imperfection used is of the type in (11.18). The axisymmetric component has wavelength $2\lambda_C = 0.602R$ and amplitude $\omega_o = 0.05\%$. The purely axisymmetric response is included in each of the two figures. The second bifurcation check

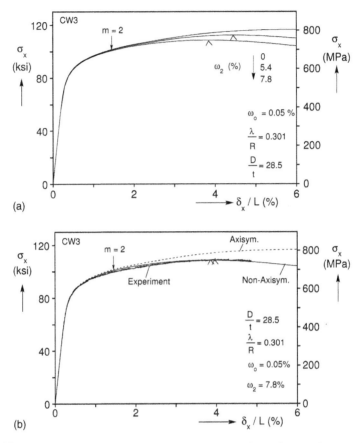

Figure 11.28 (a) Stress–displacement response for three values of non-axisymmetric imperfection ω_2 for $D/t = 28.5$. (b) Comparison of non-axisymmetric axial stress–displacement response with experimental one for $D/t = 28.5$.

yielded a non-axisymmetric mode of $m = 2$, which is seen to occur at the relatively small strain of approximately 1.45%. Three-dimensional results for a non-axisymmetric imperfection corresponding to this mode with amplitude ω_2 of 5.4% and 7.8% are included in Figure 11.28(a). The two responses separate from the axisymmetric one in a nearly tangential manner in the neighborhood of the bifurcation point. Each response subsequently attains a limit load. As ω_2 is increased, the limit load occurs at a smaller average strain. Furthermore, the limit strains are lower than the one for the case with the purely axisymmetric imperfection. What is striking is that for this to happen the values of ω_2, although still a small fraction of the wall thickness, must be two orders of magnitude larger than ω_o. When both ω_o and ω_2 are small and of the same order of magnitude, the limit strain is governed by ω_o.

Figure 11.28(b) shows a comparison of the non-axisymmetric response that best agrees with the corresponding experimental one. For $\omega_o = 0.05\%$, this corresponds to a non-axisymmetric imperfection with amplitude of 7.8%. Despite the idealizations made by the assumed periodicity and symmetries, the calculated response is seen to be in very good agreement with the measured one. The limit strain is also very well captured by the

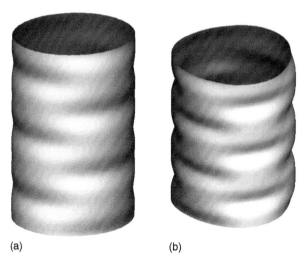

(a)   (b)

Figure 11.29   Non-axisymmetric buckling modes: (a) $m=2$ and (b) $m=3$.

analysis. A deformed configuration of a section of a shell that includes three axial waves and an $m=2$ imperfection is shown in Figure 11.29(a). The 90° rotation of the $m=2$ mode between adjacent $2\lambda_C$ long sectors of the shell is clearly illustrated. Compression beyond the calculated limit load will lead to localization of one of the axial waves, leading to a pattern like the one shown in Figure 11.12. Further compression will result in concertina folding with the folds 90° out of phase, as shown in Figure 11.4(b).

A second set of results for a tube with $D/t=43.3$ (Exp. CW14) is shown in Figure 11.30. The axisymmetric imperfection has wavelength of $2\lambda_C=0.492R$ and amplitude $\omega_o=0.05\%$. The purely axisymmetric response bifurcates with the $m=3$ mode at an average strain of 1.05%. Non-axisymmetric responses for ω_3 of 4% and 6% are shown in Figure 11.30(a). The trend is similar to that of the results for the thicker shell in Figure 11.28 although the limit loads occur at much smaller average strains. The case $\omega_3=6\%$ is compared to the corresponding measured response in Figure 11.30(b). The response as well as the position of the limit load are well captured by the analysis. The $m=3$ mode is seen fully developed in Figure 11.29(b). The rendering includes 5 full waves with adjacent sectors rotating by 60°. Following the limit load, one of the sectors will localize in the manner illustrated in Figure 11.13. Continued compression will lead to the type of folding seen in Figure 11.4(c).

Similar simulations have been conducted for several of the other tests. For the predictions to match the experimental values of $\bar{\varepsilon}_L$, the amplitudes of the two imperfections had to be selected individually. Each cylinder has geometric and load imperfections introduced by the setup that are unique to the test. Since neither of these are known to the accuracy required, imperfection amplitudes had to be estimated. At the same time, the fact that the data in Figure 11.15 follow a specific trend with relatively small scatter indicates that the imperfections may also have some underlying trends. We saw in Figure 11.22 and in the ensuing discussion that consideration of non-axisymmetric imperfections is required for both qualitative and quantitative reproduction of the experimental results.

With this as background, Bardi *et al.* [11.15] conducted an extensive parametric study aimed at establishing the functional dependence of ω_o and ω_m that reproduces the $\bar{\varepsilon}_L - D/t$

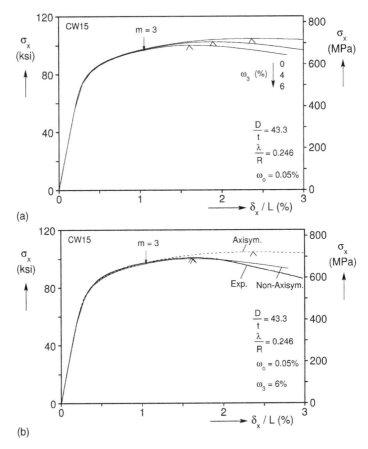

(a)

(b)

Figure 11.30 (a) Stress–displacement response for three values of non-axisymmetric imperfection ω_3 for $D/t = 43.3$. (b) Comparison of non-axisymmetric axial stress–displacement response with experimental one for $D/t = 43.3$.

experimental trend. Optimal results were obtained for $\omega_o = 0.05\%$ and

$$\omega_m = \frac{0.0008}{m^3} \left(\frac{D}{t}\right)^2. \tag{b}$$

The performance of this scheme is illustrated in Figure 11.31, where predictions for ω_2 and ω_3 are compared to the experimental results. For $D/t < 39$, $m = 2$ predictions are lower while $m = 3$ becomes preferred for $D/t > 39$. Thus, the predictions agree with the experimental data both in the magnitude of $\bar{\varepsilon}_L$ and in the mode of collapse. Finally, the two sets of predictions are assembled into one by choosing the lower one in each of the two D/t regimes. The resultant "best prediction" is compared to the experimental results in log–log scales in Figure 11.32. The predictions are now in very good agreement with the experimental results for the whole range of D/t considered.

The robustness of this scheme was tested by varying ω_o and the constant multiplier in Eq. (b). ω_o tends to shift the best fit response upwards. Provided it is kept below 0.1%, its value does not significantly alter the "best prediction" in Figure 11.32. If $\omega_o > 0.1\%$ then

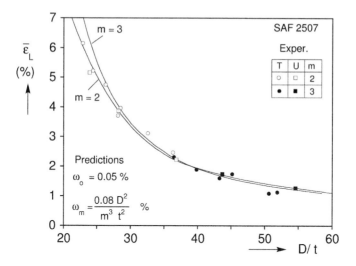

Figure 11.31 Average limit stains vs. D/t: comparison of experiments and predictions for $m = 2$ and $m = 3$.

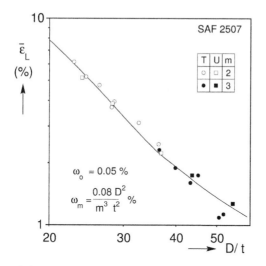

Figure 11.32 Average limit stains vs. D/t: comparison of experiments with "best" predictions.

the constant of proportionality must be increased for the predictions to be in reasonable agreement with experiments.

In summary, for the predictions to match the experimental trend of the $\bar{\varepsilon}_L - D/t$ results quantitatively as well as in the mode of buckling, the amplitude of the non-axisymmetric imperfection ω_m had to be assigned the particular functional dependence in Eq. (b). That imperfection sensitivity is dependent on D/t, and that it should increase as m increases is intuitively acceptable. As to how much this particular functional dependence is governed by the material properties of our tubes and by the imperfections introduced by the specimen fabrication and the experimental setup, is not known. In other words, until Eq. (b) is tested

Table 11.4   Three stress–strain responses adopted in parametric study.

E Msi (GPa)	σ_y ksi (MPa)	σ_o ksi (MPa)	n
28.2 (194.5)	80.0 (552)	76.95 (530.7)	9
28.2 (194.5)	85.0 (586)	82.70 (570.3)	15
28.2 (194.5)	88.0 (607)	86.70 (597.9)	30

against experiments on metals with different stress–strain responses, its universality must be questioned.

11.6   PARAMETRIC STUDY

The critical and limit states are now studied parametrically. In an effort to establish the effect of hardening on these instabilities, the material is assumed to yield isotropically and to obey the Ramberg–Osgood stress–strain response. Three hardening parameters (n) of 9, 15 and 30 are considered. The study is limited to axisymmetric buckling modes. The elastic modulus is kept constant and σ_y is assigned the values listed in Table 11.4, so that the responses cross around a strain of 1.2%. The yield stresses also vary to some degree, taking the values listed in the table. Tubes with D/t values ranging from 20 to 60 were analyzed.

The critical values $\{\sigma_C, \varepsilon_C, \lambda_C\}$ are plotted against D/t in Figure 11.33. Three sets of results are presented in each plot that correspond to the three values of n. The critical stress is seen in Figure 11.33(a) to be highest for $n = 9$ and lowest for $n = 30$. Figure 11.33(b) shows the corresponding λ_C/R vs. D/t plots. They follow the square root dependence on t/D implied by Eq. (11.4a). At the same time, increasing n results is an increase in λ_C/R.

Because of the nature of the stress–strain responses adopted, the critical strains follow powerlaw relationships with D/t and appear as linear in the log–log plot in Figure 11.33(c). The results illustrate the strong influence of the hardening exponent on ε_C. The results for $n = 30$ are significantly lower than the corresponding ones for $n = 9$.

Figure 11.33(c) includes values of the average strains at the calculated limit load for strictly axisymmetric imperfections with $\omega_o = 0.2\%$. As expected, the level of $\bar{\varepsilon}_L$ is significantly higher than the corresponding ε_C for all three hardening parameters. Once again, reduction in the hardening of the material (increase in n) reduces $\bar{\varepsilon}_L$. Indeed, the difference between $\bar{\varepsilon}_L$ for $n = 9$ and 30 is strikingly large. For example, for $D/t = 30$ and $n = 9$, $\bar{\varepsilon}_L$ is 3.17%. By contrast, for $n = 30$, $\bar{\varepsilon}_L$ is only 1.50%.

Similar results were also generated for stress–strain responses of the same hardening but with lower yield stresses. This was done by changing the length of the linear part of the stress–strain response. The effect on ε_C is relatively small. For example, in the case of $D/t = 30$ and $n = 15$, when the response is shifted downwards by 10 and 20 ksi (69 and 138 MPa), the critical strain increases by approximately 4% and 8%, respectively. Although this change depends on both n and D/t, it is of secondary importance and will not be further pursued here. $\bar{\varepsilon}_L$ is also influenced by the material yield stress. However, its effect is again relatively secondary.

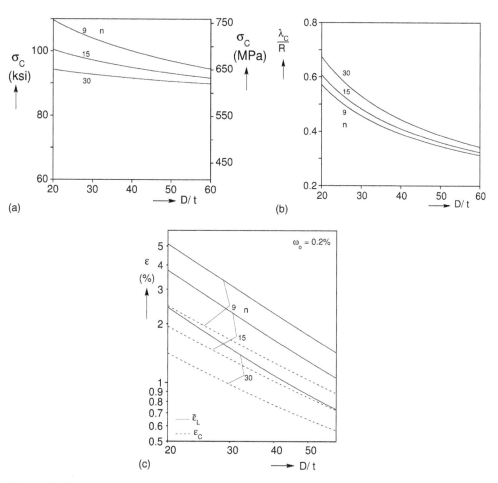

Figure 11.33 Effect of hardening parameter n on: (a) the critical stress, (b) on λ_C, and (c) on the critical and limit strains for axisymmetric buckling.

The sensitivity of ε_C and $\bar{\varepsilon}_L$ to material hardening raises the question as to how hot finished steel shells buckle and collapse. Upon yielding, the material response of such steels usually exhibits Lüders banding (see Section 13.1.2). This material instability usually extends over strains of 1–3%. At the time of the writing of this chapter, the effect of the Lüders' strain on the problem was still not well understood. However, preliminary work points to the following: the presence of Lüders bands generally complicates the onset of wrinkling and can affect $\bar{\varepsilon}_L$. Lüders banding is an inhomogeneous deformation. During its development, parts of a loaded structure, such as a compressed cylinder, can be in the relatively undeformed phase while other parts become fully deformed to the Lüders strain. Sections that undergo Lüders straining will usually wrinkle simultaneously. The amplitude of the wrinkles grows gradually until the whole domain is strained to the Lüders strain. If the pipe D/t is low enough, the wrinkles do not cause failure at this stage. Instead, as the material enters the hardening stage, they continue to grow, and eventually lead to a limit load instability similar to what has been reported here. In this case, the limit load is governed mostly by the hardening of the stable part of the response. Accordingly, the trend of the present results may be applicable. For thinner shells, collapse can be initiated at strains within the Lüders strain, in which case the present analysis is not applicable. Clearly this subject is complex and requires further investigation and understanding.

11.7 SUMMARY AND RECOMMENDATIONS

Cylinders thick enough to undergo plastic compression experience a cascade of events eventually leading to failure, that usually manifests as localized axisymmetric or non-axisymmetric folding. At some strain level in the plastic regime, the cylinder develops uniform axisymmetric wrinkling. Under continued compression the wrinkles grow stably, gradually reducing the axial rigidity of the structure. This reduction in axial rigidity eventually leads to a limit load instability. Beyond the limit load, deformation localizes. The limit load can therefore be designated as the limit state of the structure. Subsequent events can be usually tracked under displacement controlled loading.

In the case of materials with relatively healthy hardening, such as the SAF 2507 used in the experiments of [11.10], the initial axisymmetric mode reverts to a non-axisymmetric mode before the limit load is attained. Typically, the mode consists of a few ($\sim$2–5) circumferential waves. In this case, the buckling mode grows gradually, and induces a limit load instability that becomes the limit state of the structure. It is important to note that for both axisymmetric and non-axisymmetric collapse, the onset of wrinkling and the limit state can be separated by significant levels of strain. For example, in the tests presented, the separation strain was about 0.25% for $D/t = 51.9$ and more than 5% for $D/t = 22.9$. This difference can be exploited in the design of displacement-controlled applications.

The onset of wrinkling can be easily established using the classical plastic bifurcation check presented in Section 11.3 that is based on deformation theory of plasticity. In addition to the critical stress and strain, this yields the wavelength of axisymmetric wrinkling. It has been demonstrated that anisotropy can strongly influence the critical state, including the wavelength.

The evolution of wrinkling and the ensuing limit state can be simulated by modeling a λ_C long periodic domain that is assigned an initial imperfection corresponding to the axisymmetric buckling mode, as described in Section 11.4. The model outlined is discretized by adopting kinematically admissible displacement functions. Its performance can be reproduced by discretizing the structure with axisymmetric finite elements, provided the same imperfection characteristics are used. The flow theory of plasticity with isotropic hardening suffices for such calculations. Once again, anisotropy can influence the calculated limit strain and should be incorporated in the plasticity subroutine used.

The possibility of bifurcation of the axisymmetric solution to a non-axisymmetric buckling mode requires use of a custom bifurcation check like the one outlined in Section 11.4.6. Once the mode of buckling is identified, a non-axisymmetric model like the one described in Section 11.5 can be used to simulate the evolution of such buckling modes. In this model, the $2\lambda_C$ long domain is assigned imperfections with an axisymmetric and a non-axisymmetric component. This model will yield the limit state associated with this buckling mode. The accuracy of such predictions is strongly influenced by how representative the two imperfection amplitudes ω_o and ω_m are of the real structure. A general guideline on how to choose these amplitudes is not presently available.

It has been demonstrated that, among other factors, $\bar{\varepsilon}_L$ is influenced by the hardening of the material. If stress–strain data are not available at the design stage, a conservative approach is to assume that the material exhibits relatively low hardening (e.g., $n = 30$ in a Ramberg-Osgood response) along with the design yield stress. Calculate the critical state, and then run an axisymmetric simulation with a relatively high value of ω_o ($\sim$0.1–0.2%; or adopt amplitudes representative of expected imperfections if available).

REFERENCES

11.1. Klever, F.J., Palmer, A.C. and Kyriakides, S. (1994). Limit-state design of high-temperature pipelines. *Proc. 13th International Conference on Offshore Mechanics and Arctic Engineering*, Vol. V, Houston, TX, February 1994, pp. 77–92.

11.2. Yun, H.D. and Kyriakides, S. (1990). On the beam and shell modes of buckling of buried pipelines. *Soil Dyn. Earthq. Eng.* **9**, 179–193.

11.3. Murray, D.W. (1997). Local buckling, strain localization, wrinkling and postbuckling response of a line pipe. *Eng. Struct.* **19**, 360–371.

11.4. Wilkie, S.A., Dobranko, R.M. and Fladager, S.J. (2000). Case history of local wrinkling of a pipeline. *Proc. ASME International Pipeline Conference*, Vol. 2, Calgary, Alberta, Canada, October 1–5, 2000, pp. 917–922.

11.5. Oswell, J.M., Hanna, A.J., Dobranko, R.M. and Wilkie, S.A. (2000). Instrumentation and geotechnical assessment of local pipe wrinkling on the Norman Wells pipelines. *Proc. ASME International Pipeline Conference*, Vol. 2, Calgary, Alberta, Canada, October 1–5, 2000, pp. 923–930.

11.6. Peek, R. (2000). Axisymmetric wrinkling of cylinders with finite strain. *ASCE J. Eng. Mech.* **126**, 455–461.

11.7. Bardi, F.C., Yun, H.D. and Kyriakides, S. (2003). On the axisymmetric progressive crushing of circular tubes under axial compression. *Int. J. Solid. Struct.* **40**, 3137–3155.

11.8. Lee, L.H.N. (1962). Inelastic buckling of initially imperfect cylindrical shells subject to axial compression. *J. Aeronaut. Sci.* **29**, 87–95.

11.9. Batterman, S.C. (1965). Plastic buckling of axially compressed cylindrical shells. *AIAA J.* **3**, 316–325.

11.10. Bardi, F.C. and Kyriakides, S. (2006). Plastic buckling of circular tubes under axial compression. Part I experiments. *Int. J. Mech. Sci.* **48**, 830–841.

11.11. Bijlaard, P.P. (1949). Theory and tests on the plastic stability of plates and shells. *J. Aeronaut. Sci.* **16**, 529–541.

11.12. Hutchinson, J.W. (1974). Plastic buckling. *In Advances in Applied Mechanics*, **14**, Ed. C.S. Yih. Academic Press, NY, pp. 67–144.

11.13. Kyriakides, S., Bardi, F.C. and Paquette, J.A. (2005). Wrinkling of circular tubes under axial compression: effect of anisotropy. *ASME J. Appl. Mech.* **72**, 301–305.

11.14. Gellin, S. (1979). Effect of an axisymmetric imperfection on the plastic buckling of an axially compressed cylindrical shell. *ASME J. Appl. Mech.* **46**, 125–131.

11.15. Bardi, F.C., Kyriakides, S. and Yun, H.D. (2006). Plastic buckling of circular tubes under axial compression. Part II analysis. *Int. J. Mech. Sci.* **48**, 842–854.

11.16. Koiter, W.T. (1963). The effect of axisymmetric imperfections on the buckling of cylindrical shells under axial compression. *Pro. Kon. Ned. Ak. Wet.* **B66**, 265–279.

11.17. Bushnell, D. (1976). BOSOR5-Program for buckling of elastic–plastic complex shells of revolution including large deflections and creep. *Comput. Struct.* **6**, 221–239.

11.18. Tvergaard, V. (1983a). Plastic buckling of axially compressed circular cylindrical shells. *Thin Wall. Struct.* **1**, 139–163.

11.19. Needleman, A. and Tvergaard, V. (1982). Aspects of plastic post-buckling behavior. In *Mechanics of Solids*, The Rodney Hill 60th Anniversary Volume, Eds. H.G. Hopkins and M.J. Sewell. Pergamon Press, Oxford, pp. 435–498.

11.20. Ju, G.-T. and Kyriakides, S. (1992). Bifurcation and localization instabilities in cylindrical shells under bending: Part II predictions. *Int. J. Solid. Struct.* **29**, 1143–1171.

11.21. Vaze, S.P. and Corona, E. (1997). Response and stability of square tubes under bending. *ASME J. Appl. Mech.* **64**, 649–657.

11.22. Tvergaard, V. (1983b). On the transition from a diamond mode to an axisymmetric mode of collapse in cylindrical shells. *Int. J. Solid. Struct.* **19**, 845–856.

11.23. Mikkelsen, L.P. (1995). Elastic–viscoplastic buckling of circular cylindrical shells under axial compression. *Eur. J. Mech., A/Solid.* **14**, 901–920.

12

Combined Internal Pressure and Axial Compression

Offshore as well as onshore pipelines in operation are under internal pressure. The scenarios for developing axial compression outlined in Chapter 11 also apply to a pressurized pipeline. The general features of plastic buckling under axial compression and internal pressure are similar to those of pure axial loading, presented in Figure 11.2. The cylinder first wrinkles at an increasing load. The wrinkle amplitude grows, leading eventually to a limit load instability. A pipeline will fail by localized collapse at this strain, and as a result this constitutes a limit state. The biaxial state of stress lowers the axial stress levels of the various critical events described in Figure 11.2 but, as will be demonstrated, has a smaller effect on the corresponding strains. In addition, the pressure has a stabilizing effect on the axisymmetric mode, making a switch to non-axisymmetric modes more difficult. These features of the problem are first illustrated experimentally. The formulation for predicting the onset of plastic wrinkling is then developed, followed by a study of how wrinkles grow, localize and lead to collapse.

12.1 COMBINED AXIAL COMPRESSION–INTERNAL PRESSURE EXPERIMENTS

Lee [12.1] reported results from buckling experiments on cylindrical shells under combined internal pressure and axial compression. The experiments were conducted on tubes made of a soft aluminum alloy (Al-3003-0). The ends of the tubes were clamped, and consequently the onset of wrinkling was masked and was not reported. The axial stress at the onset of collapse was measured for different loading paths.

Combined loading experiments on line pipe were reported by Murray [12.2]. He used pipes of various L/D ratios with D/t s of 64.3 and 50.6. The majority of the tests involved combined compression, bending and internal pressure. The tests were particularly aimed at understanding the behavior of the pipe well past the onset of local collapse.

Lee's results did not allow for the establishment of the onset of wrinkling, and Murray's experiments dealt with more complex loadings. We thus will concentrate on the conduct and the analysis of a more recent set of experiments involving SAF 2507 super-duplex tubes compressed to failure under fixed levels of internal pressure [12.3]. The tests were designed to approximate this loading as encountered in a long pipeline.

12.1.1 Experimental Set-Up

The test set-up is the same as the one in Figure 11.5 but has been extended to allow internal pressurization of the specimens. The extended set-up, including the pressurizing system, is shown schematically in Figure 12.1. The specimen cavity was sealed with O-ring seals located on the solid end-plugs. In addition, a solid metal insert was placed in the cavity to reduce the volume of the pressurizing fluid. The pressure was monitored by an electrical pressure transducer and by mechanical dial gages. The cavity was filled with hydraulic fluid and pressurized by a stand-alone, closed-loop control, pressure intensifier. The intensifier operates on standard 3,000 psi (207 bar) hydraulic power and has a capacity of 10,000 psi (690 bar). In the present experiments, it was operated under pressure control (see Figure 12.1).

The test specimens were machined out of SAF 2507 super-duplex stock as described in Section 11.2. Wrinkles formed by compression in the presence of internal pressure have a longer wavelength than those formed under pure compression. For this reason, the test section of the combined loading test specimens was 5 in (127 mm – $L/R \approx 4.6$) long, while their overall length was 13 in (330 mm). The rest of the geometric characteristics were similar to those described in Section 11.2.

In a typical experiment, the pressure was increased to the required level and was held constant. The specimen was then axially compressed at a displacement rate that produced a strain rate of approximately 2×10^{-5} s^{-1} in the test section. Periodically, the loading was paused for a few seconds to perform axial scans of the specimen. The scans were

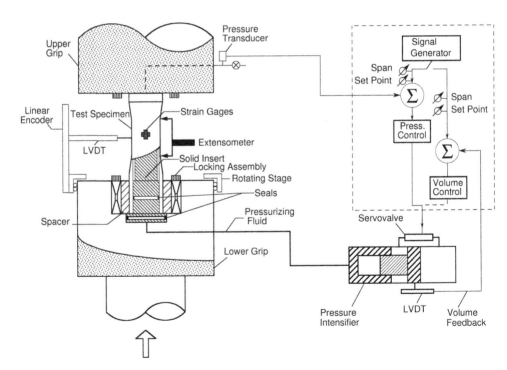

Figure 12.1   Schematic of combined compression and internal pressure test set-up.

subsequently used to identify the onset of wrinkling and to establish the evolution of the wrinkles. Typically, one or two of the wrinkles tended to localize, leading to a limit load.

12.1.2 Experimental Results

Sets of tubes of two D/t values (28.3 and 39.8) were compressed at different levels of internal pressure until a limit load instability developed. The parameters of each set of tests and the main results are summarized in Tables 12.1 and 12.2. Pressure is normalized by the yield pressure, defined as

$$P_o = S_\theta \sigma_o \left(\frac{t}{R} \right). \tag{12.1}$$

The properties of the tube stock used are listed in Table 11.2. A typical axial stress-shortening $(\sigma_x - \delta_x/L)$ response for a tube with $D/t = 28.3$, tested at a fixed pressure of 3,010 psi (208 bar or $P = 0.439 P_o$) is shown in Figure 12.2(a). Figure 12.2(b) shows a set of axial scans taken in the course of the test at the positions marked on the response with solid bullets. The response is similar to the one presented in Figure 11.6(a) for pure axial compression. The first effect of the combined loading is that the material yields at a lower axial stress, causing a lowering of the inelastic part of the response. In the early scans, the test section deformed nearly uniformly. Later scans reveal that small bulges formed at the two ends of the test section. Although the linear tapers next to the test section reduced the stress concentrations at the discontinuities, they did not eliminate them completely. Thus, at higher compression strains, small bulges formed at these locations. Wrinkles gradually began to appear in the rest of the test section as well. It was determined that the onset of wrinkling (ε_C) occurred between strains of 1.20% and 1.40% (points marked with "$\downarrow$" on response). The stress corresponding to these strains is designated as σ_C. A second major effect of internal pressure is an increase in the wrinkle wavelength. For the case shown in Figure 12.2, the wavelength was 1.5 times the value at zero pressure (see Table 12.1).

As compression progressed, the wrinkle amplitude grew, causing a reduction in the stiffness of the specimen. At higher strains, the growth of the two end bulges accelerated, resulting in a limit load at an average strain of $\bar{\varepsilon}_L = 3.50\%$. The test was terminated soon after the limit load. Interestingly, the wrinkling pattern remained axisymmetric throughout the test.

A total of six tests were performed for this D/t value. The recorded stress–displacement responses are shown in Figure 12.3. The geometric variables and the critical parameters of these tests appear in Table 12.1. Pressure lowers the response, and as a consequence, the critical and limit stresses decrease. This is demonstrated in Figure 12.4(a), where the critical stress is plotted against pressure. By contrast, the critical strain is seen to remain relatively unchanged in Figure 12.4(b). The corresponding wrinkle half-wavelengths are plotted against pressure in Figure 12.4(c). The wavelength is seen to increase nearly linearly with pressure, almost doubling in value when going from $P = 0$ to $P = 0.688 P_o$. Figure 12.5 shows photographs of two specimens at the completion of the tests. One was tested at a relatively low pressure $(P = 0.147 P_o)$, and the second at a relatively high pressure $(P = 0.688 P_o)$. The wrinkles as well as the bulges at the ends of the test sections can be seen in the photograph. The difference in wrinkle wavelength between the two is quite striking.

Table 12.1 Specimen parameters and critical variables measured in axial buckling under internal pressure experiments ($D/t = 28.3$).

Exp No.	Tube No.	D in (mm)	t in (mm)	$\dfrac{D}{t}$	$t_{min} - t_{max}$ in (mm)	P psi (bar)	σ_C ksi (MPa)	$\% \, \varepsilon_C$	$\dfrac{\lambda_C}{R}$	σ_L ksi (MPa)	$\% \, \bar{\varepsilon}_L$
IPC5	SAF4	2.2555 (57.29)	0.0802 (2.04)	28.13	0.0798–0.0807 (2.03–2.05)	0	96.3–97.5 (664–672)	1.21–1.41	0.280–0.363	109.8 (757)	3.71
IPC2	SAF4	2.2549 (57.27)	0.0801 (2.04)	28.16	0.0791–0.0810 (2.01–2.06)	1,011 (69.7)	94.5–96.8 (652–668)	1.45–1.74	0.317–0.386	106.1 (732)	3.72
IPC1	SAF4	2.2521 (57.20)	0.0781 (1.98)	28.82	0.0775–0.0787 (1.97–2.00)	2,008 (138.5)	90.1–93.1 (621–642)	1.18–1.51	0.382–0.446	101.2 (698)	3.03
IPC3	SAF4	2.2552 (57.28)	0.0797 (2.02)	28.28	0.0786–0.0803 (2.00–2.04)	3,010 (207.6)	82.2–84.3 (567–581)	1.20–1.40	0.441–0.529	95.7 (660)	3.50
IPC4	SAF4	2.2557 (57.29)	0.0800 (2.03)	28.21	0.0792–0.0806 (2.01–2.05)	4,000 (275.9)	75.2–76.9 (519–530)	1.37–1.57	0.455–0.584	87.8 (606)	3.69
IPC6	SAF4	2.2556 (57.29)	0.0801 (2.04)	28.15	0.0794–0.0808 (2.02–2.05)	4,744 (327.2)	64.2–66.1 (443–456)	1.07–1.27	0.547–0.653	80.7 (557)	4.01

Table 12.2 Specimen parameters and critical variables measured in axial buckling under internal pressure experiments ($D/t = 39.8$).

Exp No.	Tube No.	D in (mm)	t in (mm)	$\dfrac{D}{t}$	$t_{min} - t_{max}$ in (mm)	P psi (bar)	σ_C ksi (MPa)	$\% \, \varepsilon_C$	$\dfrac{\lambda_C}{R}$	σ_L ksi (MPa)	$\% \, \bar{\varepsilon}_L$
IPC11	SAF5	2.2112 (56.16)	0.0555 (1.41)	39.84	0.0546–0.0567 (1.39–1.44)	0	97.4–99.2 (672–684)	0.86–1.01	0.241–0.306	105.1 (725)	1.89
IPC10	SAF5	2.2092 (56.11)	0.0544 (1.38)	40.62	0.0537–0.0553 (1.36–1.41)	932 (64.3)	94.0–95.4 (648–658)	0.89–1.05	0.320–0.363	97.5 (672)	1.31
IPC7	SAF5	2.2117 (56.18)	0.0555 (1.41)	39.85	0.0548–0.0564 (1.39–1.43)	1,852 (127.7)	86.8–88.8 (599–612)	0.74–0.88	0.390–0.449	93.9 (648)	1.52
IPC8	SAF5	2.2121 (56.19)	0.0558 (1.42)	39.65	0.0551–0.0564 (1.40–1.43)	2,772 (191.2)	80.5–82.4 (555–568)	0.78–0.96	0.429–0.474	86.8 (599)	1.59
IPC9	SAF5	2.2114 (56.17)	0.0555 (1.41)	39.85	0.0540–0.0572 (1.37–1.45)	3,543 (244.3)	66.1–67.8 (456–468)	0.73–0.87	0.587–0.660	72.0 (497)	1.49
IPC12	SAF5	2.2137 (56.23)	0.0567 (1.44)	39.04	0.0555–0.0579 (1.41–1.47)	3,996 (275.6)	60.3–61.6 (416–425)	0.75–0.90	0.566–0.672	68.2 (470)	1.89

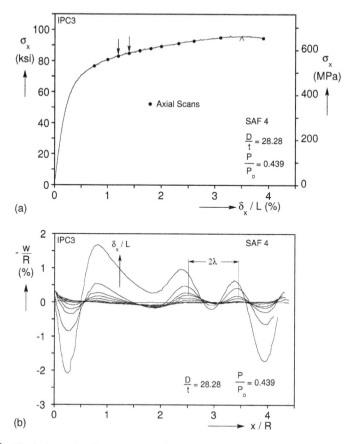

(a)

(b)

Figure 12.2 Typical results for a compression under internal pressure test: (a) recorded axial stress-shortening response and (b) axial scans showing evolution of wrinkles in test section.

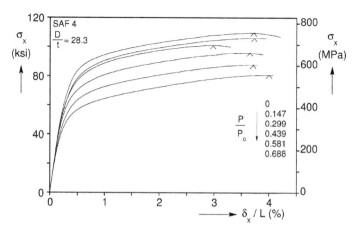

Figure 12.3 Axial stress-shortening responses recorded at different values of internal pressure for tubes with $D/t = 28.3$.

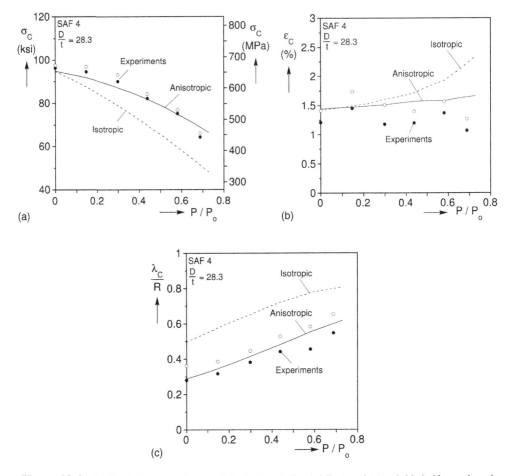

Figure 12.4   (a) Critical stress, (b) critical strain (onset of wrinkling) and (c) wrinkle half-wavelength as a function of internal pressure. Experiments and predictions for $D/t = 28.3$.

Figure 12.6(a) shows a plot of the limit stress against pressure. σ_L decreases with pressure, and is somewhat higher than the wrinkling stress. Figure 12.6(b) shows a plot of the average strain corresponding to limit stress ($\bar{\varepsilon}_L$) against pressure. $\bar{\varepsilon}_L$ is seen to increase modestly with pressure. The corresponding ε_C values are included in the figure for comparison. They are seen to be significantly lower than $\bar{\varepsilon}_L$. Finally, it is noted that in the five pressure experiments, the wrinkled tubes remained axisymmetric up to collapse. By contrast, the pure compression experiment in the set buckled into a non-axisymmetric mode ($m = 2$).

The results from the tests for $D/t = 39.8$ are broadly similar. Figure 12.7 shows the axial stress-shortening responses recorded at six different pressure levels. The main parameters are listed in Table 12.2. The critical stresses, strains and half-wavelengths are plotted against pressure in Figure 12.8. The critical stress drops with pressure, whereas the critical strain remains relatively unaffected. The wrinkle wavelength exhibits a similar increase with pressure to that observed in the previous set of results. The limit stresses and

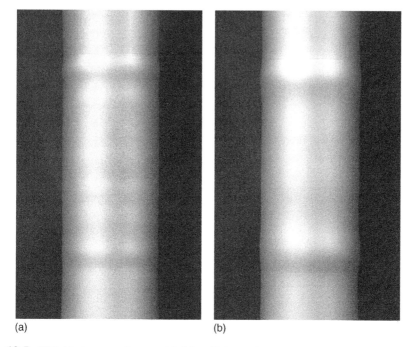

Figure 12.5   Wrinkled test specimens with $D/t = 28.3$ tested at (a) $P = 0.147P_o$ and (b) $P = 0.688P_o$.

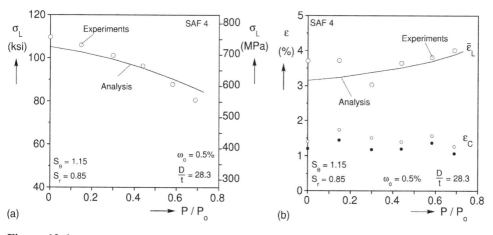

Figure 12.6   Limit stress (a) and average strain (b) as functions of internal pressure for $D/t = 28.3$. Experiments and predictions.

strains are plotted against pressure in Figure 12.9. The stress drops as pressure increases, whereas the strain increases mildly. $\bar{\varepsilon}_L$ is significantly higher than ε_C, but the difference is smaller than for the lower D/t case. It is also worth noting that both strains have much lower values than the thicker tubes in Figure 12.6. Some scatter observed in $\bar{\varepsilon}_L$ is probably caused by initial imperfections. The pressurized tubes buckled axisymmetrically for this set also, whereas the one pure compression tube buckled in a non-axisymmetric manner some time before the limit stress was reached.

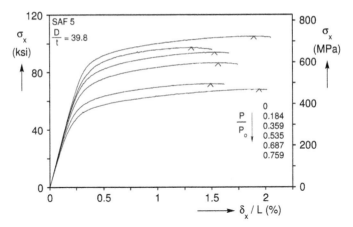

Figure 12.7 Axial stress-shortening responses recorded at different values of internal pressure for tubes with $D/t = 39.8$.

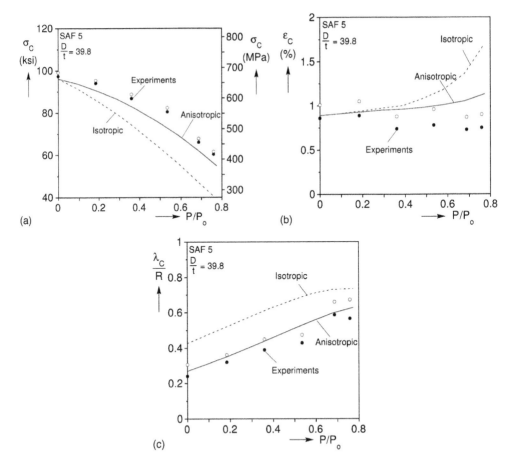

Figure 12.8 (a) Critical stress, (b) critical strain (onset of wrinkling) and (c) wrinkle half-wavelength as a function of internal pressure. Experiments and predictions for $D/t = 28.3$.

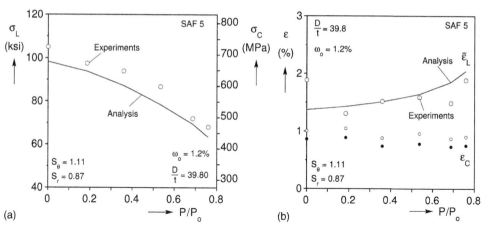

Figure 12.9 Limit stress (a) and strain (b) as functions of internal pressure for $D/t = 39.8$. Experiments and predictions.

12.2 ONSET OF AXISYMMETRIC WRINKLING

12.2.1 Formulation

The bifurcation into axially uniform wrinkling follows the same steps as those developed in Section 11.3. A circular cylinder of radius R and wall thickness t is under a state of biaxial stress given by

$$N_{xxo} = -\sigma t \quad \text{and} \quad N_{\theta\theta o} = PR = \sigma_\theta t. \tag{a}$$

The onset of buckling is represented by the same incremental equilibrium equations as pure compression:

$$\dot{N}'_{xx} = 0, \tag{b}$$

$$\dot{M}''_{xx} - \frac{\dot{N}_{\theta\theta}}{R} = -N_{xxo}\dot{w}'' = \sigma t \dot{w}''.$$

The corresponding kinematical relations are

$$\dot{\varepsilon}^o_{xx} = \dot{u}', \dot{\varepsilon}^o_{\theta\theta} = \frac{\dot{w}}{R} \quad \text{and} \quad \dot{\kappa}_{xx} = -\dot{w}''. \tag{c}$$

The instantaneous stress–strain relations are

$$\begin{Bmatrix} \dot{\sigma}_x \\ \dot{\sigma}_\theta \end{Bmatrix} = \begin{bmatrix} C_{11} & C_{12} \\ C_{12} & C_{22} \end{bmatrix} \begin{Bmatrix} \dot{\varepsilon}_x \\ \dot{\varepsilon}_\theta \end{Bmatrix}. \tag{d}$$

The buckling mode is again

$$\tilde{w} = a\cos\frac{\pi x}{\lambda} \quad \text{and} \quad \tilde{u} = b\sin\frac{\pi x}{\lambda}. \tag{12.2}$$

Substituting (12.2) into the equations above results in the following expression for the critical half-wavelength and stress (see Section 11.3 and [12.1])

$$\lambda_C = \pi \left[\frac{C_{11}^2}{12(C_{11}C_{22} - C_{12}^2)} \right]^{1/4} (Rt)^{1/2} \tag{12.3a}$$

and

$$\sigma_C = \left[\frac{C_{11}C_{22} - C_{12}^2}{3} \right]^{1/2} \left(\frac{t}{R} \right), \tag{12.3b}$$

where $[C_{\alpha\beta}]$ are the incremental deformation theory moduli (13.26) or the corresponding anisotropic version (13.28). The difference from purely axial compression enters through the equivalent stress, which in this case takes the value

$$\sigma_e = \left(\sigma_x^2 - \sigma_x\sigma_\theta + \sigma_\theta^2 \right)^{1/2} \quad \text{where } \sigma_x = \frac{-F}{2\pi Rt} \quad \text{and} \quad \sigma_\theta = \frac{PR}{t}. \tag{12.4}$$

(Here F is the compressive axial load as shown in Figure 11.16). Note that the state of stress and σ_e depend only on the values of $(\sigma_x, \sigma_\theta)$ and not on the loading path. This implies that the first bifurcation stress can be established by simple substitution of (12.4) into (12.3). By contrast, the prebuckling deformation of the cylinder depends on the path followed in the $(\sigma_x - \sigma_\theta)$ plane. Thus, unless the path is radial, the strains at the critical state must be evaluated incrementally through the flow theory equations (13.20) or the corresponding anisotropic version (13.21(b)). For constant pressure compression, (13.20) reduce to

$$\dot{\varepsilon}_x = \frac{1}{E} \left[1 + Q(\sigma_x - \sigma_\theta)^2 \right] \dot{\sigma}_x$$

and   (e)

$$\dot{\varepsilon}_\theta = \frac{1}{E} \left[-\nu + Q(2\sigma_x - \sigma_\theta)(2\sigma_\theta - \sigma_x) \right] \dot{\sigma}_x.$$

12.2.2 Predictions

A sample of axial stress–strain results corresponding to various values of internal pressures for a $D/t = 28.3$ tube is shown in Figure 12.10. The anisotropic flow theory was used, along with the properties of SAF4 in Table 11.2. As in the experiments, pressure reduces the yield stress in the axial direction. Therefore, as the pressure increases, the responses are seen to trace progressively lower axial stresses. The point at which the structure bifurcates into axial wrinkling is marked on each response with the symbol $\wedge$. The bifurcation stresses were evaluated from Eq. (12.3) using the anisotropic incremental deformation theory equations (13.28). Once σ_C is established, the corresponding strain is evaluated from the flow theory by first internally pressurizing and then incrementing the axial stress to this value.

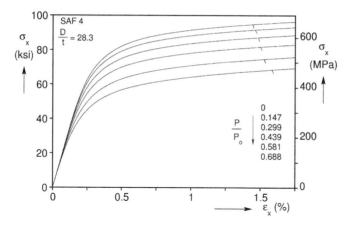

Figure 12.10 Axial stress–strain responses for a uniformly deforming cylinder compressed at different pressure levels with bifurcation points marked.

The critical states $\{\sigma_C, \varepsilon_C, \lambda_C\}$ calculated for $D/t = 28.3$ are compared to the experimental results in Figure 12.4. The results from the anisotropic material properties track the experimental stress, strains and wrinkle wavelengths quite well. The critical stress decreases with pressure, the critical strain remains relatively unaffected, and λ_C increases with pressure. The predictions assuming the material to yield isotropically ($S_r = S_\theta = 1$) deviate from the experimental values for all three variables. For biaxial loading, the anisotropy affects the whole response, not just the onset of wrinkling. Critical state predictions for the tubes with $D/t = 39.8$ are included in Figure 12.8. The anisotropic material results agree with the measured variables well, while the isotropic material results do not.

12.3 EVOLUTION OF WRINKLING

The evolution of wrinkling and the onset of axisymmetric collapse can be established by axially compressing a uniformly wrinkled cylinder in the presence of internal pressure. This can be achieved by a simple extension of the formulation presented in Section 11.4. An internal pressure work term is added to the PVW in (11.13) so that it becomes

$$2\pi R \int_0^{\lambda_C} \left\{ N_{xx} \delta \varepsilon_{xx}^o + N_{\theta\theta} \delta \varepsilon_{\theta\theta}^o + M_{xx} \delta \kappa_{xx} \right\} dx = 2\pi R P \int_0^{\lambda_C} \delta w \, dx, \tag{12.5}$$

where λ_C is the critical half-wavelength of axisymmetric wrinkles yielded by (12.3(a)). The effect of pressure is also accounted for in the flow theory constitutive equations through the appropriate state of stress. The rest of the formulation, including the imperfection adopted, the discretization and incremental solution procedure, remain the same as in Section 11.4.

A sample of results corresponding to one of the tests with D/t of 28.15 is shown in Figure 12.11(a). The pressure is $0.69P_o$, while the material model includes the anisotropy parameters for SAF4 in Table 11.2. Results for three imperfection amplitudes

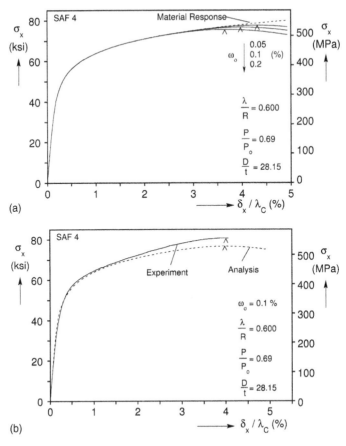

(a)

(b)

Figure 12.11 (a) Axial stress–displacement responses of a cylinder with axisymmetric imperfections of different amplitudes for $P = 0.69P_o$. (b) Comparison of an axisymmetric wrinkling response with corresponding experimental response.

$\omega_o = \{0.05, 0.1, 0.2\}\%$ are shown. The homogeneous deformation response, drawn with a dashed line, is also included. As mentioned above, internal pressure lowers the inelastic part of the response, and the imperfection causes a limit load instability. Increasing the imperfection amplitude reduces the limit strain. Figure 12.11(b) shows a direct comparison of the experimental response with a response calculated using $\omega_o = 0.1\%$. The predicted limit strain agrees very well with the experimental value. It is worth noting that the amplitude of ω_o required for this matching is rather small.

In such idealized calculations, the amplitude of the imperfection is a parameter that encompasses several of the factors that affect the experimental limit strain. These include the finiteness of the tube tested, the effect of the discontinuity at the ends (albeit reduced), a small eccentricity of the applied load, small thickness variations, etc. For this reason, and for the purpose of comparing experiments with predictions, the imperfection amplitude of each of the two sets of experiments was selected based on the best overall performance. Thus, for the $D/t = 28.3$ experiments, ω_o was chosen to be 0.5%. The axial stress-shortening responses calculated with this imperfection for the six test pressures

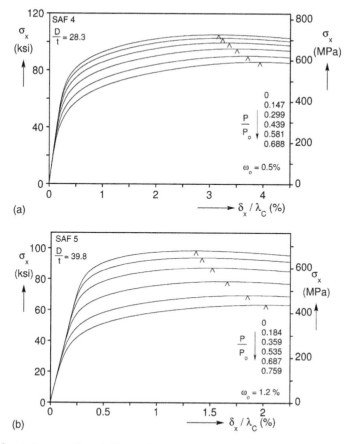

Figure 12.12 Axisymmetric wrinkling axial stress–displacement responses for various pressures. (a) $D/t = 28.3$ and (b) $D/t = 39.8$.

are shown in Figure 12.12(a). The imperfection wavelength used in each case is the one yielded by the bifurcation check algorithm. Pressure lowers the response, but simultaneously causes the limit stress to occur at progressively higher strain. The limit stresses and strains are compared to the experimental values in Figure 12.6. The predicted limit stresses are seen to be in good agreement with the measured values. The trend of the corresponding average strains is in agreement with that of the experiments for higher values of pressure. For $P = 0$ and $0.147P_o$ the predictions are lower than the measured values. The reader is reminded that for $P = 0$, the cylinder failed in a non-axisymmetric buckling mode not captured by the present analysis.

For the $D/t = 39.8$ experiments, $\omega_o = 1.2\%$ was adopted, along with the λ_C predicted for each case. The calculated $\sigma_x - \delta_x$ responses for the six test pressures are shown in Figure 12.12(b). The trend is similar to that of the thicker tubes, but the limit point occurs at much lower strain levels. The predicted values of σ_L and $\bar{\varepsilon}_L$ are compared to the measured values in Figure 12.9. The limit stress follows the experimental trend quite well. $\bar{\varepsilon}_L$ is also in good agreement with the experiments, except in the case of $P = 0$. In this test, the cylinder again failed by non-axisymmetric buckling not captured by the present analysis.

In summary, the inelastic response and limit state of relatively thick cylinders compressed axially in the presence of internal pressure are reproduced well by an axisymmetric model. For the present experiments, appropriate modeling of the plastic anisotropy in the tubes played an important role in this successful performance of the analysis.

12.4 PARAMETRIC STUDY

The parametric study of the critical and limit states conducted in Chapter 11 is extended to the problem of combined compression and internal pressure. The material is assumed to yield isotropically and is assigned the three Ramberg-Osgood stress–strain responses with the properties listed in Table 11.4 (see Figure 12.13). Tubes with D/t values ranging from 20 to 60 were analyzed. The tubes were compressed to failure at pressure levels of 0, $0.5P_o$ and $0.7P_o$. The critical values $\{\sigma_C, \varepsilon_C, \lambda_C\}$ are plotted against D/t in Figures 12.14(a), 12.14(b) and 12.14(c). Nine sets of results are presented in each plot corresponding to the three values of n and the three pressure levels. Internal pressure lowers the axial stress and results in a corresponding reduction in σ_C, in the same manner as seen in the experiments (e.g., Figure 12.4(a)). σ_C is highest for $n = 9$ and lowest for $n = 30$. Because of the nature of the stress–strain responses adopted, the critical strains follow powerlaw relationships with D/t and appear as linear in the log–log plot in Figure 12.14(b). Internal pressure causes a modest increase in ε_C. Even more importantly, the results illustrate the strong influence of the hardening exponent on ε_C. The results for $n = 30$ are significantly lower than the corresponding ones for $n = 9$. Figure 12.14(c) shows the corresponding λ_C vs. D/t plots. Once again, the results illustrate that pressure increases λ_C. At the same time, increasing n results in an increase in λ_C.

The limit strain was also studied parametrically by considering strictly axisymmetric imperfections. The half-wavelength adopted in each case corresponds to the critical value λ_C. The amplitude of the imperfection used had a constant value of $\omega_o = 0.2\%$. The calculated values of $\bar{\varepsilon}_L$ are plotted against D/t in Figure 12.15 in log–log scales. As

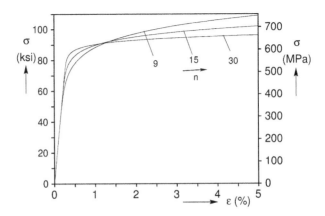

Figure 12.13   Three stress–strain responses adopted in the parametric study.

expected, for each pressure level $\bar{\varepsilon}_L$ is significantly higher than the corresponding ε_C in Figure 12.14(b). In addition, pressure has the effect of increasing $\bar{\varepsilon}_L$ for all three materials. Even more importantly, as was the case for ε_C, reduction in the hardening of the material (increase in n) reduces $\bar{\varepsilon}_L$. The limit strain is also influenced by the material yield stress. However, its effect is relatively secondary. Thus, for a fixed D/t and loading, material hardening is an important variable for $\bar{\varepsilon}_L$.

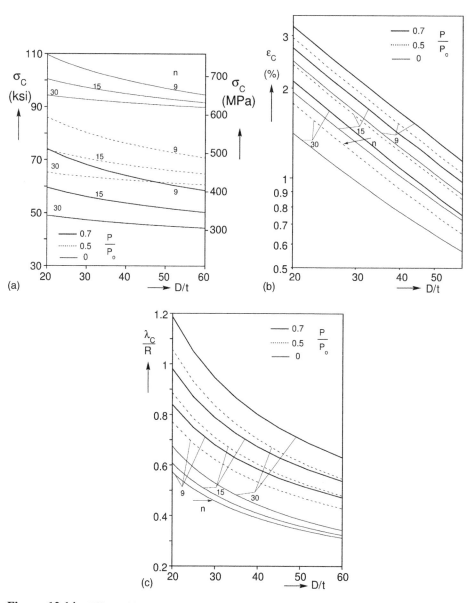

Figure 12.14 Effect of hardening parameter n on: (a) the critical stress, (b) the critical strain and (c) the wrinkle wavelength, for three pressure levels.

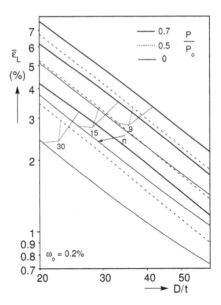

Figure 12.15   Effect of hardening parameter n on the average limit strains for three pressure levels.

12.5   SUMMARY AND RECOMMENDATIONS

The inelastic response as well as the limit state of long cylinders compressed in the presence of internal pressure is similar to that of cylinders loaded under pure compression. Internal pressure interacts plastically with axial compression. For compression under a fixed internal pressure, this interaction results in a lowering of the axial stress–strain response. J_2-type plasticity with isotropic hardening, or its anisotropic counterpart (see Chapter 13), can adequately capture this interaction.

At some plastic strain level, the cylinder develops uniform axisymmetric wrinkling. Under continued compression the wrinkles grow stably, gradually reducing the axial rigidity of the structure. This reduction in axial rigidity eventually leads to a limit load instability. Beyond the limit load, deformation localizes. The limit load is designated as the limit state of the structure. In the experiments of [12.3] on SAF 2507 stainless steel, all pressurized cylinders remained axisymmetric to the end of the test, past the limit load. By contrast, the pure compression tests developed non-axisymmetric buckling modes (see Chapter 11).

The stress at the onset of wrinkling and the corresponding wavelength can be easily established using the classical plastic bifurcation check presented in Section 12.2 that is based on deformation theory of plasticity. By contrast, if the stress path is not proportional, as was the case for the tests presented, the corresponding strain must be evaluated using the flow theory of plasticity. Anisotropy influences all three variables, and consequently, the anisotropic versions of these constitutive models must be used if it is present in the structures analyzed. For the material used in the tests presented, the critical stress decreased with internal pressure, whereas the critical strain remained relatively unaffected. The wrinkle wavelength generally increases with pressure.

The evolution of wrinkling and the resultant limit state can be established by modeling a λ_C long periodic domain that is assigned an initial imperfection corresponding to the axisymmetric buckling mode, as described in Section 12.3. The model outlined is discretized by adopting kinematically admissible displacement functions. Its performance can be reproduced by discretizing the structure with axisymmetric finite elements instead, provided the same imperfection characteristics and constitutive models are used. The flow theory of plasticity with isotropic hardening suffices for such calculations. Once again, anisotropy can influence the calculated limit strain, and should be incorporated in the plasticity subroutine used if required by the application.

For typical line pipe steels and D/t values in the range of 20–60, the limit strain is usually significantly larger than the critical strain. This difference can be exploited in strain-controlled applications. In stress-controlled applications, the critical stress can be adopted as a conservative design criterion. The limit strain generally increases with pressure. In addition, $\bar{\varepsilon}_L$ is influenced by the hardening of the material, as it tends to decrease as the hardening decreases (larger n). If stress–strain data are not available at the design stage, a conservative approach is to assume that the material exhibits relatively low hardening (e.g., $n = 30$ in a Ramberg–Osgood response) along with the design yield stress. Calculate the critical state and subsequently run an axisymmetric simulation with a relatively high value of ω_o ($\sim 0.5\%$; adopt amplitudes representative of expected imperfections if available).

REFERENCES

12.1. Lee, L.H.N. (1961). Inelastic buckling of cylindrical shells under axial compression and internal pressure. *Developments in Mechanics, Proc. 7th Midwestern Mechanics Conference* Vol. 1, Plenum Press, New York, pp. 190–202.

12.2. Murray, D.W. (1997). Local buckling, strain localization, wrinkling and postbuckling response of a line pipe. *Eng. Struct.* **19**, 360–371.

12.3. Paquette, J.A. and Kyriakides, S. (2006). Plastic buckling of tubes under axial compression and internal pressure. *Int. J. Mech. Sci.* **48**, 855–867.

13

Elements of Plasticity Theory

Plasticity plays a pivotal role in the development of most subjects covered in this book series. To assist the reader, this chapter presents some of the basic ideas behind incremental metal plasticity. A brief outline of phenomenological aspects of elastic–plastic behavior is presented at the outset, followed by a cursory review of the main stress tensor concepts required in later developments. The classical formulation of J_2 flow theory with isotropic hardening, the main model used in the book, is described next in some detail. This is followed by a brief exposé of anisotropic plasticity suitable for modeling pipe yield anisotropy. Three nonlinear kinematic hardening models used in special problems involving reverse loading are outlined next. Throughout the book, bifurcation checks in the plastic range are performed using the incremental version of the J_2 deformation theory or its anisotropic counterpart. For this reason, a section is devoted to summarizing the basic equations of this theory. This chapter is aimed at readers with some familiarity with plasticity. The uninitiated reader is referred to more complete treatments of the subject for an in-depth understanding (e.g., see list of books in references).

13.1 PRELIMINARIES

13.1.1 Aspects of Uniaxial Behavior

We consider a uniaxial tension test on a typical structural metal. The test specimen is cylindrical, with a test section of initial length L_o and initial and deformed cross sectional areas A_o and A, respectively. Figure 13.1 shows two views of the recorded force-elongation ($F - \Delta L$) response. Force is converted into engineering stress, $\sigma = F/A_o$, and the elongation into engineering strain, $\varepsilon = \Delta L/L_o$. The response in compression is essentially the same for smaller values of strain but differs at high strains. The response exhibits a linear region (OA in Figure 13.1(a)) with a corresponding image on the compression side. This is the *linearly elastic regime*, and its slope is the *Young modulus* (E) of the material. Beyond A, the response enters the *elastic-plastic regime* and becomes nonlinear. The slope of the response decreases rapidly initially, forming a transition knee that usually ends at a strain smaller than 1%. For higher strains the slope changes at a much slower rate. Unloading from any point beyond A, say along BC, follows a linear trajectory with a slope E. In the process, the elastic part of the strain (ε^e) is recovered, whereas *plastic* deformation (ε^p) is permanent. For example, when the bar is fully unloaded, the

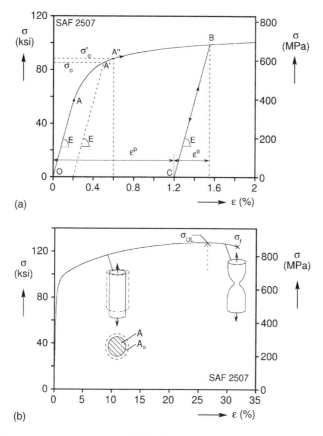

Figure 13.1 Stress–strain response of a SAF 2507 stainless steel: (a) small strain regime with two yield stress definitions and (b) large strain regime showing the ultimate and failure stresses.

plastic strain is

$$\varepsilon^p = \varepsilon - \frac{\sigma}{E}. \tag{a}$$

The area under the response OABC is unrecoverable energy (*plastic work*). Reloading follows essentially the same path (CB), and the original nonlinear trajectory is rejoined as if the unloading had not occurred (approximately). The stress level at which this occurs is significantly higher than the stress at the initial elastic limit. Such a material response is called *strain* or *work hardening*, as the apparent stress level at which plastic action takes place increases with strain or plastic work.

The elastic limit (A) will be the initial boundary between the elastic and plastic regimes for all plasticity models that follow. The *yield stress* is the boundary between the two regimes for the purposes of structural design. It is at a stress level higher than the elastic limit and is defined by convention. The most common definition of yield stress (σ_o) corresponds to the stress at a strain offset of 0.002, as shown in Figure 13.1(a) (point A'). The API definition of yield stress (σ_o'), widely used in the oil and gas industry, corresponds to the stress at a total strain of 0.005 (point A'').

Two useful measures of the nonlinearity of a stress–strain response are the *tangent modulus* (E_t), its instantaneous slope, and the *secant modulus* (E_s), each expressed as follows:

$$E_t = \frac{d\sigma}{d\varepsilon} \quad \text{and} \quad E_s = \frac{\sigma}{\varepsilon}. \tag{13.1a}$$

Using (a),

$$\frac{d\varepsilon^p}{d\sigma} = \left[\frac{1}{E_t} - \frac{1}{E}\right] \quad \text{and} \quad \frac{\varepsilon^p}{\sigma} = \left[\frac{1}{E_s} - \frac{1}{E}\right]. \tag{13.1b}$$

Along with elongation, a tensile test causes lateral contraction of the cross section. The transverse strain (ε_t) is related to the axial one through *Poisson's ratio* $v = -\varepsilon_t/\varepsilon$. In the elastic regime, steels have Poisson's ratios in the range of 0.25–0.3. For structural metals, the plastic part of the strain is essentially *incompressible* (volume preserving or that $v = 0.5$). This difference between the elastic and plastic Poisson's ratios usually requires that the strain be decomposed into its elastic and plastic parts.

At higher values of strain, the force–displacement response becomes softer and eventually a maximum value is reached as shown in Figure 13.1(b). At this point, the specimen starts to develop a neck. The neck extends over a length of about 2–3 times the lateral dimension of the cross section, while the rest of the specimen remains cylindrical. Since beyond the load maximum the deformation in the test section is inhomogeneous, this part of the response no longer represents just the material behavior. The diameter of the neck gets progressively smaller while the recorded load decreases, and at some stage the specimen fails by fracture. The stress corresponding to the load maximum is known as the *ultimate stress* $(\sigma_{UL}$, also known as *tensile strength*) while the corresponding strain is denoted by ε_{UL}. Their values depend on alloy content and on the processing. For pipeline steels, the ultimate strain is usually higher than 15%. The ultimate stress is typically in the range $1.08 < \sigma_{UL}/\sigma_o < 1.33$ $(0.75 < \sigma_o/\sigma_{UL} < 0.93$; yield-to-ultimate strain ratio is more commonly quoted for line pipe). The "strain" at failure (ε_f) is often quoted as a measure of *ductility*. It is usually based on the net elongation of the part of the test section of specified original length (typically 2 inches) that contains the neck. Since the length of the necked zone changes with the dimensions of the cross section, this measure of ductility is not a suitable material variable. A more consistent way of measuring and comparing ductility from tests using different specimen geometries is the *reduction of area* at failure (smallest cross sectional area of neck/initial cross sectional area). If, however, the specimen geometry is kept constant, ε_f can be a useful tool for comparing the ductility of steels as their alloy composition and processing are varied.

The stress–strain responses of most structural metals, including line-grade carbon steels and stainless steels, exhibit some sensitivity to the rate of loading (e.g., the *strain rate* $(\dot{\varepsilon})$. At room temperature the sensitivity is relatively small and can usually be neglected for processes producing strain rates $\dot{\varepsilon} < 10^{-2}\,\text{s}^{-1}$. (Note that uniaxial tests like the one in Figure 13.1 are usually conducted at $10^{-4} < \dot{\varepsilon} < 10^{-3}\,\text{s}^{-1}$.) For higher strain rates and for temperatures higher than about 150°C, the strain rate effects can become significant for some applications. The rate dependence is then established experimentally and modeled through a rate- and temperature-dependent plasticity theory (viscoplasticity).

Figure 13.2 illustrates the response following unloading from a plastic state (B) and reverse loading (BCD). The new elastic limit at C is seen to be at a stress level that is much lower than that corresponding to compression of the virgin material (stress at initial

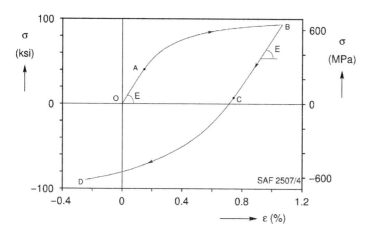

Figure 13.2   Stress–strain response showing initial loading followed by unloading and reverse loading.

elastic limit). Indeed, the apparent "strengthening" along AB from tension results in a corresponding reduction in compressive strength (*Bauschinger effect*). Equally important is the significant difference between the shapes of the initial monotonic response OAB and BCD. The former exhibits a more abrupt transition from elastic to elastic–plastic behavior than the latter. This difference must be accounted for in problems involving reverse loading. Continuation of reverse loading to the same compressive strain level as the strain at B, followed by a second unloading and tensile loading usually closes the loop, generating a symmetric *hysteresis*.

Structural metal stress–strain responses are commonly represented through powerlaw relationships. The three-parameter Ramberg–Osgood expression given by

$$\varepsilon = \frac{\sigma}{E}\left[1 + \frac{3}{7}\left(\frac{\sigma}{\sigma_y}\right)^{n-1}\right] \tag{13.2}$$

is widely used in the book because it facilitates parametric studies of physical results. A procedure for fitting measured responses through its three parameters appears in Appendix C.

13.1.2 Discontinuous Yielding

The transition from elastic to plastic deformation of hot-finished, low-carbon steel is often characterized by a material instability known as *Lüders strain*. The macroscopic effect of the instability is inhomogeneous deformation. Figure 13.3 shows a typical stress–displacement response recorded in a uniaxial test on a X60 line-grade carbon steel. Localized plastic deformation starts at σ_U (*upper yield stress*) with a sudden drop in stress. Under displacement control, Lüders deformation spreads along the length of the specimen (see Figure 13.3(a)), while the stress remains essentially constant (σ_L, *lower yield stress*). When the whole test section has been deformed to the Lüders strain ($\Delta\varepsilon_L \approx 2.67\%$), the material hardens and the specimen reverts back to homogeneous deformation. For such a response, the yield stress is assigned the value of the plateau stress σ_L (60.7 ksi – 418 MPa).

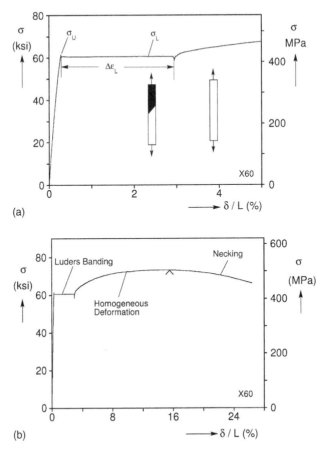

Figure 13.3 Stress–strain response of an X60 steel exhibiting Lüders banding: (a) small strain regime and (b) straining to failure.

Figure 13.3(b) shows the continuation of this X60 steel response until just before failure of the specimen. The material hardens from a strain of about 2.97% to 15.5%, and the specimen deforms homogeneously. It reaches an ultimate stress of 73.22 ksi (505 MPa, $\sigma_o/\sigma_{UL} = 0.829$). Beyond this point the specimen necks. (In this test the strain was measured with a 1.0 in (25.4 mm) gage length extensometer. The neck occurred within this gage length.)

The inhomogeneous nature of the deformation was illustrated in [13.1] using tensile tests on strips made from similar type steels. The strips were coated with a brittle lacquer that shattered at a strain in the range of 1–2%. Figure 13.4 shows a set of photographs of the surface of the strip taken at constant time intervals during the test. Lighter color represents relatively deformed material (shattered coating), and the dark color relatively undeformed material. The two regions are separated by a relatively sharp inclined front. The front is seen to propagate from left to right at nearly constant speed (see [13.1]).

The initial stress peak (*upper yield stress σ_U*), as well as the level and extent of the stress plateau (*lower yield stress σ_L* and $\Delta\varepsilon_L$, respectively) depend on the grain size (the

Figure 13.4   Photographic sequence showing evolution of Lüders strain in a tension test on an X70 steel strip (from [13.1]).

finer the grain, the higher the values of σ_U, σ_L and $\Delta\varepsilon_L$) and on alloy content. They are also sensitive to temperature and to the rate of loading [13.2].

Two main theories of the microscopic events behind the instability have survived scientific scrutiny. Cottrell and Bilby [13.3] attributed the upper yield stress to the pinning of dislocations by carbon and nitrogen atoms which naturally tend to form "atmospheres" around them. They postulated that initial yielding requires a higher stress in order to pull the dislocations out of their atmospheres. Once released, the dislocations can be moved by a lower stress. The pinning effect of interstitial impurities is widely accepted for several reasons including the reappearance of Lüders strain following mild heat treatment that allows these atoms to migrate and re-pin the dislocations (*strain aging*). The second theory, due to Johnston and Gilman [13.4], attributes the load drop to multiplication of dislocations. As their number increases, the stress required to move them decreases.

The localized nature of Lüders deformation can have undesirable repercussions in structures deformed into the plastic range. These are often avoided by cold-finishing a pipe to a degree that exhausts the Lüders strain. For smaller diameter pipe, this is achieved by stretching. For larger diameter seamless pipe, it is achieved by cold reduction of the diameter of the pipe by passing it through three or four roll reducers (see Chapter 3).

Recently, it has been demonstrated that discontinuous Lüders deformation can be modeled by adopting a non-monotonic (*up–down–up*) stress–strain response calibrated to measured results such as those in Figure 13.3 [13.1, 13.5]. In such attempts, the microscale material instability is assumed to be transferred in an average way to the continuum level. When implemented in finite element models with fine meshes, the characteristic scale becomes the size of the elements.

13.1.3 Multiaxial Behavior

Let the *stress tensor* $\sigma = \sigma^T$ have principal stresses $(\sigma_1, \sigma_2, \sigma_3)$ and corresponding principal directions (n_1, n_2, n_3) related through

$$(\sigma - \sigma_i I)n_i = 0. \tag{13.3}$$

The stress tensor *invariants* (I_1, I_2, I_3) in terms of general and principal stresses are given by

$$I_1 = \sigma_{ii} = \sigma_1 + \sigma_2 + \sigma_3,$$

$$I_2 = \frac{1}{2}(\sigma_{ii}\sigma_{jj} - \sigma_{ij}\sigma_{ji}) = \sigma_1\sigma_2 + \sigma_2\sigma_3 + \sigma_3\sigma_1, \tag{13.4}$$

$$I_3 = \det|\sigma| = \sigma_1\sigma_2\sigma_3.$$

The *deviatoric stress* tensor is defined as

$$s_{ij} = \sigma_{ij} - \frac{1}{3}\sigma_{kk}\delta_{ij}. \tag{13.5}$$

The principal deviatoric stresses and directions are, respectively

$$s_i = \sigma_i - \frac{I_1}{3} \quad \text{and} \quad n_i, \ i = 1, 2, 3. \tag{13.6}$$

The invariants of s are

$$J_1 = s_{kk} = 0,$$

$$J_2 = \frac{1}{2}s_{ij}s_{ij} = \frac{1}{2}(s_1^2 + s_2^2 + s_3^2), \tag{13.7}$$

$$J_3 = \det|s| = \frac{1}{3}s_{ij}s_{jk}s_{ki} = s_1s_2s_3.$$

13.1.4 Yield Criteria

An important characteristic of metal plasticity is that, to a first approximation, yielding is unaffected by hydrostatic states of stress. If we also invoke that yielding is initially isotropic, and that the material yields at the same stress levels in tension and compression, a general expression for a yield criterion is

$$f(J_2, J_3^2) = \text{const.} \tag{13.8}$$

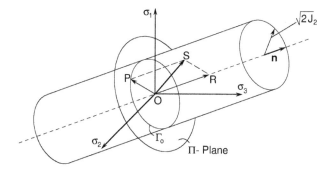

Figure 13.5 Stress tensor decomposition into hydrostatic (**OP**) and deviatoric (**OR**) components (principal stress space).

These characteristics afford a helpful geometrical representation of the yield surface in stress space which, without any loss of generality, will be represented by the three principal stress components. Let **OS** be the vector representing the state of stress $(\sigma_1, \sigma_2, \sigma_3)^T$ on the yield surface (see Figure 13.5). It can be readily decomposed into vector **OR** along the direction $n = \frac{1}{\sqrt{3}}(1, 1, 1)^T$ and vector **OP** in the plane with normal n. $\mathbf{OR} = \frac{1}{3}I_1 n$ is the hydrostatic component of σ and $\mathbf{OP} = (s_1, s_2, s_3)^T$ is the deviatoric stress vector with amplitude $|\mathbf{OP}| = (s_1^2 + s_2^2 + s_3^2)^{1/2} = \sqrt{2J_2}$. The plane containing **OP** is known as the Π-plane.

The yield function (13.8) corresponds to a cylindrical surface with axis n in this space (Figure 13.5). In view of its insensitivity to I_1, it is sufficient to consider its intersection, Γ_o, with the Π-plane. Isotropy and equivalence of tensile and compressive stress states on the yield surface dictate that the shape of Γ_o repeat every 30° (Hill [13.32], p. 18). Furthermore, it can be shown that the surface must be convex [13.6–13.8].

a. The von Mises and Tresca Yield Criteria

In 1913 von Mises postulated [13.9] that *yielding occurs when J_2 reaches a critical value*; or in the words of Hencky [13.10], *when the distortional energy reaches a critical value*. The criterion can be written in several forms:

$$J_2 = \frac{1}{2}s \cdot s = k^2,$$

$$J_2 = \frac{1}{6}[(\sigma_1 - \sigma_2)^2 + (\sigma_2 - \sigma_3)^2 + (\sigma_3 - \sigma_1)^2] = k^2, \tag{13.9a}$$

$$J_2 = \frac{1}{6}[(\sigma_{11} - \sigma_{22})^2 + (\sigma_{22} - \sigma_{33})^2 + (\sigma_{33} - \sigma_{11})^2 + 6(\sigma_{12}^2 + \sigma_{23}^2 + \sigma_{31}^2)] = k^2.$$

Calibrated to a uniaxial test with yield stress σ_o, $k = \sigma_o/\sqrt{3}$. The criterion leads to a circular cylindrical surface in the $(\sigma_1, \sigma_2, \sigma_3)$ space. Its intersection with the Π-plane is a circle of radius $\sqrt{\frac{2}{3}}\sigma_o$ (Figure 13.6(a)). For plane stress states, (13.9a) reduces to

$$[\sigma_{11}^2 - \sigma_{11}\sigma_{22} + \sigma_{22}^2 + 3\sigma_{12}^2]^{1/2} = \sigma_o. \tag{13.9b}$$

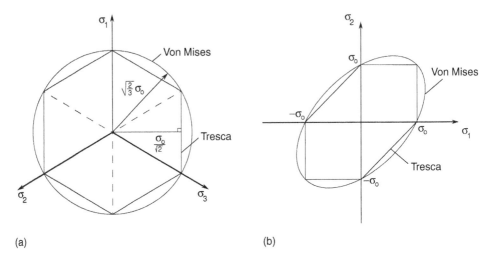

Figure 13.6   The von Mises and Tresca yield surfaces in: (a) the Π-plane and (b) the $\sigma_1 - \sigma_2$ plane.

The principal stress version of (13.9b) reduces to the ellipse shown in Figure 13.6(b). An alternative yield criterion due to Tresca (1864) is based on the postulate that *yielding occurs when the maximum shear reaches a critical value*, which can be written as

$$\max\left\{\left|\frac{\sigma_1 - \sigma_2}{2}\right|, \left|\frac{\sigma_2 - \sigma_3}{2}\right|, \left|\frac{\sigma_3 - \sigma_1}{2}\right|\right\} = \kappa. \qquad (13.10)$$

Calibrated to the same uniaxial test, $\kappa = \sigma_o/2$. This yield cylinder is hexagonal with the cross section shown in Figure 13.6(a). In the case of plane stress ($\sigma_3 = 0$), the yield surface reduces to the polygon shown in Figure 13.6(b). Both criteria are accepted in structural design. The simplicity of Tresca's criterion makes it attractive in analysis, while the continuous nature of the von Mises yield function has made it the dominant candidate for use in subsequent yielding (flow rules).

b. Anisotropic Yielding of Thin-Walled Tubes

Forming processes through which seamless tubular components are manufactured induce yield anisotropy to the finished product. Anisotropy affects the collapse pressure as well as other limit states important in structural design and must be included in modeling. Hill's [13.11, 13.32] quadratic anisotropic yield function captures these effects adequately for the needs of these problems. Reduced to plane stress in polar cylindrical coordinates (x, θ, r), it can be stated as follows

$$\left[\sigma_x^2 - \left(1 + \frac{1}{S_\theta^2} - \frac{1}{S_r^2}\right)\sigma_x\sigma_\theta + \frac{1}{S_\theta^2}\sigma_\theta^2 + \frac{1}{S_{x\theta}^2}\sigma_{x\theta}^2\right]^{1/2} = \sigma_{ox}, \qquad (13.11)$$

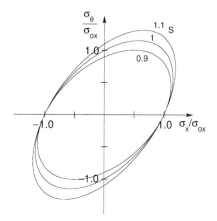

Figure 13.7   Effect of anisotropy variable S on the yield surface in the $\sigma_x - \sigma_\theta$ plane.

where

$$S_\theta = \frac{\sigma_{o\theta}}{\sigma_{ox}}, \quad S_r = \frac{\sigma_{or}}{\sigma_{ox}} \quad \text{and} \quad S_{x\theta} = \frac{\sigma_{ox\theta}}{\sigma_{ox}}.$$

Here σ_{ox}, $\sigma_{o\theta}$ and σ_{or} are the yield stresses in the corresponding directions and $\sigma_{ox\theta}$ is the yield stress under pure shear. A methodology for determining these is described in Appendix B [13.12]. Several of the problems of interest do not involve shear. If in addition σ_{or} is not available, the simplifying assumption $S_r = S_\theta \equiv S$ often suffices. The effect of the anisotropy variable S on the yield surface in the $(\sigma_x, \sigma_\theta)$ plane is illustrated in Figure 13.7. $S = 1$ corresponds to isotropic yielding.

13.2   INCREMENTAL PLASTICITY

13.2.1   The Flow Rule

A defining characteristic of metal plasticity is path dependence. Thus, constitutive equations must be formulated incrementally. Incremental plasticity models are defined through a *flow rule* that relates the strain increment to the stress increment. It depends on the current stress state on the current yield surface (YS), and on rules that describe the evolution of the yield surface with history. Although working incremental elastic–plastic models preceded 1950, Hill [13.32] and Drucker [13.6–13.8] are responsible for showing that:

1. *yield-surfaces are convex*,
2. *the instantaneous plastic strain increment at σ is normal to the yield surface*,
3. *the strain increments are linearly related to the stress increments*.

Without going into details, these characteristics imply that the simplest flow rule has the form

$$d\varepsilon_{ij}^p = \frac{1}{H} \left(\frac{\partial f}{\partial \sigma_{mn}} d\sigma_{mn} \right) \frac{\partial f}{\partial \sigma_{ij}}, \tag{13.12}$$

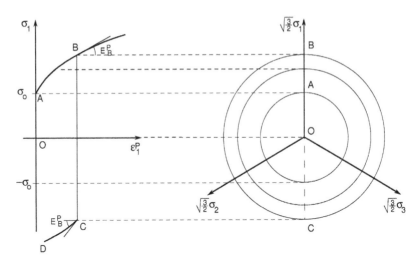

Figure 13.8   Isotropic hardening and its effect on reverse loading.

where f is the current YS (e.g., (13.8)), and H is a scalar function of σ and of the loading history. Under these circumstances $d\varepsilon_{ii}^p = 0$ (*incompressibility* of plastic strains).

A part of every strain increment is *recoverable*, or *elastic*, $(d\varepsilon^e)$ and obeys the usual isotropic elastic relationships

$$d\varepsilon_{ij}^e = \frac{1}{E}\left[(1 + v)d\sigma_{ij} - vd\sigma_{kk}\delta_{ij}\right].$$ (13.13)

The total strain increment is then given by

$$d\varepsilon_{ij} = d\varepsilon_{ij}^e + d\varepsilon_{ij}^p .$$ (13.14)

13.2.2   J$_2$ Flow Theory with Isotropic Hardening

In *isotropic hardening* plasticity, the YS stays centered at the origin and grows any time plastic action takes place, so that σ stays on the YS. During growth, it maintains its original shape, as shown schematically in Figure 13.8 for a uniaxial test. Its growth depends strictly on the magnitude of the equivalent stress and thus

$$f(\sigma) = \sigma_{e\,max}.$$ (13.15)

This is a rather restrictive evolution law for the YS, that often does not represent actual material behavior well. Thus, for example, on unloading from a plastic state such as point B in Figure 13.8, the material remains elastic until point C is reached with a compressive stress equal to $|\sigma_B|$. Because B and C have the same stress levels, their plastic moduli are the same, resulting in a discontinuity at C. Actual metals yield much earlier (closer to $2\sigma_o$ from B) and in a smooth manner, as clearly seen in Figure 13.2. The degree of inaccuracy introduced by this assumption depends on the hardening of the material, with the discrepancy increasing as the hardening increases. It should be pointed out, however, that in many problems that experience primarily loading, this deficiency of the model plays only a minor role. Thus, despite many developments in the subject over the last 25 years, isotropic hardening plasticity remains widely used due to its simplicity.

If we assume that initial and subsequent yield surfaces are Mises-type, (13.15) becomes

$$f = \sigma_e = \sqrt{3J_2(s)} = \left(\frac{3}{2}s \cdot s\right)^{1/2} = \sigma_{e\,\max},$$
(13.16a)

where σ_e represents the magnitude of the state of stress that is known as the *equivalent stress*. A scalar measure of the plastic strain increment which is work-compatible to σ_e, called *equivalent plastic strain increment*, is given by

$$d\varepsilon_e^p = \left(\frac{2}{3}d\varepsilon^p \cdot d\varepsilon^p\right)^{1/2}$$
(13.16b)

(i.e., $\sigma_e d\varepsilon_e^p = dW^p$). Specializing the flow rule to this yield function results in

$$d\varepsilon_{ij}^p = \frac{1}{H}\frac{9}{4\sigma_e^2}(s \cdot d\sigma)s_{ij}.$$
(13.17)

H can be evaluated by calibrating (13.17) to a test. For a uniaxial test,

$$H = \frac{d\sigma_1}{d\varepsilon_1^p}.$$

Combining the elastic and plastic strain increments results in

$$d\varepsilon_{ij} = \frac{1}{E}\left\{\left[(1+v)d\sigma_{ij} - v d\sigma_{kk}\delta_{ij}\right] + 9Q(s \cdot d\sigma)s_{ij}\right\}, \quad \text{or} \quad d\varepsilon = Dd\sigma$$
(13.18a)

where

$$Q = \frac{1}{4\sigma_e^2}\left(\frac{E}{E_t(\sigma_e)} - 1\right).$$

This inverts to

$$d\sigma = Cd\varepsilon \text{ where } C_{ijkl} = \frac{E}{1+v}\left\{\frac{1}{2}(\delta_{ik}\delta_{jl} + \delta_{il}\delta_{jk}) + \frac{v}{1-2v}\delta_{ij}\delta_{kl} - \frac{9Qs_{ij}s_{kl}}{1+v+6Q\sigma_e^2}\right\}.$$
(13.18b)

For plane stress, (13.16a) reduces to

$$f = \left[\sigma_x^2 - \sigma_x\sigma_\theta + \sigma_\theta^2 + 3\sigma_{x\theta}^2\right]^{1/2} = \sigma_{e\,\max}.$$
(13.19)

Explicit, incremental strain–stress relationships for plane stress can be written as follows:

$$\left\{ \begin{array}{c} d\varepsilon_x \\ d\varepsilon_\theta \\ d\varepsilon_{x\theta} \end{array} \right\} = D \left\{ \begin{array}{c} d\sigma_x \\ d\sigma_\theta \\ d\sigma_{x\theta} \end{array} \right\}$$

where   (13.20a)

$$D = \frac{1}{E} \left[\begin{array}{cc} 1 + Q(2\sigma_x - \sigma_\theta)^2 & -\nu + Q(2\sigma_x - \sigma_\theta)(2\sigma_\theta - \sigma_x) \\ -\nu + Q(2\sigma_x - \sigma_\theta)(2\sigma_\theta - \sigma_x) & 1 + Q(2\sigma_\theta - \sigma_x)^2 \\ 3Q(2\sigma_x - \sigma_\theta)\sigma_{x\theta} & 3Q(2\sigma_\theta - \sigma_x)\sigma_{x\theta} \end{array} \right.$$

$$\left. \begin{array}{c} 6Q(2\sigma_x - \sigma_\theta)\sigma_{x\theta} \\ 6Q(2\sigma_\theta - \sigma_x)\sigma_{x\theta} \\ 1 + \nu + 18Q\sigma_{x\theta}^2 \end{array} \right].$$

The plane stress version of the inverted Eqs. (13.18b) can now be written as follows [13.13]:

$$d\sigma_{\alpha\beta} = \hat{C}_{\alpha\beta\gamma\delta} d\varepsilon_{\gamma\delta}, \quad (\alpha, \beta, \gamma, \delta) = 1, 2$$

where   (13.20b)

$$\hat{C}_{\alpha\beta\gamma\delta} = C_{\alpha\beta\gamma\delta} - \frac{C_{\alpha\beta33}C_{\gamma\delta33}}{C_{3333}}.$$

In the case of anisotropic yielding of the type represented by (13.11), Eq. (13.19) becomes

$$f = \left[\sigma_x^2 - \left(1 + \frac{1}{S_\theta^2} - \frac{1}{S_r^2}\right) \sigma_x\sigma_\theta + \frac{1}{S_\theta^2}\sigma_\theta^2 + \frac{1}{S_{x\theta}^2}\sigma_{x\theta}^2 \right]^{1/2} = \sigma_{e\,\text{max}} \qquad (13.21a)$$

(i.e., the anisotropy remains the same during growth of the YS). A work-compatible plastic equivalent strain increment is given in Appendix B. Following the same procedure as above results in the following anisotropic version of the constitutive matrix in (13.20):

$$D_a = \frac{1}{E} \left[\begin{array}{cc} 1 + Q(2\sigma_x - \beta\sigma_\theta)^2 & -\nu + Q(2\sigma_x - \beta\sigma_\theta)(2\alpha\sigma_\theta - \beta\sigma_x) \\ -\nu + Q(2\sigma_x - \beta\sigma_\theta)(2\alpha\sigma_\theta - \beta\sigma_x) & 1 + Q(2\alpha\sigma_\theta - \beta\sigma_x)^2 \\ Q(2\sigma_x - \beta\sigma_\theta)\gamma\sigma_{x\theta} & Q(2\alpha\sigma_\theta - \beta\sigma_x)\gamma\sigma_{x\theta} \end{array} \right.$$

$$\left. \begin{array}{c} Q(2\sigma_x - \beta\sigma_\theta)2\gamma\sigma_{x\theta} \\ Q(2\alpha\sigma_\theta - \beta\sigma_x)2\gamma\sigma_{x\theta} \\ 1 + \nu + 2Q\gamma^2\sigma_{x\theta}^2 \end{array} \right] \qquad (13.21b)$$

where

$$\alpha = \frac{1}{S_\theta^2}, \quad \beta = \left(1 + \frac{1}{S_\theta^2} - \frac{1}{S_r^2}\right), \quad \gamma = \frac{1}{S_{x\theta}^2}, \quad \text{and} \quad Q = \frac{1}{4\sigma_e^2}\left(\frac{E}{E_t(\sigma_e)} - 1\right).$$

13.3 THE DEFORMATION THEORY OF PLASTICITY

The so-called *deformation theory of plasticity* was an early attempt (Henky [13.10], Nadai [13.14], Il'yushin [13.15]) to represent the nonlinearity associated with plastic deformations through *total* stress–strain relationships. This in essence results in nonlinear elastic relationships, with the additional constraint that the inelastic part be incompressible. Plastic deformation is path-dependent which, of course, cannot in general be captured through total theories. As a result, with two exceptions, deformation theories are not used for elastic–plastic problems. The first exception is for problems with loadings that result in proportional (or nearly proportional, Budiansky [13.16]) stress paths. For such stress paths, the J_2 incremental relationships (13.17) integrate out to the corresponding J_2 deformation theory ones. The convenience of the total expressions allows algebraic manipulation and the derivation of closed-form solutions for several practical problems.

The second class of problems where the J_2 deformation theory is preferred is in elastic–plastic bifurcation buckling calculations. Bifurcation from a plastic state is by its nature a non-proportional increment. Thus, it is naturally expected that it should be best represented by flow theories. Despite this, it has been known for more than 50 years that the use of flow theories (most notably the J_2 flow theory) results in significant overprediction of the critical stress and even more so the strains (e.g. [13.13, 13.17]). By contrast, use of the J_2 deformation theory often leads to better agreement between experiment and analysis. The reasons for this unorthodox state of affairs are complex and not yet resolved. Possible contributors pointed out over the years are:

(a) Plastic deformation may be causing local distortion (even the development of corners) of the YS that is not captured by simple J_2 flow rule with isotropic hardening.
(b) Experimentally, bifurcation is not an instantaneous event and, as a result, is difficult to establish with accuracy.
(c) Actual structures always have imperfections, which "smooth" the process and affect the recorded "onset" of instability.

There may indeed be additional reasons for this unorthodox state of affairs.

In the work presented in this book, it was consciously decided that, lacking a better alternative, deformation theory will be adopted for all plastic bifurcation calculations presented. Flow theory will be used for all other purposes.

13.3.1 The J_2 Deformation Theory

In this most commonly used deformation theory, the plastic strain components are proportional to the deviatoric stress components, with the scalar function of proportionality depending strictly on J_2 (or equivalently on $\sigma_e = \sqrt{3J_2}$) as follows:

$$\varepsilon_{ij}^p = g(J_2)s_{ij}. \tag{13.22a}$$

This has the properties that $\varepsilon_{ii}^p = 0$ and is invariant to coordinate transformation. The function $g(J_2)$ can be determined from any simple proportional loading test. When calibrated to a uniaxial test (13.22a) becomes

$$\varepsilon_{ij}^p = \frac{3}{2}\left[\frac{1}{E_s(\sigma_e)} - \frac{1}{E}\right]s_{ij}. \tag{13.22b}$$

It is easy to show that (13.22b) is scalarized by adopting the measure of *equivalent plastic strain*:

$$\varepsilon_e^p = \left(\frac{2}{3}\boldsymbol{\varepsilon}^p \cdot \boldsymbol{\varepsilon}^p\right)^{1/2}. \tag{13.23}$$

The model is completed by adding the elastic components of strain to (13.22)

$$\varepsilon_{ij} = \frac{1+\nu}{E}\sigma_{ij} - \frac{\nu}{E}\sigma_{kk}\delta_{ij} + \frac{3}{2}\left[\frac{1}{E_s} - \frac{1}{E}\right]s_{ij}. \tag{13.24a}$$

This is often written as

$$\varepsilon_{ij} = \frac{1+\nu_s}{E_s}\sigma_{ij} - \frac{\nu_s}{E_s}\sigma_{kk}\delta_{ij} \quad \text{where} \quad \nu_s = \frac{1}{2} + \frac{E_s}{E}\left(\nu - \frac{1}{2}\right). \tag{13.24b}$$

13.3.2 Incremental J_2 Deformation Theory

In bifurcation checks the incremental form of (13.24) is required, which can be shown to be

$$d\varepsilon_{ij} = \frac{1}{E}\left[(1+\nu)d\sigma_{ij} - \nu d\sigma_{kk}\delta_{ij}\right] + \frac{3}{2}\left(\frac{1}{E_s} - \frac{1}{E}\right)ds_{ij} + \frac{9}{4}\left(\frac{1}{E_t} - \frac{1}{E_s}\right)\frac{(s \cdot d\sigma)}{\sigma_e^2}s_{ij},$$

or

$$\tag{13.25a}$$

$$d\boldsymbol{\varepsilon} = \boldsymbol{D}_d d\boldsymbol{\sigma}.$$

This inverts to $d\boldsymbol{\sigma} = \boldsymbol{C}_d\, d\boldsymbol{\varepsilon}$ where

$$[C_{ijkl}]_d = \frac{E}{1+\nu+h}\left\{\frac{1}{2}(\delta_{ik}\delta_{jl} + \delta_{il}\delta_{jk}) + \frac{3\nu+h}{3(1-2\nu)}\delta_{ij}\delta_{kl} - \frac{h's_{ij}s_{kl}}{1+\nu+h+\frac{2}{3}h'\sigma_e^2}\right\}$$

$$\tag{13.25b}$$

with

$$h = \frac{3}{2}\left(\frac{E}{E_s} - 1\right) \quad \text{and} \quad h' \equiv \frac{dh}{dJ_2}.$$

Reducing (13.25a) to plane stress results in

$$\left\{\begin{array}{c} d\varepsilon_x \\ d\varepsilon_\theta \\ d\varepsilon_{x\theta} \end{array}\right\} = \boldsymbol{D}_d\left\{\begin{array}{c} d\sigma_x \\ d\sigma_\theta \\ d\sigma_{x\theta} \end{array}\right\},$$

where

$$\tag{13.26}$$

$$\boldsymbol{D}_d = \frac{1}{E_s}\begin{bmatrix} 1 + q(2\sigma_x - \sigma_\theta)^2 & -\nu_s + q(2\sigma_x - \sigma_\theta)(2\sigma_\theta - \sigma_x) \\ -\nu_s + q(2\sigma_x - \sigma_\theta)(2\sigma_\theta - \sigma_x) & 1 + q(2\sigma_\theta - \sigma_x)^2 \\ 3q(2\sigma_x - \sigma_\theta)\sigma_{x\theta} & 3q(2\sigma_\theta - \sigma_x)\sigma_{x\theta} \end{bmatrix}$$

$$\begin{array}{c} 6q(2\sigma_x - \sigma_\theta)\sigma_{x\theta} \\ 6q(2\sigma_\theta - \sigma_x)\sigma_{x\theta} \\ 1 + \nu_s + 18q\sigma_{x\theta}^2 \end{array}$$

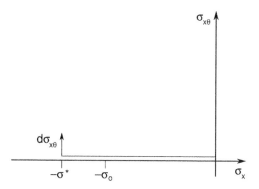

Figure 13.9 Bifurcation from a uniaxial state to one involving shear also.

where

$$q = \frac{1}{4\sigma_e^2}\left(\frac{E_s}{E_t} - 1\right).$$

An illustrative example of the difference between the use of J_2 flow and deformation theories in bifurcation calculations is the following. Consider a structure initially loaded plastically to a uniform uniaxial stress $\sigma_x = -\sigma^*$. At some stage, bifurcation buckling results in a deformation involving additional components of stress such as, for example, a shear stress increment, $d\sigma_{x\theta}$, as shown in Figure 13.9. From the J_2 flow theory Eqs. (13.20) the effective shear modulus of this increment is the elastic one (G), i.e.,

$$d\sigma_{x\theta} = 2G d\varepsilon_{x\theta} = \frac{E}{1+\nu}d\varepsilon_{x\theta}.$$

By contrast, the J_2 deformation theory Eqs. (13.26) result in

$$d\sigma_{x\theta} = 2G_r d\varepsilon_{x\theta} = \frac{2G}{1 + 3G\left(\dfrac{1}{E_s} - \dfrac{1}{E}\right)}d\varepsilon_{x\theta}.$$

$G_r \leq G$ and is known as the *reduced shear modulus* of deformation theory.

13.3.3 Anisotropic Deformation Theory

Anisotropy, when present, will affect the onset of plastic bifurcation as it does subsequent events [13.18]. If deformation theory is used for bifurcation checks, anisotropy must be accounted for. The corresponding formulation to the anisotropic flow theory in Eqs. (13.21) is developed by assuming that there exists a complementary strain energy density function $U^c(\sigma_e)$ such that the plastic strain is given by

$$\varepsilon_{ij}^P = \frac{\partial U^c}{\partial \sigma_{ij}} = \frac{\partial U^c}{\partial \sigma_e}\frac{\partial \sigma_e}{\partial \sigma_{ij}} = \left(\frac{1}{E_s} - \frac{1}{E}\right)\sigma_e\frac{\partial \sigma_e}{\partial \sigma_{ij}} \tag{13.27a}$$

where

$$
\sigma_e = \left[\sigma_x^2 - \left(1 + \frac{1}{S_\theta^2} - \frac{1}{S_r^2} \right) \sigma_x \sigma_\theta + \frac{1}{S_\theta^2} \sigma_\theta^2 + \frac{1}{S_{x\theta}^2} \sigma_{x\theta}^2 \right]^{1/2}. \qquad (13.27b)
$$

The incremental version of (13.27a) required in bifurcation calculations is developed as in [13.16]. When specialized to the "plane" stress problems of interest here, it can be written as follows

$$
\left\{ \begin{array}{c} d\varepsilon_x \\ d\varepsilon_\theta \\ d\varepsilon_{x\theta} \end{array} \right\} = \boldsymbol{D}_{da} \left\{ \begin{array}{c} d\sigma_x \\ d\sigma_\theta \\ d\sigma_{x\theta} \end{array} \right\}
$$

where

$$
\boldsymbol{D}_{da} = \frac{1}{E_s} \left[\begin{array}{ccc} 1 + q(2\sigma_x - \beta\sigma_\theta)^2 & -\hat{v}_s + q(2\sigma_x - \beta\sigma_\theta)(2\alpha\sigma_\theta - \beta\sigma_x) & \\ -\hat{v}_s + q(2\sigma_x - \beta\sigma_\theta)(2\alpha\sigma_\theta - \beta\sigma_x) & \alpha + \dfrac{E_s}{E}(1 - \alpha) + q(2\alpha\sigma_\theta - \beta\sigma_x)^2 & \\ q(2\sigma_x - \beta\sigma_\theta)\gamma\sigma_{x\theta} & q(2\alpha\sigma_\theta - \beta\sigma_x)\gamma\sigma_{x\theta} & \end{array} \right.
$$

$$
\left. \begin{array}{c} q(2\sigma_x - \beta\sigma_\theta)2\gamma\sigma_{x\theta} \\ q(2\alpha\sigma_\theta - \beta\sigma_x)2\gamma\sigma_{x\theta} \\ \dfrac{\gamma}{2} + \left(1 + v - \dfrac{\gamma}{2} \right) \dfrac{E_s}{E} + 2q\gamma^2\sigma_{x\theta}^2 \end{array} \right]
$$

$$
(13.28)
$$

and

$$
\alpha = \frac{1}{S_\theta^2}, \quad \beta = \left(1 + \frac{1}{S_\theta^2} - \frac{1}{S_r^2} \right), \quad \gamma = \frac{1}{S_{x\theta}^2}, \quad q = \frac{1}{4\sigma_e^2} \left(\frac{E_s(\sigma_e)}{E_t(\sigma_e)} - 1 \right)
$$

$$
\text{and} \quad \hat{v}_s = \frac{\beta}{2} + \frac{E_s}{E} \left(v - \frac{\beta}{2} \right).
$$

13.4   NONLINEAR KINEMATIC HARDENING

Comparison of Figures 13.2 and 13.8 demonstrates that isotropic hardening is incapable of capturing the Bauschinger effect associated with unloading and reverse loading from a plastic state. For such loading histories, material hardening is modeled by allowing the YS to follow the stress vector by translation. In the simplest approach, known as *kinematic hardening*, the size and shape of the YS are kept constant. If a Mises-type yield function is adopted, then

$$
f(\boldsymbol{\sigma} - \boldsymbol{\alpha}) = \left[\frac{3}{2}(\boldsymbol{s} - \boldsymbol{a}) \cdot (\boldsymbol{s} - \boldsymbol{a}) \right]^{1/2} = \sigma_0. \qquad (13.29)
$$

Here $\boldsymbol{\alpha}$ is the variable that tracks the position of the center of the YS and $\boldsymbol{a}$ is its deviator. A key question in such models is how to relate the position of the YS to the plastic

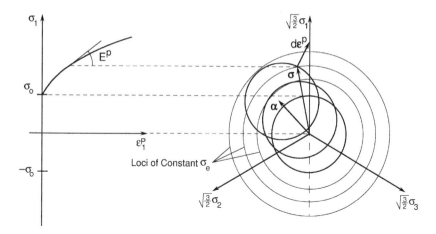

Figure 13.10 Evolution of yield surface in the Drucker–Palgen nonlinear kinematic hardening model. Isotropic surfaces are loci of constant plastic modulus.

modulus. A second important aspect is how the YS translates in stress space. Following are three models which approach the plastic modulus issue in different ways, while the *hardening rule* is incrementally decoupled from it. The models differ in complexity and exhibit different advantages and disadvantages. They have been chosen amongst other candidate models because of their relative simplicity and demonstrated ability to adequately represent aspects of mechanical behavior of tubular structures.

13.4.1 The Drucker–Palgen Model [13.19]

This is by far the simplest nonlinear kinematic hardening model. Its simplicity is rooted in the assumption that the plastic modulus H is strictly a function of σ_e,

$$H = H(\sigma_e), \quad \sigma_e = \left(\frac{3}{2} s \cdot s \right)^{1/2}. \tag{13.30}$$

In other words, loci of equal σ_e, such as the concentric circles centered at the origin of the Π-plane shown in Figure 13.10, have a constant plastic modulus (Eq. (13.30) is a more general version of the stricter choice of H proposed in [13.19], where $H = H(\sigma_e^{2N})$. This is an oversimplification of actual behavior, which can be problematic for more complex cyclic loading histories (e.g. [13.20]). However, for relatively simple loading histories, the benefits of the model's simplicity can outweigh any inaccuracies resulting from its use. Incorporating (13.29) and (13.30), for given σ and α, the flow rule (13.12) becomes

$$d\varepsilon_{ij}^p = \frac{1}{H(\sigma_e)} \frac{9}{4\sigma_o^2} [(s - a) \cdot d\sigma](s_{ij} - a_{ij}) \tag{13.31}$$

$(a_{ij} = \alpha_{ij} - \frac{1}{3}\alpha_{kk}\delta_{ij})$. Thus, $d\varepsilon^p$ is influenced by the current position of the YS. When calibrated to a uniaxial stress–strain response, H becomes

$$H = \frac{d\sigma_1}{d\varepsilon_1^p}.$$

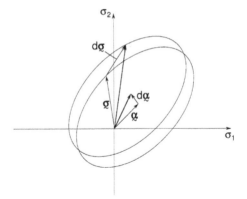

Figure 13.11   Schematic representation of the consistency condition in kinematic hardening
plasticity.

For problems involving reverse loading, the calibration should be based on the stable
hysteresis response (e.g., BCD in Figure 13.2).

Once $d\varepsilon^p$ is evaluated, the YS is translated according to an appropriate hardening rule,
which in general can be expressed as

$$d\boldsymbol{\alpha} = d\mu\boldsymbol{v}, \quad \boldsymbol{v} \cdot \boldsymbol{v} = 1, \tag{13.32}$$

where $\boldsymbol{v}$ is the unit vector in the direction of translation and $d\mu$ is the amount of translation.
The parameter $d\mu$ is evaluated by requiring that the vector $(\boldsymbol{\sigma} + d\boldsymbol{\sigma})$ be on the YS at its new
position of $(\boldsymbol{\alpha} + d\boldsymbol{\alpha})$ as shown in Figure 13.11 (*consistency condition*). For infinitesimal
increments, this condition is

$$\frac{\partial f}{\partial \sigma_{ij}} d\alpha_{ij} = \frac{\partial f}{\partial \sigma_{kl}} d\sigma_{kl}. \tag{13.33}$$

Once $d\boldsymbol{\alpha}$ is also calculated, the variables are updated and the process is repeated for the
next loading increment.

Implemented in the manner described, the model affords the user freedom of choice
of the hardening rule. The choice of hardening direction can play an important role in
predictions of the model [13.21–13.24]. Following are the various rules found to be useful
in different problems.

Kinematic Hardening Rules

- *The Ziegler Rule* [13.25]

$$v_{ij} = \frac{(\sigma_{ij} - \alpha_{ij})}{|\sigma_{ij} - \alpha_{ij}|}. \tag{13.34a}$$

- *The Phillips Rule* [13.26]

$$v_{ij} = \frac{d\sigma_{ij}}{|d\sigma_{ij}|}. \tag{13.34b}$$

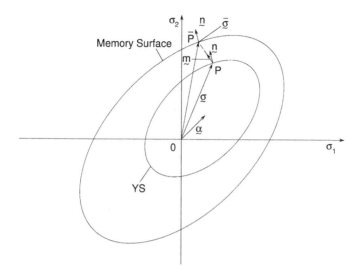

Figure 13.12 The Mroz hardening direction applied to the yield surface and a memory surface [13.21].

- *The Armstrong–Frederick Rule* [13.27]

$$v_{ij} = \frac{[(1-k)(\sigma_{ij} - \alpha_{ij}) - k\alpha_{ij}]}{|(1-k)(\sigma_{ij} - \alpha_{ij}) - k\alpha_{ij}|}. \tag{13.34c}$$

where k is a calibration constant.
- *The Mroz Rule* [13.28]
This rule is for models with two or more nesting (non-intersecting) surfaces. In the case of two surfaces, the YS and a second surface that contains it, the YS translates along the direction $\mathbf{P\overline{P}}$ in Figure 13.12, given by

$$v_{ij} = \frac{(\overline{\sigma}_{ij} - \sigma_{ij})}{|\overline{\sigma}_{ij} - \sigma_{ij}|}. \tag{13.34d}$$

Here $\overline{\sigma}$ is the image σ on the outer surface (i.e., $\overline{P}$ on outer surface has the same normal at point P on the YS – see (13.36b)). The rule ensures that the two surfaces come into contact tangentially. In the case of the Drucker–Palgen model a second surface can represent the largest state of stress in history (*memory surface* [13.21]). Its center is at the origin and grows whenever the YS comes in contact with it (see also Section 13.4.3).

13.4.2 The Dafalias–Popov Two-Surface Model

The Dafalias–Popov model [13.29, 13.30] is one of the most effective and flexible model frameworks for complex loading histories. Here we outline the version of the model applicable to cyclically stable materials that undergo relatively few loading cycles. A unique feature of the model is the method used to evaluate the plastic modulus. In the uniaxial setting represented in Figure 13.13 by the stable hysteresis bcde, the plastic

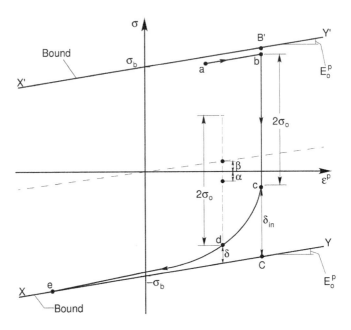

Figure 13.13   Uniaxial stress–strain variables in the Dafalias–Popov model.

modulus at point d depends on the stress variables δ and δ_{in}. Both are distances measured from line XY, called *the bound*, which is the tangent to the stress–plastic strain response at a large value of strain (point e in this case). Thus, δ is the distance of point d from the bound, and δ_{in} is the distance of the last elastic state, point c, from the same line. The plastic modulus H is related to these variables as follows:

$$H(\delta, \delta_{in}) = E_o^P + h\left(\frac{\delta}{\delta_{in} - \delta}\right), \tag{13.35}$$

where E_o^P is the modulus of the bound and h is a calibration constant which, in the simplest case, is evaluated through a one-point fit of an experimental stress–strain response as outlined in Appendix E. A second bounding line X'Y' is drawn parallel to XY as shown in the figure.

In the multiaxial setting, the YS bc is represented by (13.29). B'C becomes a *bounding surface* (BS) that encloses the YS and is defined by

$$F(\overline{\sigma} - \beta) = \left[\frac{3}{2}(\overline{s} - b) \cdot (\overline{s} - b)\right]^{1/2} = \sigma_b. \tag{13.36a}$$

Here σ_b is the size of the BS, $\overline{\sigma}$ is the congruent point on the BS to σ on the YS, β is the center of the BS, and $\overline{s}$ and b are, respectively, their deviators. The two surfaces are geometrically similar, and as a result points P and $\overline{P}$ are congruent when they have the same normals as shown in Figure 13.14. Thus, the two points are related through

$$(\overline{\sigma} - \beta) = \frac{\sigma_b}{\sigma_o}(\sigma - \alpha). \tag{13.36b}$$

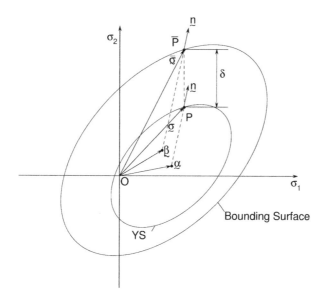

Figure 13.14 Yield and bounding surfaces and associated variables of the Dafalias–Popov model.

The scalar δ is generalized as follows (see Figure 13.14)

$$\delta = [(\bar{\sigma} - \sigma) \cdot (\bar{\sigma} - \sigma)]^{1/2}. \tag{13.37}$$

a. Evolution of the Yield Surface

The YS translates in stress space according to a chosen hardening rule, which in general is expressed as (13.32). The amount of translation $d\mu$ is chosen so that the consistency condition (13.33) is satisfied. The hardening directions in (13.34) can also be used in the current model.

b. Evolution of the Bounding Surface

The translation of the BS is coupled to that of the YS as follows:

1. *YS and BS Not in Contact*

$$d\beta_{ij} = d\alpha_{ij} - dM m_{ij}, \tag{13.38a}$$

where

$$m_{ij} = \frac{(\bar{\sigma}_{ij} - \sigma_{ij})}{|\bar{\sigma}_{ij} - \sigma_{ij}|}, \tag{13.38b}$$

and

$$dM = \left(1 - \frac{E_o^p}{H}\right)\left(\frac{d\sigma_{ij} n_{ij}}{m_{kl} n_{kl}}\right). \tag{13.38c}$$

In (13.38c), n is the unit normal to the YS at the current stress point.

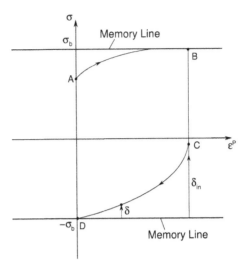

Figure 13.15   Uniaxial stress–strain variables in the Tseng–Lee model.

2. *YS and BS in Contact*
   In this case

$$d\beta_{ij} = d\xi \nu_{ij}, \tag{13.39a}$$

where

$$\nu_{ij} = \frac{(\bar{\sigma}_{ij} + d\sigma_{ij} - \beta_{ij})}{|\bar{\sigma}_{ij} + d\sigma_{ij} - \beta_{ij}|}. \tag{13.39b}$$

$d\xi$ is evaluated by requiring that a consistency condition for the BS be satisfied. The YS moves so as to remain in tangential contact with the BS at the current stress point.

13.4.3 The Tseng–Lee Two-Surface Model

This is an alternate two-surface model with the difference that the outer surface is now a *memory surface* (MS) [13.31]. It encloses the YS (13.29) and stays centered at the origin as shown in Figure 13.12. The MS is represented by

$$F(\sigma) = \left[\frac{3}{2} s \cdot s\right]^{1/2} = \sigma_b. \tag{13.40}$$

Its size, σ_b, represents the maximum stress state in history and grows isotropically any time its current value is exceeded. Thus, during the initial loading the flow rule is based on the MS and

$$d\varepsilon_{ij}^p = \frac{1}{H} \left(\frac{\partial F}{\partial \sigma_{mn}} d\sigma_{mn}\right) \frac{\partial F}{\partial \sigma_{ij}}, \tag{a}$$

where

$$H = \frac{d\sigma_1}{d\varepsilon_1^p}$$

is evaluated from the monotonic stress–strain response (AB in Figure 13.15). During this phase of the loading, the YS stays attached to the MS at the current stress point. It detaches on the first reverse loading that induces plastic deformation and translates according to the Mroz direction (13.38b – along $\mathbf{P\overline{P}}$ Figure 13.12) but with the simpler definition of $\bar{\sigma}$ given by

$$\bar{\sigma} = \frac{\sigma_b}{\sigma_o}(\sigma - \alpha). \tag{13.41}$$

Now the flow rule (13.12) is based on the yield function (13.29) and the plastic modulus H is evaluated as follows:

$$H(\delta, \delta_{in}) = E_m^p \left[1 + h\left(\frac{\delta}{\delta_{in} - \delta} \right) \right]. \tag{13.42}$$

In this case, E_m^p is the plastic modulus of the MS at σ_b and δ is a stress measure from the current plastic state to the MS, which in the uniaxial setting is represented by two horizontal lines located at the maximum stress level reached in history $\pm\sigma_b$. Similarly, δ_{in} is the value of δ at the last elastic state (see Figure 13.15). h is a fit parameter of the monotonic (AB) or stable hysteresis response (BCD) determined as outlined in Appendix F. In the multiaxial setting, δ is again generalized according to (13.37).

As the YS translates, the Mroz hardening direction rule ensures that as $\sigma_e \to \sigma_b$ the two surfaces come into tangential contact. When σ_b is exceeded, the MS becomes once more the operative surface and the YS follows, remaining in contact at the current stress point. The simpler evolution of the MS is an advantage over the previous model, which makes for a more robust numerical implementation.

REFERENCES

13.1. Miller, J.E. and Kyriakides, S. (2000). On the propagation of Lüders bands in steel strips. *ASME J. Appl. Mech.* **67**, 645–654.

13.2. Hall, E.O. (1970). *Yield Point Phenomena in Metals and Alloys.* Plenum Press, New York.

13.3. Cottrell, A.H. and Bilby, B.A. (1948). Dislocation theory of yielding and strain ageing of iron. *Proc. Physical Society* **62/I-A**, 49–62.

13.4. Johnston, W.G. and Gilman, J.J. (1959). Dislocation velocities, dislocation densities, and plastic flow in Lithium Fluoride crystals. *J. Appl. Phys.* **30**, 129–144.

13.5. Aguirre, F., Kyriakides, S. and Yun, H.D. (2004). Bending of steel pipes with Lüders bands. *Int. J. Plast.* **20**, 1199–1225.

13.6. Drucker, D.C. (1950). Some implications of work-hardening and ideal plasticity. *Quart. Appl. Math.* **7**, 411–418.

13.7. Drucker, D.C. (1952). A more fundamental approach to plastic stress–strain relations. *Proc. 1st US Congress of Applied Mechanics*, ASME, 487–491.

13.8. Drucker, D.C. (1960). Plasticity. In *Structural Mechanics*, J.N. Goodier and N.J. Hoff, Ed., Pergamon Press, London, pp. 407–454.

13.9. von Mises, R. (1913). Mechanik der festen Körper im plastisch-deformablen Zustand. *Nachrichten von der Koniglichen Gesellschaft derWissenschaften zu Göttingen, Mathematisch-Physikalische Klasse*, 582–592.

13.10. Hencky, H. (1924). Zur theorie plastischer deformationen und der hierdurch im material hervorgerufenen nachspannungen. *Zeitschrift für Angewandte Mathematik und Mechanik* **4**, 323–334.

13.11. Hill, R. (1948). A theory of the yielding and plastic flow of anisotropic metals. *Proc. Royal Society* **A 193**, 281–297.

13.12. Kyriakides, S. and Yeh, M.-K. (1988). Plastic anisotropy in drawn metal tubes. *ASME J. Eng. Ind.* **110**, 303–307.

13.13. Hutchinson, J.W. (1974). Plastic buckling. In, *Advances in Applied Mechanics*, Vol. 14, Ed. C.S. Yih, Academic Press, New York, pp. 67–144.

13.14. Nadai, A. (1923). Der beginn des fliessvorganges in einem tordierten stab. *Zeitschrift fur Angewandte Mathematik und Mechanik* **3**, 442–454.

13.15. Il'yushin, A.A. (1945). Relations between the theory of St. Venant–Levy–Mises and the theory of small elastic–plastic deformations. *Prikladnaia Matematika i Mekhanika* **9**, 207–218.

13.16. Budiansky, B. (1959). A reassessment of deformation theories of plasticity. *ASME J. Appl. Mech.* **26**, 259–264.

13.17. Batdorf, S.B. (1949). Theories of plastic buckling. *J. Aeronaut. Sci.* **16**, 405–408.

13.18. Kyriakides, S., Bardi, F.C. and Paquette, J.A. (2005). Wrinkling of circular tubes under axial compression: effect of anisotropy. *ASME J. Applied Mech.* **72**, 301–305, 2005.

13.19. Drucker, D.C. and Palgen, L. (1981). On stress–strain relations suitable for cyclic and other loading. *ASME J. Appl. Mech.* **48**, 479–485.

13.20. Hassan, T. and Kyriakides, S. (1992). Ratcheting in cyclic plasticity, Part I: Uniaxial behavior. *Int. J. Plast.* **8**, 91–116.

13.21. Shaw, P.-K. and Kyriakides, S. (1985). Inelastic analysis of thin-walled tubes under cyclic bending. *Int. J. Solids Struct.* **21**, 1073–1100.

13.22. Hassan, T., Corona, E. and Kyriakides, S. (1992). Ratcheting in cyclic plasticity, Part II: Multiaxial behavior. *Int. J. Plast.* **8**, 117–146.

13.23. Corona, E., Hassan, T. and Kyriakides, S. (1996). On the performance of kinematic hardening rules in predicting a class of biaxial ratcheting histories. *Int. J. Plast.* **12**, 117–145.

13.24. Kyriakides, S., Corona, E. and Miller, J.E. (2004). Effect of yield surface evolution on bending induced cross sectional deformation of thin-walled sections. *Int. J. Plast.* **20**, 607–618.

13.25. Ziegler, H. (1959). A modification of Prager's hardening rule. *Quart. Appl. Math.* **17**, 55–65.

13.26. Phillips, A. and Lee, C.W. (1979). Yield surfaces and loading surfaces: experiments and recommendations. *Int. J. Solids Struct.* **15**, 715–729.

13.27. Armstrong, P.J. and Frederick, C.O. (1966). *A Mathematical Representation of the Multiaxial Bauschinger Effect*. Berkley Nuclear Laboratories, R&D Department Report No. RD/B/N/73.

13.28. Mroz, Z. (1967). On the description of anisotropic workhardening. *J. Mech. Phys. Solids* **15**, 163–175.

13.29. Dafalias, Y.F. and Popov, E.P. (1975). A model of nonlinearly hardening materials for complex loading. *Acta Mech.* **21**, 173–192.

13.30. Dafalias, Y.F. and Popov, E.P. (1976). Plastic internal variables formalism of cyclic plasticity. *ASME J. Appl. Mech.* **43**, 645–651.

13.31. Tseng, N.T. and Lee, G.C. (1983). Simple plasticity model of two-surface type. *ASCE J. Eng. Mech.* **109**, 795–810.

Books in Plasticity

13.32. Hill, R. (1950). *The Mathematical Theory of Plasticity*. Oxford University Press, New York.

13.33. Mendelson, A. (1968). *Plasticity: Theory and Applications*. McMillan, New York.

13.34. Katchanov, L.M. (1971). *Foundations of the Theory of Plasticity*. North-Holland Publishing Co., Amsterdam.

13.35. Chakraparty, J. (1987). *Theory of Plasticity*. McGraw-Hill, New York.

13.36. Chen, W.F. and Han, D.J. (1988). *Plasticity for Structural Engineers*. Springer-Verlag, New York.

13.37. Lubliner, J. (1990). *Plasticity Theory*. McMillan, New York.

13.38. Khan, A.S. and Huang, S. (1995). *Continuum Theory of Plasticity*. John Wiley, New York.

Appendix A
Mechanical Testing

Mechanical testing is an essential part of quality control and certification of pipeline production. It is most often conducted at the pipe mill. A minimum amount of testing is required by design codes. Additional testing to match the needs of a specific project is often specified by the customer. This section outlines the four main categories of mechanical testing commonly performed: tension and compression tests for establishing the material stress–strain properties, Charpy and other impact tests for toughness, hardness tests for welds and heat-affected zones and the split-ring test for estimating bending-type residual stresses in the pipe.

A.1 TENSILE AND COMPRESSIVE MATERIAL STRESS–STRAIN RESPONSES

A.1.1 Tension Tests

The main mechanical properties of pipe materials are established by tensile testing. Because the properties along the axis and transverse to it can differ, distinction should be made between axial or *longitudinal* specimens and circumferential or *transverse* specimens. Full thickness, dog-bone shaped axial specimens are usually relatively easy to extract. ASTM A370-03a [A.1] (Figure 3) and API SP 5L [A.2] (Figure 4) provide guidelines as to preferred specimen dimensions. For smaller diameter and thinner wall pipes, a specimen width (w) between $1.5\,t$ and $2\,t$ and a gage length (L_g) of at least $5w$ can be used as an alternate guideline for convenience (see Figure A.1). Saw cutting followed by machining to size should be the preferred method of extracting the specimens from the pipe. If flame cutting is used, then the heat-affected zone should not be closer than 4 in (100 mm) from the edges of the specimen.

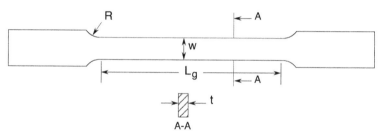

Figure A.1 Typical flat tensile specimen geometry.

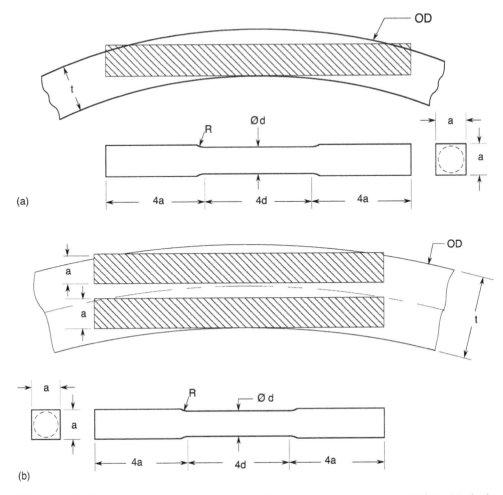

Figure A.2 Transverse round test specimens and locations in the pipe cross section: (a) single specimen and (b) two specimens.

Transverse specimens can be more difficult to extract, especially from smaller diameter and thinner wall pipes. For this reason, API SP 5L [A.2] allows testing of specimens extracted from a transverse sector of pipe that are then flattened. This, of course, alters the measured stress–strain response significantly. The changes induced depend strongly on the diameter and thickness of the pipe and, to some degree, on the manufacturing process. Correcting the measured data for the flattening is difficult and beyond the normally available means at most steel mills. For this reason, flattening of specimens should be avoided. On the other hand, during production, flattened specimens can be useful for comparing transverse properties within a heat and between heats for quality control of pipes of the same dimensions.

The best alternative is a transverse specimen machined out of the thickness of the pipe, as shown in Figure A.2(a). Whenever possible, the specimen should encompass most of the pipe wall thickness. For pipes known to have significant gradients in properties through the thickness, two specimens can be extracted side-by-side, as shown in Figure A.2(b). Round test section specimens is a convenient option, although specimens of other shapes are also acceptable. Clearly, following this option does not allow use of standard-dimensioned specimens. Instead, specimen dimensions will be pipe-specific. For round specimens, the test section should be at least 4 diameters long, while other dimensions will depend on the method of gripping. These can be scaled to match those in Figure 5 of [A.1] (see also Figure 4 of [A.2]).

Testing is performed in a standard testing machine (either servo-hydraulic or screw-type) under displacement control. The rate of loading should be chosen such as to induce a strain rate in the range of $10^{-4} < \dot{\varepsilon} \leq 10^{-3}\,\mathrm{s}^{-1}$ to the test section. The strain should be measured with an extensometer mounted in the test section (Figure A.3). A 1- or 2-inch gage length extensometer with a strain range of 40–50% is preferable. The addition of strain gages (two placed diametrically opposite to each other) is recommended when very accurate measurements of the elastic modulus and of the initial yielding behavior are required. Gage marks can be scribed on the test section (usually at 1 or 2 inch spacing) to allow crude estimates of elongation at failure. The load is measured by a calibrated load cell that is usually an integral part of the testing facility.

A variety of gripping mechanisms are available, which also determine the specimen design. Hydraulic grips with either round or flat specimen wedge inserts are the most versatile and easy to use. Figure A.3 shows a round specimen with square ends inside hydraulic grips with flat grip wedges. This arrangement is useful for both tensile and compressive tests. Mechanical, vise-type grips are equally effective, but are usually limited to smaller cross section specimens.

Typical stress–strain data from a steel with a continuous transition from elastic to plastic behavior are shown in Figure 13.1. The elastic modulus (E), the 0.2% strain offset yield stress (σ_o), and the alternative yield stress definition corresponding to the stress at a strain of 0.5% ($\sigma_o' \equiv$ *yield strength*, API) are marked in Figure 13.1(a). The response at higher strains, including the onset and growth of necking as well as the failure of the neck, is shown in Figure 13.1(b). The maximum value of engineering stress recorded is the ultimate stress ($\sigma_{UL} \equiv$ *tensile strength*).

Figure 13.3 shows stress–strain data from an X60 carbon steel that exhibits discontinuous yielding. In this case, the yield stress is defined as the stress plateau associated with Lüders bands (σ_L). Once the Lüders bands have been exhausted, the response hardens and becomes similar to the one in Figure 13.1.

Tensile tests are usually continued past necking of the test section and are terminated when the specimen fails. Necking often occurs outside the extensometer gage length, and in other cases the extensometer range is exceeded. In such cases, the crosshead displacement can provide an estimate of the elongation of the gage length. A crude estimate of the elongation at failure can be obtained from the change in length between scribe marks, originally 2 inches apart, that encompass the necked zone. A more dependable estimate of ductility is the reduction in area of the smallest cross section within the neck.

A.1.2   Compression Tests

The tensile and compressive stress–strain responses of hot-finished tubular steel products are similar, at least for small strains. Tensile tests are thus sufficient to determine the mechanical properties. Cold-formed or cold-finished products (e.g., UOE pipe) can exhibit differences between tensile and compressive properties. Transverse compressive

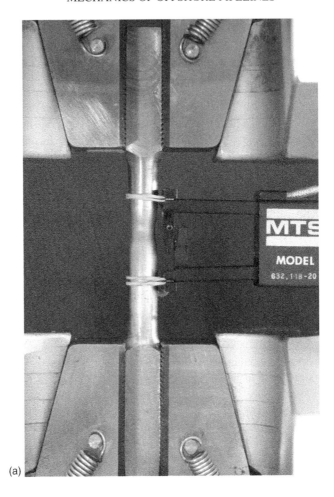

(a)

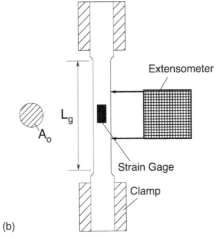

(b)

Figure A.3   Round test specimen in hydraulic grips: (a) photograph and (b) schematic.

properties are then required in order to design such pipes against collapse. Compressive uniaxial testing is more difficult because the specimen can buckle before sufficient material response is secured. Buckling can be delayed, however, by making the test section shorter. The specimen dimensions in Figure A.2 will usually allow a buckling-free test up to a strain of at least 1%. An additional requirement is nearly perfect grip alignment. Because of through thickness variation of properties, some bending of the specimen is unavoidable. The effect of bending must be negated by mounting two extensometers diametrically opposite to each other (two strain gages also suffice) and averaging their outputs.

A.2 TOUGHNESS

Toughness describes the ability of a material to resist rapid fracture. It is usually established by impacting a pre-notched specimen with an object traveling at a known velocity. Tough materials such as steels undergo significant plastic deformation during impact and dissipate a significant amount of energy prior to rupture. Rupture involves nucleation, growth and coalescence of voids and is known as *ductile* fracture. The fracture surface is usually rough with sheared regions. By contrast, in *brittle* fracture the material often cleaves along preferred crystallographic planes, leaving behind a crystalline fracture surface appearance. Brittle cracks absorb much less energy.

Body-centered cubic steels show a marked temperature dependence in their yield stress. Since the velocity of dislocations is proportional to the stress, at low temperature dislocations move at a higher velocity and have a higher chance of forming a crack by dislocation coalescence [A.3]. The possibility of nucleation of cracks at twins and at carbide inclusions also increases at lower temperatures. These and other low temperature effects, combined with high strain rates, result in a transition from ductile to brittle fracture as the temperature decreases. The consequence of this transition is a marked decrease in the energy absorbed by a dynamic crack. The Charpy V-notch impact test is the most common way of establishing the ductile-to-brittle transition temperature.

A.2.1 Charpy V-Notch Impact Test (CVN)

The Charpy impact test is the most common test for measuring the fracture toughness of materials. Its popularity is based on the ease with which it can be performed and the relatively simple equipment required. A notched specimen of specified dimensions is broken by a single blow of a freely-swinging pendulum. The pendulum is released from a fixed height (see Figure A.4). By knowing the mass of the pendulum, the kinetic energy at impact is determined. The machine measures the height reached by the pendulum after it strikes the specimen. The difference between the start and finish heights gives the energy absorbed by the broken specimen. Charpy machines used for testing steel usually have energy capacities in the range of 220–300 ft-lb (300–400 J).

The standard Charpy specimen is 55 mm long with a 10 mm square cross section. A 2 mm deep, 45° V-notch is machined at mid-span. The notch has a specified radius of 0.25 mm. The specimen is simply supported as shown in Figure A.5, and is struck on the opposite side of the notch by a striker attached to the pendulum.

Tests are performed at different temperatures, often down to −60°C, in order to establish the transition temperature from ductile to brittle failure. Above the transition temperature,

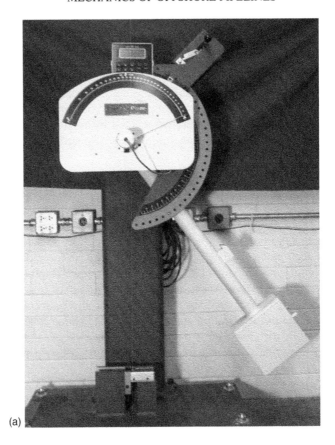

(a)

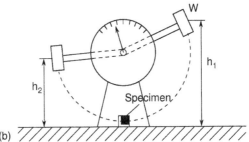

(b)

Figure A.4   (a) A Charpy impact testing machine. The test specimen is at the base of the machine while the striker is attached to the swinging pendulum. (b) Impact hammer height before and after impacting the specimen.

impacted specimens fail in a ductile manner. This involves microvoid coalescence, which requires relatively large amounts of energy. The fracture surface is usually rough, with clear signs of significant shear deformation. At lower temperatures, closer to the transition temperature, specimens fail in a brittle (usually cleavage) manner and have a crystalline fracture surface appearance. Brittle fractures absorb less energy. A typical impact energy (C_V)–temperature (T) plot for a low carbon steel is shown in Figure A.6. The drop in

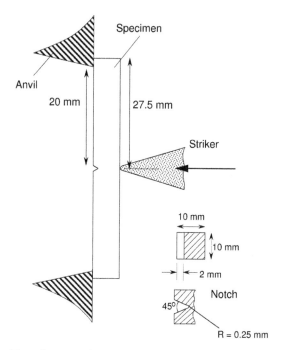

Specimen

Anvil

20 mm

27.5 mm

Striker

10 mm

10 mm

2 mm

Notch

45°

R = 0.25 mm

Figure A.5 Charpy V-notch test specimen geometry and set up.

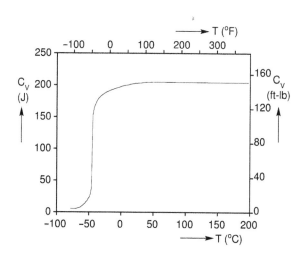

Figure A.6 Charpy V-notch impact energy as a function of temperature for a low carbon steel (C = 0.11 wt%) [A.4].

impact energy around −40°C indicates the transition from ductile to brittle fracture. In this case, C_V drops from about 200 J at 50°C and above down to 10 J at −70°C and lower temperatures.

The test is only an indicator of the material toughness and of the transition temperature. The energy absorbed depends on specimen and notch geometry, as well as on the setup. If these are kept constant (e.g., as per [A.1]), the measured energy absorbed can be a useful tool for comparing the toughness, as the alloying and processing of steel is varied.

When evaluating line pipe steels, specimens should be cut along the transverse direction. C_V in the range of 25–75 J (18–55 ft-lb) at a temperature in the range of 0°C to −40°C (32°F to −40°F) is typically specified for transverse specimens. Modern steels usually outperform these guidelines (e.g., see Table 3.3). The toughness is usually higher along the axial direction for both seamless and seam-welded pipe. This is because of preferential orientation of grains during rolling. Because of this difference, requirements are higher for axial specimens. Special attention should also be paid to both longitudinal and girth welds, as they, as well as the heat-affected zones, can have a different (lower) C_V.

A.2.2   Drop-Weight Tear Test (DWTT)

The drop-weight tear test is an alternative to the Charpy V-notch test, often conducted in addition to it. The test involves a 12×3 in (305×76 mm) full thickness or 0.75 in (19 mm – the smaller of the two) specimen extracted from the transverse direction of the pipe. The specimen is flattened and either a V-notch or a Chevron notch is placed at mid-span (see API RP 5L3 [A.5]). The specimen is simply supported at the ends and is impacted by a steel hammer of specified mass that is dropped from a specified height. The specimen fractures dynamically and the fracture surfaces are evaluated. The area of the cross section that has undergone shear fracture instead of cleavage fracture is estimated and reported as a fraction (percent) of the total area.

It has been shown that the sheared area fraction is related to the speed of crack propagation in gas lines. If the DWTT sheared area is above 40%, the speed of propagation is lower than the decompression speed of about 450 m/s and the crack is arrested. Thus, for gas lines, a DWTT shear area of 80–85% at the lowest operating temperature is usually specified.

A.3   HARDNESS TESTS

Hardness is most commonly defined as the resistance of a material to indention. A hard (diamond or hardened steel) cone or pyramid-shaped indenter is pressed into the material with a known force, causing plastic deformation, as shown in Figure A.7. The indention diameter or depth is then a measure of the material hardness. Hardness measurements approximately represent the material strength. They are thus used to test narrow zones of material otherwise hard to analyze, such as welds and heat-affected zones. There are several different types of indention tests (Brinell, Vickers, Rockwell), which differ mainly in the shape and size of the indenter.

The Rockwell test is perhaps the most widely used. It uses a small cone-shaped diamond indenter, called the brale, with a slightly rounded end and an included angle of 120° (a steel ball of prescribed diameter is used for softer materials). The cone is first pressed into the material at a minor load of 10 kg and the indention depth is recorded. The major load is then applied, increasing the indention depth (value depends on the scale being used, e.g., 100 kg for Rockwell B and 150 kg for Rockwell C). The load is reduced back to the minor level and the difference in indention depth (Δh, mm) is measured by a dial

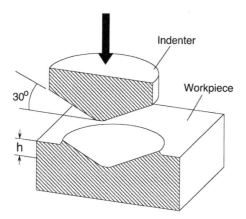

Figure A.7 Rockwell hardness indenter and test specimen.

gage on the machine. The hardness number is calculated as follows

$$HRX = M - \frac{\Delta h}{0.002}.$$ (A.1)

where X represents the scale, and $M = 100$ for diamond indenters (A, C and D scales) and 130 for steel indenters (B, E, M, R, etc. scales). Accordingly, a decrease in Δh of 0.002 mm increases the hardness by one unit.

The Vickers hardness test is an alternative to Rockwell that is used more widely in Europe. In this test, a small diamond indenter with a pyramidal shape is pressed into the surface of the material. The resulting impression is observed under a microscope and recorded. The loads involved range between 1 and 100 kg. The hardness is calculated as follows

$$HVX = \frac{2P \sin \theta}{d^2}.$$ (A.2)

where X and P are the applied load in kg, 2θ is the angle between two opposite sides of the pyramid ($136°$), and d is the average length in mm of the two diagonals of the nearly square indention.

Conversions of HRX numbers to other hardness numbers, such Vickers (HV) and Brinell (HB), appear in Tables 2–6 in [A.1]. Included are the corresponding tensile strengths. These should be considered to be approximate, as they can vary with material hardening and other factors.

A.4 RESIDUAL STRESSES

In most major manufacturing processes, pipe is finished by cold forming, which often induces residual stress fields. The mechanical behavior of pipe under external pressure and bending is influenced to some degree by circumferential bending stresses. The amplitude of this particular residual stress field is determined in a simple pipe ring splitting test [A.6]. A section of the pipe (one-third to one-half a diameter in length is sufficient) is cut along a generator with a saw. The ring either springs open or closes. If it opens, the gap

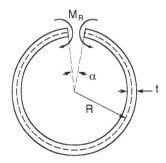

Figure A.8   Schematic of a ring that springs open upon splitting.

is measured; if it closes, then a second cut is made, removing a sliver of the ring, and the amount of closing is measured.

Simple opening up or closing of the ring can be related to a bending-type residual stress field, as shown in Figure A.8. Strength of materials considerations, based on the usual thin wall assumptions, yields the following expression for the amplitude of the bending stress field

$$|\sigma_R| = \frac{Et}{4\pi R}\alpha. \tag{A.3}$$

where α is the included angle of the opening or closing (for rings longer than $0.1D$ replace E in (A.3) by $E/(1 - \nu^2)$. For very thick pipes ($D/t \sim 10$ or less), a more accurate expression can be derived from elasticity (Eq. (20) [A.7]; see also [A.8]). Such measurements performed on seamless X42 and X65 pipes produced σ_R/σ_o in the range of 0.1–0.45. Similar amplitudes have also been measured in small diameter seamless SS-304 tubes.

REFERENCES

A.1. ASTM A370-03a (2003). *Standard Test Methods and Definitions for Mechanical Testing of Steel Products.*

A.2. API Specification 5L (2004). *Specification for Line Pipe*, 43rd Edition. American Petroleum Institute. Washington, DC.

A.3. Honeycomb, R.W.K. (1982). *Steels: Microstructure and Properties*. Edward Arnold, London.

A.4. Boyer, H.E. and Gall, T.L. (1985). *Metals Handbook*. American Society for Metals. Metals Park, Ohio.

A.5. API Recommended Practice 5L3 (1996). *Recommended Practice for Conducting Drop-Weight Tear Tests on Line Pipe*, 3rd Edition. American Petroleum Institute.

A.6. Frame, W.M. (1938). Casing setting depths are not assured by physical properties of the steel. *API Drilling and Production Practice*, pp. 323–352.

A.7. Yeh, M.-K. and Kyriakides, S. (1986). On the collapse of inelastic thick-walled tubes under external pressure. *ASME J. Energy Resour. Technol.* **108**, 35–47.

A.8. Timoshenko, S. and Goodier, J.N. (1970). *Theory of Elasticity*, 3rd Edition, Section 29. McGraw-Hill. New York.

Appendix B
Plastic Anisotropy in Tubes

Forming processes commonly used to manufacture tubes and pipes often induce some yield anisotropy in the finished products. Such anisotropy affects the collapse pressure and other limit states of pipelines. The anisotropy must thus be determined and included in the modeling of the various limit states. Hill-type anisotropy [B.1] is usually adequate for modeling the yield characteristics of tubes. For commonly-used pipe dimensions, a state of "plane" stress usually suffices. The yield function in terms of polar cylindrical coordinate stress components can be written as

$$f = \sigma_e = \left[\sigma_x^2 - \left(1 + \frac{1}{S_\theta^2} - \frac{1}{S_r^2} \right) \sigma_x \sigma_\theta + \frac{1}{S_\theta^2} \sigma_\theta^2 + \frac{1}{S_{x\theta}^2} \sigma_{x\theta}^2 \right]^{1/2}, \qquad \text{(B.1)}$$

where $S_\theta = \sigma_{o\theta}/\sigma_{ox}$, $S_r = \sigma_{or}/\sigma_{ox}$, $S_{x\theta} = \sigma_{ox\theta}/\sigma_{ox}$, $\{\sigma_{ox}, \sigma_{or}, \sigma_{o\theta}\}$ are the yield stresses in the respective directions (axial, through-thickness and circumferential), and $\sigma_{ox\theta}$ is the yield stress under pure shear.

Four experiments are required to establish the anisotropy constants in (B.1). For larger diameter and relatively thick tubes, the mechanical properties in the three normal directions can be measured directly on test coupons extracted from each direction. The shear anisotropy requires an independent shear test. For smaller diameter tubes, the following experimental procedure has been found to be effective [B.2].

B.1 ANISOTROPY TESTS

Two of the material constants can be evaluated directly by measuring the uniaxial stress–strain response in the axial and circumferential directions. In the axial direction, the response is measured from a uniaxial test on an axial coupon cut along the axis of the tube. (For small diameter tubes the whole tube can be pulled axially instead.) In the circumferential direction, the response can be best measured using a "lateral" internal pressure test on a section of the tube. In this test, the specimen experiences only circumferential stress. The anisotropy constant S_r can be evaluated by conducting any combined stress test involving a radial path in the $\sigma_x - \sigma_\theta$ plane. For convenience, we choose for this a hydrostatic pressure inflation test in which the ratio of the circumferential to the axial stress is 2. For the last two tests, a pressurizing system coupled with a closed-loop control testing machine is used.

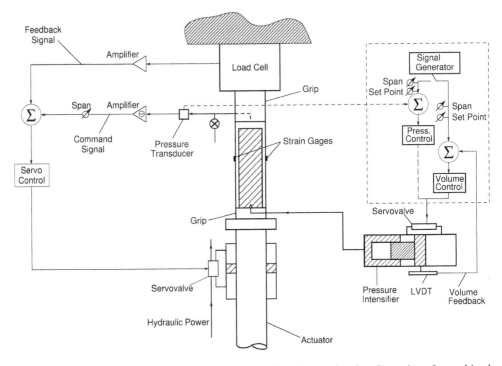

Figure B.1   Experimental set up for establishing plastic anisotropy in tubes. It consists of a combined
internal pressure-axial loading test facility.

Figure B.1 shows schematically such a system. The specimen is sealed at both ends
using grips compatible with available testing machine accessories and mounted in a standard servo hydraulic testing machine (custom-made grips are usually required). The axial
and circumferential strains in the test section are measured with strain gages mounted
on the surface of the tube. The specimen is filled with a fluid such as hydraulic oil
and subsequently pressurized. In the case shown, the pressurizing unit consists of a
10,000 psi (690 bar) pressure intensifier that operates on standard 3,000 psi (200 bar)
hydraulic power. It has its own independent closed-loop control system and is run under
volume control. The pressure is monitored via a pressure transducer whose output is
amplified so that at full scale it gives the standard 10 V. The testing machine is run in load
control and must have the option of accepting an external command signal. The pressure,
axial force, and strains are recorded on a data acquisition system for later processing.

B.1.1   Lateral Pressure Test

Pure lateral pressure is achieved by providing an axial compressive load to equilibrate the
pressure load at the ends of the tube (PA_i, where A_i is the internal cross sectional area of
the tube). The output of the pressure transducer, suitably amplified through an inverting
amplifier, is used as the command signal for the axial servo controller. As the pressure
in the tube is gradually increased, the actuator moves to keep the axial force at $-PA_i$. In

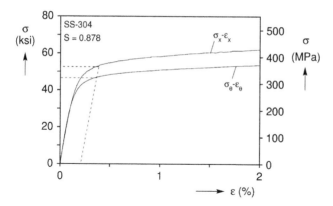

Figure B.2   Stress–strain responses in the axial and circumferential directions of an SS304 tube that illustrate anisotropic yielding.

this fashion, the axial force due to the pressure is reacted by the testing machine and the tube experiences stresses of $\sigma_x = 0$ and $\sigma_\theta = PR/t$.

A common anisotropy present in drawn tubes is the type shown in Figure B.2. The stress–strain response in the circumferential direction has a somewhat lower yield stress than the one in the axial direction, while the shape of the responses is similar. The case shown was conducted for the purposes of calculating the collapse pressure of a particular tube. As mentioned in Section 4.3, in such cases the assumption that $S_r = S_\theta \equiv S$ suffices, and the shear anisotropy is not required. For this case, S was found to be 0.878. For pipes that buckle plastically, S has a proportional impact on the collapse pressure.

B.1.2   Hydrostatic Pressure Test

For this case no external command signal is required from the pressure transducer. The machine is placed in load control with the axial load set at zero. As the pressure increases, the actuator moves, keeping the machine applied force at zero. The axial force due to the pressure is reacted by the tube, and as a result the stresses that develop are now $\sigma_x = PR/2t$ and $\sigma_\theta = PR/t$. These are used in (B.1) to evaluate $\sigma_e(P)$ with $S_r = S_\theta = 1$ initially. The work compatible equivalent plastic strain increment is established by requiring the $dW^p = \sigma_e d\varepsilon_e^p = \sigma_{ij} d\varepsilon_{ij}^p$ (see also [B.3]). For this biaxial loading it is given by

$$
d\varepsilon_e^p = \sqrt{2} \left[\frac{\dfrac{1}{S_\theta^2}(d\varepsilon_x^p)^2 + (d\varepsilon_\theta^p)^2 + \left(1 + \dfrac{1}{S_\theta^2} - \dfrac{1}{S_r^2}\right) d\varepsilon_x^p d\varepsilon_\theta^p}{\dfrac{1}{S_\theta^2} + \dfrac{1}{S_r^2} - \dfrac{1}{2} - \dfrac{1}{2}\left(\dfrac{1}{S_\theta^2} - \dfrac{1}{S_r^2}\right)^2} \right]^{1/2}.
\tag{B.2}
$$

The measured strains are used in (B.2), again initially with $S_r = S_\theta = 1$. The resultant $\sigma_e - \varepsilon_e^p$ measured in an SAF 2507 super-duplex tube is shown in Figure B.3(a). Included are the corresponding responses measured in the axial and circumferential directions. This type of tube was cold finished, and as a result the yield stress in the circumferential direction is higher than in the axial direction. It was found that $S_\theta = 1.14$ brings the two

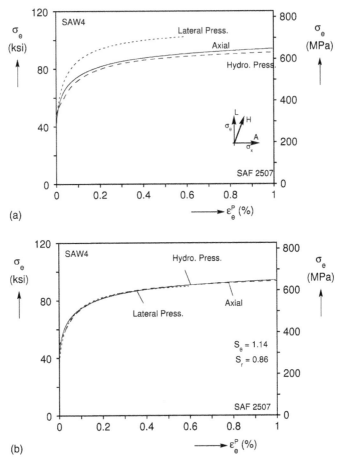

Figure B.3　(a) Equivalent stress–plastic strain responses from three tests on SAF2507 exhibiting anisotropy. (b) Coalescence of the three responses after the anisotropy parameters in (B.1) and (B.2) have been appropriately selected.

uniaxial responses together. With S_θ fixed at this value in (B.1) and (B.2), S_r was varied until at $S_r = 0.86$ all three responses coalesced as shown in Figure B.3(b).

B.1.3　Torsion Test

The parameter $S_{x\theta}$, when required, can be evaluated in an additional test involving pure shear. For smaller diameter pipes this can be a pure torsion test on a section of pipe. For pure shear loading, the equivalent stress and equivalent plastic strain increments are, respectively

$$\sigma_e = \frac{\sigma_{x\theta}}{S_{x\theta}} \quad \text{and} \quad d\varepsilon_e^p = 2S_{x\theta}d\varepsilon_{x\theta}^p, \tag{B.3}$$

where $S_{x\theta} = 1/\sqrt{3}$ when the material yields isotropically. $S_{x\theta}$ is chosen by matching the equivalent stress–plastic strain response produced by (B.3) to the uniaxial response.

REFERENCES

B.1. Hill, R. (1948). A theory of the yielding and plastic flow of anisotropic metals. *Proc. Roy. Soc.* A **193**, 281–297.

B.2. Kyriakides, S. and Yeh, M.-K. (1988). Plastic anisotropy in drawn metal tubes. *ASME J. Eng. Ind.* **110**, 303–307.

B.3. White, G.N. and Drucker, D.C. (1950). Effective stress and effective strain in relation to stress theories of plasticity. *J. Appl. Phys.* **21**, 1013–1021.

Appendix C
The Ramberg–Osgood
Stress–Strain Fit

Structural metal stress–strain responses can often be represented through a power-law relationship. One of the most commonly used relationships is the three-parameter Ramberg–Osgood expression

$$\varepsilon = \frac{\sigma}{E}\left[1 + \frac{3}{7}\left(\frac{\sigma}{\sigma_y}\right)^{n-1}\right], \tag{C.1}$$

where E, σ_y and n are fit parameters. These can be determined from a measured stress–strain response as follows:

(a) E is the slope of the linearly elastic part of the response.
(b) σ_y is the stress at the intersection of the stress–strain response and a line through the origin with a slope of $0.7E$ (see Figure C.1).
(c) An approximate value of n is obtained from the slope of the linear part of the plot of $\log\left(\varepsilon - \frac{\sigma}{E}\right)$ vs. $\log(\sigma)$.
(d) Plot this first estimate of the fit and compare it directly with the experimental response. Adjust σ_y and n (usually slightly) iteratively until the best fit is achieved.

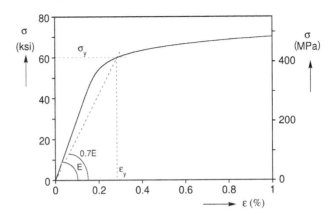

Figure C.1 Ramberg–Osgood construction lines.

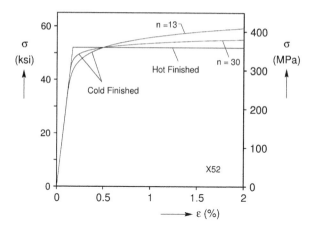

Figure C.2 Three X52 generic stress–strain responses.

Table C.1 X52 stress–strain parameters.

E Msi (GPa)	σ_o' ksi (MPa)	σ_y ksi (MPa)	n	σ_o ksi (MPa)	σ_{el} ksi (MPa)	E' ksi (MPa)
30	52.0	45.97	13	50.07	32.18	–
(207)	(359)	(317)		(345)	(222)	
30	52.0	49.43	30	51.15	35.0	–
(207)	(359)	(341)		(353)	(241)	
30	52.0	–	–	52.0	52.0	50
(207)	(359)			(359)	(359)	(345)

It is not always possible to obtain a close fit for small values of strain, say on the order of 0.5%, and for larger values simultaneously. Therefore, it is best to restrict the fit to the part of the response relevant to the problem at hand. If the strain range of interest is large, the response can be fitted with (C.1) up to some strain level and with another fit (a straight line often works) for larger values.

Generic X52 Stress–Strain Responses

Several parametric studies conducted in this book involve the three X52 line grade steel stress–strain responses shown in Figure C.2. Two are continuous and have the Ramberg–Osgood parameters listed in Table C.1. The third is bilinear with a small post-yield modulus (E') that is meant to represent hot-finished pipe. The stress at a strain of 0.005 (σ_o') is 52 ksi (359 MPa) for all three. The stress at a strain offset of 0.002 (σ_o) used in this book for normalizations is also listed. The stress at the elastic limit (σ_{el}) is a variable used in numerical plasticity calculations. A Poisson's ratio of $\nu = 0.3$ is used with these stress–strain responses.

Appendix D
Sanders' Circular Cylindrical
Shell Equations

Several viable nonlinear shell theories are available for circular cylindrical shells. Versions of these theories that correspond to the same level of approximation differ from each other in small ways that do not usually affect the solution. In this book we have adopted mainly Sanders' [D.1] shell equations with the assumptions that mid-surface strains are small and rotations are "small but finite." The strain–displacement and equilibrium equations are presented here.

a. Strain–Displacement Equations

Consider a circular cylindrical shell of radius R and wall thickness t. Define a cylindrical coordinate system $\{x, \theta, r\}$ and let $\{u, v, w\}$ be the corresponding displacements. The membrane and bending components of the strains in terms of the displacements are

$$\varepsilon_{xx}^o = \frac{\partial u}{\partial x} + \frac{1}{2}\left(\frac{\partial w}{\partial x}\right)^2 + \frac{1}{2}\phi^{2*},$$

$$\varepsilon_{\theta\theta}^o = \frac{1}{R}\frac{\partial v}{\partial \theta} + \frac{w}{R} + \frac{1}{2}\left[\frac{v^{**}}{R} - \frac{1}{R}\frac{\partial w}{\partial \theta}\right]^2 + \frac{1}{2}\phi^{2*},$$

$$\varepsilon_{x\theta}^o = \frac{1}{2}\left(\frac{1}{R}\frac{\partial u}{\partial \theta} + \frac{\partial v}{\partial x}\right) + \frac{1}{2R}\frac{\partial w}{\partial x}\left(\frac{\partial w}{\partial \theta} - v^{**}\right), \tag{D.1}$$

$$\kappa_{xx} = -\frac{\partial^2 w}{\partial x^2},$$

$$\kappa_{\theta\theta} = -\frac{1}{R^2}\left(\frac{\partial^2 w}{\partial \theta^2} - \frac{\partial v^{**}}{\partial \theta}\right),$$

$$\kappa_{x\theta} = -\frac{1}{2}\left[\frac{1}{R}\frac{\partial^2 w}{\partial x \partial \theta} + \left(\frac{1}{R}\frac{\partial^2 w}{\partial x \partial \theta} - \frac{1}{R}\frac{\partial v^{**}}{\partial x}\right) - \frac{1}{R}\phi^{*}\right],$$

378

where

$$\phi = \frac{1}{2}\left(\frac{\partial v}{\partial x} - \frac{1}{R}\frac{\partial u}{\partial \theta}\right).$$

The total strains are given by

$$\varepsilon_{\alpha\beta} = (\varepsilon^o_{\alpha\beta} + z\kappa_{\alpha\beta})/(A_\alpha A_\beta)^{1/2}, \quad A_1 \cong 1 \text{ and } A_2 \cong 1 + \frac{z}{R}, \quad (\alpha, \beta) \equiv (x, \theta) \text{ no sum},$$

(D.2)

where z is the through-thickness coordinate.

b. Equilibrium Equations

The force and moment intensities are given by

$$N_{\alpha\beta} = \int_{-t/2}^{t/2} \frac{A_1 A_2}{(A_\alpha A_\beta)^{1/2}} \sigma_{\alpha\beta} dz, \quad M_{\alpha\beta} = \int_{-t/2}^{t/2} \frac{A_1 A_2}{(A_\alpha A_\beta)^{1/2}} \sigma_{\alpha\beta} z dz, \quad (\alpha, \beta) \quad \text{no sum}. \quad (D.3)$$

The equilibrium equations in terms of the force and moment intensities are

$$\frac{\partial N_{xx}}{\partial x} + \frac{1}{R}\frac{\partial N_{x\theta}}{\partial \theta} - \frac{1}{2R^2}\frac{\partial M^*_{x\theta}}{\partial \theta} - \frac{1}{2R}\frac{\partial}{\partial \theta}[\phi(N_{xx} + N_{\theta\theta})]^* = 0,$$

$$\frac{\partial N_{x\theta}}{\partial x} + \frac{1}{R}\frac{\partial N_{\theta\theta}}{\partial \theta} + \frac{Q^{**}_{\theta\theta}}{R} + \frac{1}{2R}\frac{\partial M^*_{x\theta}}{\partial x} - \frac{1}{R}[\phi_2 N_{\theta\theta} + \phi_1 N_{x\theta}]^{**}$$

$$+ \frac{1}{2}\frac{\partial}{\partial x}[\phi(N_{xx} + N_{\theta\theta})]^* = 0,$$

$$\frac{\partial Q_{xx}}{\partial x} + \frac{1}{R}\frac{\partial Q_{\theta\theta}}{\partial \theta} - \frac{N_{\theta\theta}}{R} - \frac{\partial}{\partial x}[\phi_1 N_{xx} + \phi_2 N_{x\theta}] - \frac{1}{R}\frac{\partial}{\partial \theta}[\phi_1 N_{x\theta} + \phi_2 N_{\theta\theta}] + q = 0,$$

$$\frac{\partial M_{xx}}{\partial x} + \frac{1}{R}\frac{\partial M_{x\theta}}{\partial \theta} - Q_{xx} = 0,$$

(D.4)

$$\frac{\partial M_{x\theta}}{\partial x} + \frac{1}{R}\frac{\partial M_{\theta\theta}}{\partial \theta} - Q_{\theta\theta} = 0,$$

where

$$\phi_1 = -\frac{\partial w}{\partial x} \quad \text{and} \quad \phi_2 = -\frac{1}{R}\frac{\partial w}{\partial \theta} + \frac{v^{**}}{R}.$$

Two additional levels of simplification can be made when the problem allows it:

(a) Rotations about the normal to the shell are neglected by dropping terms with (•)*.
(b) The shallow-shell equations, also known as the Donnell–Mushtari–Vlasov equations (DMV), are arrived at by eliminating terms with (•)* and (•)**. In this case, characteristic deformation wavelengths must be small compared to the minimum principal radius of curvature.

REFERENCE

D.1. Sanders Jr., J.L. (1963). Nonlinear theories for thin shells. *Quart. Appl. Math.* **21**, 21–36.

Appendix E
Stress–Strain Fitting for the Dafalias–Popov Model

This model is usually used for loading histories involving loading and reverse loading. For this reason, the initial monotonic stress–strain response and the subsequent stable hysteresis are fitted separately. First convert the two responses into stress–plastic strain responses by using

$$\varepsilon^P = \varepsilon - \frac{\sigma}{E}. \tag{E.1}$$

Now follow the fitting steps below.

a. Fit of Monotonic Stress–Strain Response

The monotonic part of the response (o-a-b-c in Figure E.1(a)) is fitted as follows:

1. E_m is Young's modulus of o-a.
2. σ_{om} is the elastic limit of o-a-b-c.
3. The bounding line is constructed by drawing the tangent to a-b-c at the maximum value of strain available.
4. E_{om}^P is the slope of the bound.
5. σ_{bm} is the stress value at the point of intersection of the bound and the stress axis ($\varepsilon^P = 0$). σ_{bm} will be the radius of the bounding surface during monotonic loading.
6. Calculate the variable δ_{in} as follows

$$\delta_{in} = \sigma_{bm} - \sigma_{om}. \tag{E.2}$$

7. Select a point along a-b-c (point b in Figure E.1(a)) and measure the stress difference, δ, between it and its image on the bound. Point b is chosen such that

$$0.1 \leq \frac{\delta}{\delta_{in}} \leq 0.5. \tag{E.3}$$

8. Evaluate the plastic strain at point b, ε_b^P.

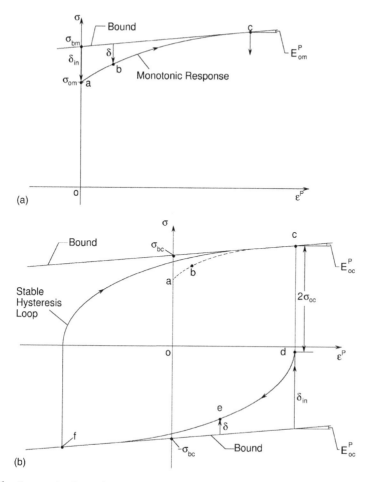

Figure E.1   Stress–plastic strain responses with definitions of Dafalias–Popov model parameters: (a) monotonic response and (b) Fit of Stable Hysteresis Response.

9. h is evaluated by substituting the values of δ and ε^P at point b into (E.4)

$$h = \frac{\delta}{\varepsilon^P} - \frac{\delta_{in}}{\varepsilon^P}\left[\ln\left(\frac{\delta}{\delta_{in}}\right) + 1\right]. \tag{E.4}$$

b.   Stable Hysteresis Stress–Strain Response

The stable hysteresis loop c-d-e-f in Figure E.1(b) is fitted in a similar fashion.

1. E_c is the elastic modulus of the hysteresis curve; that is, it is the modulus of section c-d of the unloading curve. Usually, but not always, $E_c = E_m$.

2. σ_{oc} is the radius of the yield surface calculated as follows. Identify the elastic limit of the hysteresis curve (point d in Figure E.1(b)). Then

$$\sigma_{oc} = \frac{1}{2}(\sigma_c - \sigma_d), \tag{E.5}$$

where σ_c and σ_d are the stresses corresponding to points c and d in the figure.

3. Construct the bounding line to be the tangent at point f of the response.
4. Measure E^P_{oc}, the slope of the bound.
5. Evaluate the stress σ_{bc} at the intersection of the bound and the stress axis ($\varepsilon^P = 0$). σ_{bc} will be the radius of the bounding surface.
6. The variable δ_{in} is the absolute value of the difference between the stress at point d and its image on the bound as shown in Figure E.1(b)

$$\delta_{in} = 2\sigma_{bc} - 2\sigma_{oc}. \tag{E.6}$$

7. Select a point along d-e-f (point e in Figure E.1(b)) and measure the absolute value of stress difference δ between it and its image on the bound. Point e is chosen according to (E.3).
8. Evaluate the plastic strain at points d and e.
9. h_c is evaluated as follows:

$$h_c = \frac{\delta}{\Delta\varepsilon^P} - \frac{\delta_{in}}{\Delta\varepsilon^P}\left[\ln\left(\frac{\delta}{\delta_{in}}\right) + 1\right], \tag{E.7}$$

where

$$\Delta\varepsilon^P = |\varepsilon^P_d - \varepsilon^P_e|.$$

The stress–strain parameters obtained by the procedures described above should be adjusted by trial and error until a satisfactory fit is obtained. Use the fit parameters as initial values and perform a numerical simulation of the experiment you are fitting. Compare the numerical fit of the stress–strain response with the experimental one and adjust the parameters (usually vary just h) until you obtain an optimum fit.

Appendix F
Stress–Strain Fitting for the
Tseng–Lee Model

This model is another candidate to be used for loading histories involving loading and reverse loading. The model is calibrated using uniaxial stress–strain data as described next. The monotonic part of the response (o-a-b) is fitted with the Ramberg–Osgood fit as described in Appendix C. For the results in Figure F.1, o-a-b had the following Ramberg–Osgood parameters:

$$E = 30.3 \times 10^3 \, \text{ksi}, \quad \sigma_y = 36.0 \, \text{ksi}, \quad n = 20.0, \quad \sigma_{om} = 25.3 \, \text{ksi}, \tag{a}$$

where σ_{om} is the proportional limit stress.

The material parameters for the basic hysteresis curve (b-c-d-e), E_c, σ_{oc} and h are obtained as follows:

1. E_c is the elastic modulus of the hysteresis curve; i.e., it is the slope of section b-c of the unloading curve. Usually, but not always, $E_c = E_m$.

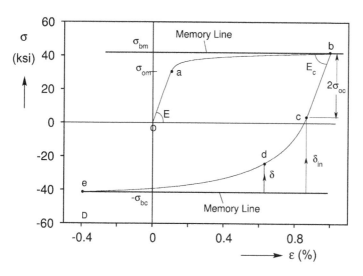

Figure F.1 Fit parameters of uniaxial monotonic and hysteresis stress–strain responses for the Tseng–Lee two-surface model.

2. σ_{oc} is the radius of the yield surface, calculated as follows. Identify the elastic limit of the hysteresis curve (point c). Then

$$\sigma_{oc} = \frac{1}{2}(\sigma_b - \sigma_c), \qquad (F.1)$$

where σ_b and σ_c are the stresses corresponding to points b and c on the response.

3. h is a shape parameter representing the hardening characteristics of the hysteresis curve. It can be evaluated as follows:

- Find the slope, E_m, of the monotonic curve at point b. The corresponding plastic modulus at point b is

$$E_m^P = \left(\frac{1}{E_m} - \frac{1}{E} \right)^{-1} \qquad (F.2)$$

- The stress at point b, σ_b, represents the current size of the memory surface.
- Construct the memory line by drawing a horizontal line through point b, i.e., at stress σ_b.
- Construct the second memory line to be the horizontal line $\sigma = -\sigma_b$.
- Evaluate δ_{in} from

$$\delta_{in} = \sigma_c + \sigma_b. \qquad (F.3)$$

- Choose a point on the unloading curve, point d, and evaluate δ as follows:

$$\delta = \sigma_d + \sigma_b. \qquad (F.4)$$

Point d is chosen such that

$$0.1 \leq \frac{\delta}{\delta_{in}} \leq 0.5. \qquad (F.5)$$

- Evaluate the plastic strains corresponding to points c and d as follows:

$$\varepsilon_c^P = \varepsilon_c - \frac{\sigma_c}{E_C} \quad \text{and} \quad \varepsilon_d^P = \varepsilon_d - \frac{\sigma_d}{E_C} \qquad (F.6)$$

where ε_c and ε_d are the strains at points c and d, respectively.

- h is evaluated by solving the following nonlinear equation

$$\frac{\delta_{in} h}{(h-1)^2} \ln \left[1 - \left(\frac{h-1}{\delta_{in} h} \right)(\delta_{in} - \delta) \right] + \frac{\delta_{in} - \delta}{h-1} = -E_m^P \left(\varepsilon_c^P - \varepsilon_d^P \right). \qquad (F.7)$$

For the curve shown in Figure F.1, the following parameters were used for the hysteresis curve

$$E_C = 28.5 \times 10^3 \text{ ksi}, \quad \sigma_{oc} = 18.0 \text{ ksi}, \quad h = 50.0. \qquad (b)$$

The stress–strain parameters obtained by the procedures described above should be adjusted by trial and error until a satisfactory fit is obtained. Use the fit parameters as initial values and perform a numerical simulation of the experiment you are fitting.

Compare the numerical fit of the stress–strain response with the experimental one, and adjust the parameters (usually vary just h) until you obtain an optimum fit.

In case you do not wish to solve the nonlinear equation (F.7) in order to evaluate h, a good initial trial value for h for steels is

$$40 \le h \le 60. \tag{c}$$

Select a value for h and then improve it by trial and error.

Appendix G
Glossary and Nomenclature

Anisotropy, Plastic   The condition where the pipe material yields at different stress levels in different directions. For example, pipelines often have different yield stresses in the axial and circumferential directions.

Barrel of Oil (bbl)   Oil volume is measured in barrels. One barrel has 42 US gal (159 l).

Bauschinger Effect   Early yielding and rounding of a stress–strain response on unloading and reverse loading from a tensile or compressive stress state.

Bifurcation Buckling   A mathematical term that describes the point at which a previously unique solution *bifurcates* into two possible solutions.

Buckling   The process of switching from a stiff and structurally sound configuration to one that is less stiff and often dangerous. *Elastic* buckling implies that the structure buckles while the material is still linearly elastic. *Plastic* buckling implies that at buckling the material is either partially or fully plasticized.

Buckle Arrestor   Circumferential stiffener placed at regular intervals along a pipeline with the purpose of arresting a propagating buckle and limiting the damage to the distance between two adjacent arrestors (usually several hundred feet apart).

Casing   Tubular structure used for lining oil wells.

Catenary Riser   A freely suspended pipe that connects a pipeline or a well on the sea floor to a facility above the surface of the sea, in the process acquiring a catenary shape.

Cathodic Protection   System that reduces the external corrosion of a pipeline by regulating the electrical potential between the steel and its environment.

Charpy V-Notch Impact Test   Common test for measuring the fracture toughness of a material.

Collapse   A structure with negative stiffness usually undergoes large deformations (collapses) without the need of additional external effort.

Collapse Pressure   The value of external pressure at which a pipe collapses.

Compliant Tower   A slender, nearly uniform frame tower that supports a deck at the top and is fixed to the seabed with piles. Its slender design makes it more compliant, and thus it deflects more than other fixed platforms, but with periods that are much longer than those of hurricane sea waves.

Conductors   Tubes partially driven into the sea floor through which wells are drilled, completed and produced.

Crack Arrestor   Circumferential stiffener placed at regular intervals along a pipeline with the purpose of arresting a running crack.

Diameter (D)   The size of the pipe is specified by the outside diameter in units of inches. The standard diameters range from 0.405 to 80 in. Diameters from 2.375 to 12.75 in can be fractional (thus an "8-inch" pipe has actual diameter of 8.625 in). From 14 inch and higher, the sizes go up by 2-inch increments. These sizes are accepted worldwide, and thus forming tools at pipe mills follow these standards.

Drop Weight Tear Test   Test method for determining the fracture toughness of heavy sections of materials.

Ductility   A measure of the extent of plastic deformation that a material can sustain. Usually reported in percent strain at failure in a tensile test within a 2-inch gage length that includes the neck.

Electric Resistance Welded Pipe (ERW)   Pipe with a longitudinal weld formed by butt welding (usually without the addition of a filler metal – Section 3.4).

Export Line   A pipeline that transports processed oil and gas between platforms or between a platform and a shore facility.

Fixed Platform   Tubular truss structure with a working deck. The platform is secured to the seabed with piles.

Flexible Pipe   Pipe with layered construction of steel strands separated by polymeric layers that results in relatively small bending stiffness.

Floating Production System (FPS)   Semi-submersible floating platform secured in place with mooring lines and anchors.

Floating Production, Storage and Offloading (FPSO) Vessel   A vessel used in place of an offshore platform that also serves as a storage and offloading facility for deepwater applications in remote areas.

Flowline   A pipeline that transports well fluids from a manifold or a sled near the wellhead to the first downstream process component such as a platform.

Fracture Toughness   Measure of material resistance to crack propagation.

Free Span   Section of pipeline that is freely suspended between two seabed shoulders.

Grade   Pipeline steels are designated as Grade A, B and X that define manufacturing and product specifications such as chemistry, fracture toughness, ultimate strength, etc. Some of Grade A and all of Grade X are followed by two or three digits that represent the minimum specified yield stress in ksi; thus, X52 is required to have a yield stress of at least 52 ksi (359 MPa).

Gravity Platform   Large reinforced concrete platforms that are held in place mainly by their own weight. They are used in large fields in water depths less than about 1,000 ft (300 m).

Hardness   Measure of the resistance of a material to indention by a spherical or pyramidal indentor (Rockwell, Vickers, Brinell scales). Hardness can be approximately related to the material yield stress and thus is a non-destructive method of measuring yield stress locally.

Heat Affected Zone (HAZ)   Part of the base metal adjacent to a weld that is altered by heat during welding.

Injection Line   A pipeline that directs fluids into a formation to support hydrocarbon production.

Integral Buckle Arrestor   A thicker section of pipe ring that is welded in a pipeline at chosen intervals with the purpose of arresting an incoming propagating buckle.

Jacket   A frustum-shaped frame structure that carries a deck at the top and is fixed to the seabed by piles.

J-Lay   A pipeline installation method where the pipe leaves the vessel in a vertical (or nearly vertical) orientation, and the suspended section acquires a "J" shape.

J-Tube   A J-shaped tube installed on a platform extending from the sea floor to the platform deck through which a pipe is pulled to form a riser.

Jumper   A short section of prefabricated tube that connects the wellhead to a sled or a manifold.

Lazy Wave Riser   Shaping the upper section of a catenary riser to form a local undulation that isolates the sagbend from motions of the host platform.

Limit Load   A load maximum in a structural response that usually indicates the onset of collapse.

Limit State   A state beyond which a structure is no longer operable.

Localization of Buckling or Collapse   Inelastic buckling modes tend to localize leading to local rather than global collapse.

Manifold   Seabed structure that connects several incoming flowlines from wells to an outgoing flowline that connects to a production facility such as a platform.

Normalizing   Process of heating a steel above the transformation temperature followed by air cooling to permit recrystallization and tempering.

Operator   Many offshore fields are co-owned by more than one company. One of the owners is designated as the operator, with the responsibility to operate the host facility and all its associated flowlines, pipelines and other facilities.

Ovality (Δ_o)   This is one measure of pipe cross section *out-of-roundness* defined as follows

$$\Delta_o = \frac{D_{max} - D_{min}}{D_{max} + D_{min}}.$$

Ovality reduces the collapse pressure, and as such, low values are desirable (see Section 4.4.3). High ovality can impede the welding of pipes to each other, and consequently must be minimized.

Overbend   The section of a pipeline that bends over the stinger of the lay barge.

Pipeline   Piping used to transport fluids on land or offshore.

Pipeline Inspection Gage (PIG) Device pulled through a pipeline for inspection as well as cleaning.

Pipe-in-Pipe System This is a system of two concentric pipes. The inner one carries hydrocarbons, usually at elevated pressure and temperature. The outer one is exposed to the water and is designed to carry external pressure. The annulus between the two pipes is either empty or contains insulation material. Pipe-in-pipe systems are usually used for thermal insulation and sometimes for mechanical protection of the inner tube.

Propagating Buckle Collapse under external pressure is usually local. If the external pressure is high enough the local collapse has the tendency to propagate and in the process flatten large sections of the structure.

Reeling A pipeline installation method where a long section of line is prewound onto a large diameter reel and subsequently installed at an offshore site by unspooling.

Remotely Controlled Vehicle (ROV) Submarine vehicles used to conduct many underwater operations ranging from observation of pipeline touchdown during installation to operation of equipment such as manifolds, etc.

Riser Tubing that connects a pipeline or a well on the sea floor to a facility above the surface of the sea.

Sagbend The section of pipe near the seabed that bends as the pipe acquires a natural catenary shape during installation.

Seamless Pipe A pipe without a seam produced by a hot working process (Section 3.3). Seamless pipes are typically made up to diameters of 16 in.

S-Lay A pipeline installation method where the pipe leaves the installation vessel in a horizontal (or nearly horizontal) orientation, and the suspended section takes an "S" shape.

Sled Anchored seabed structure that acts as the starting point for a flowline and connects to a well or manifold with a jumper tube.

Slip-On Buckle Arrestor A ring that slips over the pipeline, used to locally strengthen it against collapse. Placed periodically along the line, it is used to stop an incoming propagating buckle.

Snaking Seafloor buckling of a pipeline due to thermal expansion.

Spar A type of platform in which the deck rests on a vertical cylindrical vessel held in place with mooring lines.

Standard Cubic Foot of Gas (scf) Gas is measured in standard cubic feet, corresponding to the volume at one atmosphere and 60°F. Metric measurements are in Normal Cubic Meters (Nm^3), corresponding to the volume at one bar and 0°C.

Stinger A long, boom-like structure, rigid or articulated, that supports the pipeline for a certain distance as it comes off a lay-vessel.

Straightness This is a measure of the maximum lateral deviation of a pipe joint from a straight configuration. Typically, it is required to be below 0.2% of the length. Higher values can impede welding.

Strain Aging   Strength recovery following plastic deformation due to diffusion of soluble atoms that re-pin dislocations. Process accelerated by heating.

Strain Hardening   Increase in yield stress with plastic deformation in the direction of loading.

Submerged Arc Welding (SAW)   Electrical arc welding where the electrode is covered by a granular flux coating. The flux burns, releasing CO_2 that shields the molten weld metal.

Tensile Strength (Ultimate Stress)   This is the maximum engineering stress recorded in a uniaxial tension test. For a ductile material, it corresponds to the onset of necking.

Tensioner   Device on a lay-vessel for applying tension to the suspended pipeline.

Tension Leg Platform (TLP)   Floating platform connected to the sea floor with vertical tethers.

Thermo-Mechanical Control Processing (TMCP)   Plate rolling at controlled temperature usually followed by accelerated cooling. Results in fine-grained steel with high strength, high ductility and high toughness.

Transition Temperature   Temperature below which fracture toughness undergoes a significant reduction.

Tree   Assemblage of tubes and valves mounted at the top of a well used to regulate the flows in and out of the well.

Umbilicals   Conduits that connect facilities on a host platform with subsea facilities. They provide electrical, hydraulic, chemical injection and fiber optic connections for operating subsea facilities such as trees, manifolds, sleds, etc.

UOE Pipe (or SAW Pipe)   Pipe cold formed from plate by four steps: Crimping, U-ing, O-ing and Expansion (Section 3.6).

Upheaval Buckling   Vertical buckling of a section of buried pipeline out of its trench due to thermal expansion.

Wall Thickness (t)   For every standard pipe diameter there are several standard wall thicknesses. For example, for $D = 8.625$ in 20 standard thicknesses are available, ranging from 0.125 to 1.000 in. Selecting one of the standard thicknesses is important for seamless pipe, as they define a set of internal tools available at the pipe mill for their manufacture (see Section 3.3). By contrast, UOE pipe thicknesses other than the standard values can be easily accommodated.

Wall Thickness Eccentricity (Ξ_o)   This is one measure of wall thickness variation present primarily in seamless pipe and defined as follows

$$\Xi_o = \frac{t_{max} - t_{min}}{t_{max} + t_{min}}.$$

Large eccentricity is undesirable as it impedes welding and can adversely affect some mechanical properties.

Welded Pipe (Continuous)   This is a pipe that is formed from plate with one longitudinal seam produced by a continuous welding process. The two main categories are *UOE* (or *Submerged-Arc Welded*, SAW) pipe and *Electric Resistance Welded* (ERW) pipe. UOE pipe is made from 16 in and higher while ERW pipe diameters range from 2.375 to 24 in.

Wet Buckle   This is a buckle that leads to localized collapse of a pipeline, which in turn results in fracturing of the pipe wall, allowing water to flood the pipeline.

Yield Stress (Strength) (σ_o') This is the stress recorded in a uniaxial tensile or compressive test on the pipe material at a strain of 0.5%. This book primarily uses the more conventional yield stress definition, σ_o, that represents the stress at a strain offset of 0.2%. For materials with healthy hardening, the two definitions tend to differ by a few percent.

NOMENCLATURE

D	pipe outer diameter
D_o	pipe mean diameter ($=D-t$)
E	Young's modulus
E_t	tangent modulus
E_s	secant modulus
J_2, J_3	invariants of deviatoric stress tensor
n	Ramberg–Osgood hardening parameter
M	moment
M_o	fully plastic moment ($=\sigma_o D_o^2 t$)
P	pressure
P_C	elastic buckling pressure
P_{CO}	collapse pressure
$\hat{P}_{CO}$	calculated collapse pressure
P_o	yield pressure $\left(=2\sigma_o \dfrac{t}{D_o}\right)$
P'_o	yield pressure $\left(=2\sigma'_o \dfrac{t}{D_o}\right)$
P_{oT}	yield pressure at tension T
P_P	propagation pressure
R	cylinder mid-surface radius ($=(D-t)/2$)
$S_\theta, S_r, S_{r\theta}, S$	plastic anisotropy parameters
s	deviatoric stress tensor
T	tension
T_o	yield tension ($=\pi\sigma_o D_o t$)
t	pipe wall thickness
x, θ, r	polar cylindrical coordinates
Δ_o	initial ovality $\left(=\dfrac{D_{max}-D_{min}}{D_{max}+D_{min}}\right)$
$\varepsilon, \varepsilon_x, \varepsilon_\theta, \varepsilon_{x\theta}$	strain components
$\boldsymbol{\varepsilon}$	strain tensor
ε_C	critical (bifurcation) buckling strain
ε_L	strain at a limit load
κ	curvature
κ_1	t/D_o^2
λ_C	wrinkle wavelength
ν	Poisson's ratio
Ξ_o	pipe wall eccentricity $\left(=\dfrac{t_{max}-t_{min}}{t_{max}+t_{min}}\right)$
$\sigma, \sigma_x, \sigma_\theta, \sigma_{x\theta}$	stress components
$\boldsymbol{\sigma}$	stress tensor
σ_C	critical (bifurcation) buckling stress
σ_o	yield stress (stress at a strain offset of 0.2%)
σ'_o	API yield stress (stress at a strain of 0.5%)
σ_R	amplitude of bending-type residual stress
σ_y	Ramberg–Osgood yield stress
σ_{UL}	ultimate stress (tensile strength)

Appendix H
Units and Conversions

The *International System of Units* (*Système International-SI*) constitute the units of modern science and technology worldwide. US engineering and manufacturing in general, and the petroleum industry in particular, still use the so-called *US Customary Units*, which are by and large (but not always) based on the British Imperial Units. In this book we have tried to quote both units in most places. When quoting units from applications we typically use the units used in the country of origin. Listed below are definitions and conversion factors from US Customary to SI units.

Length:
1 meter (m) = 1,000 mm
1 inch (in) = 25.4 mm
1 foot (ft) = 0.3048 m
1 mile (mi) = 5,280 ft = 1.60934 km
1 nautical mile = 1.852 km

Mass:
1 kilogram (kg) = 1,000 g
1 pound (lb) = 1/2.20462 kg = 0.45359237 kg
1 metric tonne (t) = 1,000 kg
1 ton (long) = 2,240 lb = 1.0160469088 t

Time:
1 hour (h) = 60 minutes (min) = 3,600 seconds (s)

Temperature:
$T°F = 1.8 T°C + 32$
$T K = T°C + 273.15$

Acceleration:
1 foot/second^2 (ft/s^2) = 0.3048 m/s^2
1 inch/second^2 (in/s^2) = 0.0254 m/s^2

Density:
1 lb/ft^3 = 16.018463 kg/m^3

Energy/Work:
1 Joule (J) = 1 Nm
1 foot lbf (ft-lb) = 1.355818 J
1 inch lbf (in-lb) = 0.112985 J
1 British thermal unit = 1.05587 kJ

Force:
1 lbf = 4.4482 N
1 Newton (N or kg m/s^2) = 1/9.8066 kgf
1 N = 7.233 pdl (lb ft/s^2) = 7.233/32.174 or 0.2248 lbf
1 kip = 1,000 lbf = 4.4482 kN

Power: 1 Watt (W) = 1 Joule/second (J/s or Nm/s)
              1 foot lbf/second (ft-lb/s) = 1.355818 W
              1 British thermal unit/second (Btu/s) = 1.05587 kW
              1 horse power (550 ft-lb/s) = 745.7 W

Pressure: 1 bar = 10^5 N/m^2 (or 10^5 Pa) = 14.504 psi
              1 foot of sea water = 0.444 psi

Stress: 1 lbf/inch^2 (psi) = 6.894757 kN/m^2 (or kPa)
              1 ksi = 6.894757 MN/m^2 (or MPa)

Velocity: 1 foot/second (ft/s) = 0.3048 m/s
              1 inch/second (in/s) = 0.0254 m/s
              1 mile/hour (mi/h) = 1.60934 km/h
              1 knot = 1.852 km/h

Volume: 1 ft^3 = 0.028317 m^3
              1 m^3 = 1,000 liters (l)
              1 gallon-US (gal) = 231 in^3 = 3.785411
              1 barrel of oil (bbl) = 42 gal = 158.98731

REFERENCE

H.1. Mechtly, E.A. (1973). The International System of Units: Physical constants and conversion factors. Second Revision. NASA SP-7012.

Index

Printed and bound by CPI Group (UK) Ltd, Croydon, CR0 4YY

16/10/2024

01774872-0003